100 Advances in Polymer Science

Macromolecules: Synthesis, Order and Advanced Properties

With contributions by
K. A. Armitstead, Y. Chujo, P. Corradini,
M. Fischer, G. Goldbeck-Wood, G. Guerra,
A. Halperin, H.-H. Kausch, A. Keller,
J. P. Kennedy, B. Keszler, T. P. Lodge,
T. Q. Nguyen, T. Saegusa, M. Stamm,
M. Tirrell

With 175 Figures and 25 Tables

Springer-Verlag Berlin
Heidelberg GmbH

ISBN 978-3-662-14995-9   ISBN 978-3-540-38404-5 (eBook)
DOI 10.1007/978-3-540-38404-5

Library of Congress Catalog Card Number 61-642

© Springer-Verlag Berlin Heidelberg 1992
Originally published by Springer-Verlag Berlin Heidelberg New York in 1992.
Softcover reprint of the hardcover 1st edition 1992

Typesetting: Th. Müntzer, Bad Langensalza; Printing: Heenemann, Berlin;

02/3020-5 4 3 2 1 0 — Printed on acid-free paper

Editors

Foreword

The 100th repetition of an event is always a moment of reflection and mostly one of joy and pride. So it is with great pleasure that the Publisher and the Editors present this Jubilee Volume. Through the special composition of its contents the Editors wish to reinforce and underline the aims they have been following for the last 30 years.

Conceived in the late fifties and born in 1958 as one of the first collections dedicated to reviewing the state of the art of a rapidly developing field of science — that of polymers — Advances in Polymer Science immediately established an excellent reputation for itself. Several factors favourably assisted the birth and childhood of this series: firstly the timeliness of its inception and the need felt strongly by any active scientist to inform himself quickly and authoritatively on recent accomplishments in his own or a related field. This need was well-satisfied by the style of the collection: comprehensive articles of great topicality. The concept to let experts collect and evaluate pertinent articles — mostly including their own extensive work — proved valuable in those days of hand-compiled references; it seems to be even more useful today where computer-assisted literature searches easily provide a flood of abstracts from which it is often difficult to tell how important a full paper will be for a particular problem. In all fairness to the authors of original research papers it can be said that a good review fulfills its purpose best if it relieves its reader of the necessity to go back to each individual reference.

The second factor was very definitely the "good luck" of the Board of Editors to find expert authors. Some of the very first reviews (and many of the later ones) have become "classics":

Inhalt des 1. Bandes

*(please, note that at that time the three principal scientific languages were still
represented).*

Even in the beginning, all aspects of macromolecular science were covered:
polymerization reactions, the characterization of molecules by physicochemical
or purely physical methods in solution, in the melt, or in the solid, structure
analysis, transport phenomena, and the whole spectrum of polymer properties

and behavior have been treated in the past. Many subjects have been reviewed repeatedly (by the same or other authors) demonstrating the intention of the Editors to update constantly the information assembled in Advances of Polymer Science whenever this seemed necessary.

The present volume is a reflection of the principles maintained through 33 years of editorial work. Leading scientists have consented to give an overview over an area of particular and recent progress. The guide-line chosen for this volume:

Macromolecules: Synthesis, Order and Advanced Properties

encompasses the above-mentioned spectrum of branches of polymer science; it also expresses the philosophy of this Series and of its Board of Editors to encourage and support the interaction between chemistry, physics and technology. The 8 contributions of this volume are on chemical methods to create and modify (Keszler/Kennedy; Chujo/Saegusa; Halperin/Tirrell/Lodge), physico-chemical methods of analysis (Nguyen/Kausch), physical concepts of order (Corradini/Guerra; Armitstead/Goldbeck-Wood/Keller) and materials science aspects of solids (Fischer; Stamm). The chain molecule and its behavior is the central issue of all of these contributions.

The Publisher and the Board of Editors welcome this opportunity of the appearance of this Jubilee Volume to thank whole-heartedly the present authors and all those in the past for their important and highly appreciated efforts, for outstanding contributions and for their excellent collaboration. They are also grateful to their former colleagues in the Board, espectially to those who were active in the formative years and who put so much effort into establishing Advances in Polymer Sciences as the most cited journal in polymer science. They include J. D. Ferry, W. Kern, G. Natta, W. Prins, G. V. Schulz, A. J. Staverman, J. K. Stille, and H. Stuart.

We trust that this work will continue for the benefit and satisfaction of the scientific community, helping to improve the conditions and the effectiveness of research and contribute through the resulting progress towards meeting the serious challenges posed by modern society.

Editors   Publisher

Table of Contents

Synthesis of High Molecular Weight Poly (β-Pinene)

B. Keszler, J. P. Kennedy
Institute of Polymer Science, The University of Akron, Akron OH 44325-3909,
USA

In the course of attempts to develop conditions for the living polymerization of β-pinene (β-PIN) we discovered that the conventional "H_2O"/EtAlCl$_2$ system produces poly(β-PIN)s of heretofore unattained high molecular weights (at least up to ~40,000) and that the molecular weight can be controlled by the polymerization temperature in the −23 to −100 °C range. The log $\bar{M}_n$ versus 1/T plot is linear and yields $\Delta H_{DP} = -0.83$ kcal/mol. The repeat unit of the high molecular weight polymer is

$$-CH_2-C \overset{CH_2-CH_2}{\underset{CH_2-CH_2}{\diagdown}} CH-C(CH_3)_2-$$

which reflects isomerization polymerization. The T_g of poly(β-PIN) is 65 °C by DSC. Melt drawn fibers are amorphous by X-ray analysis.

Advances in Polymer Science, Vol. 100
© Springer-Verlag Berlin Heidelberg 1992

1 Introduction

The living carbocationic polymerization (LC$^\oplus$Pzn-mechanistic) of isobutylene has recently been described [1–3]. In view of the great structural similarity between the cationic polymerization of isobutylene and β-PIN [4] efforts have been made to find conditions for the LC$^\oplus$Pzn of β-PIN. Incentive for this research arose because of the possibility of obtaining controlled molecular weight and narrow molecular weight distribution poly(β-PIN)s. β-PIN can be readily polymerized by cationic techniques [4–8], however, the molecular weights of these products rarely exceed ~2000, indeed the highest molecular weight poly(β-PIN) described to date is ~3400 [8].

While our primary objective of developing conditions for the LC$^\oplus$Pzn of β-PIN has not been reached, a simple route to relatively high molecular weight product (at least up to 40,000) was found. Experiments described in this communication also demonstrate that the molecular weight of poly(β-PIN) can be controlled by the polymerization temperature.

2 Experimental Materials

The synthesis and purification of cumyl alcohol (CumOH), *p*-dicumyl methyl ether (DCE)) and 2-chloro-2,4,4-trimethylpentane (TMPCl), and the sources and purification of methyl chloride (MeCl), methylcyclohexane (MCHx), isobutylene have been described [9, 10]. β-Pinene (β-PIN), (Aldrich), was chromatographed over alumina (activity I, Fisher), and freshly distilled over CaH_2 under nitrogen; according to ^{1}H-NMR spectroscopy and GC analysis the purity was >99%. 2,6-Di-*tert*-butylpyridine (DtBP), (Aldrich), anhydrous *N,N*-dimethylacetamid (DMA), (Aldrich), ethylaluminum dichloride (EtAlCl$_2$), 1.0 M solution in hexanes (Aldrich), and methanol (Fisher) were used as received.

3 Procedures

Polymerizations were carried out in a dry box under nitrogen at various temperatures in large (~75 mL) test tubes agitated with turbomix. Solvents, (i.e. CH_3Cl or CH_3Cl/MCHx mixtures), Lewis base and/or DtBP if used, TMPCl, β-PIN, in this order, were charged into the reactors. After cooling to the desired temperature polymerizations were initiated by the addition of TiCl$_4$ or EtAlCl$_2$ solutions. The reactions were quenched with prechilled methanol. The conventional AMI (all monomer in) and the diagnostic IMA (incremental monomer addition) techniques have been employed (2). Detailed experimental conditions are given in the text and table or figure captions. Molecular weights were determined by a Waters High Pressure GPC instrument (Model 6000A Pump) using a series of thermostated Ultrastyragel columns (100, 500, 10^3, 10^4, 10^5 Å), a Differential Refractometer (Model 410) a UV Absorbance Detector (Model 440) and a WISP 710B Automatic Sampler. The flow rate of THF was 1 mL/min. Calibration curves were obtained with narrow molecular weight polyisobutylene standards.

The weight average molecular weights of a few representative poly(β-PIN) samples were also determined by low angle laser light scattering (Chromatix KMX-6 photometer) combined with a Waters 150 C GPC instrument using a series of thermostated Ultrastyragel columns (10^3, 10^4 Å). The polymer constans K = 1.603 E-7 and K = 1.690 E-7 for poly(β-PIN)s with $\bar{M}_{nGPC}$ = 2050 and $\bar{M}_{nGPC}$ = 6800, respectively were calculated from data obtained with a laser differencial refractometer (Chromatix KMX-16).

T_g was determined by DSC using a Du Pont (Model 9900) instrument. Fibers were drawn manually with a glass rod from heated poly(β-PIN) melts. Wide angle X-ray diffractogramms (WAXD pattern) were obtained by using a Rigaku X-ray Generator (CN4012K1).

4 Results and Discussion

4.1 Preliminary Experimentation

In view of the great structural similarity between the propagating sites in the cationic polymerization of β-PIN and isobutylene and their respective polymers (4), and our considerable experience accumulated with the $LC^{\oplus}Pzn$ of isobutylene [1–3], efforts have been made to adapt $LC^{\oplus}Pzn$ conditions found to yield living polyisobutylenes for the polymerization of β-PIN.

Thus a variety of β-PIN polymerizations have been carried out under conditions isobutylene yields living polymer. Table 1 shows the results of representative experiments.

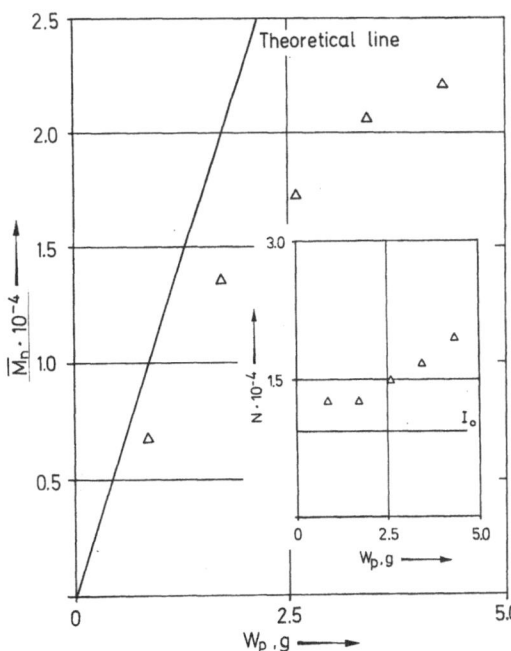

Fig. 1. Molecular weights ($\bar{M}_n$) and number of polymer chains (N) as a function of the weight of poly(β-PIN) obtained (W_p) with the TMPCl/EtAlCl$_2$/DtBP system ([M] = 0.25 mol/L, [EtAlCl$_2$] = [TMPCl] = 4×10^{-3} mol/L, MeCl/MCHx = 50/50, -80 °C, IMA conditions)

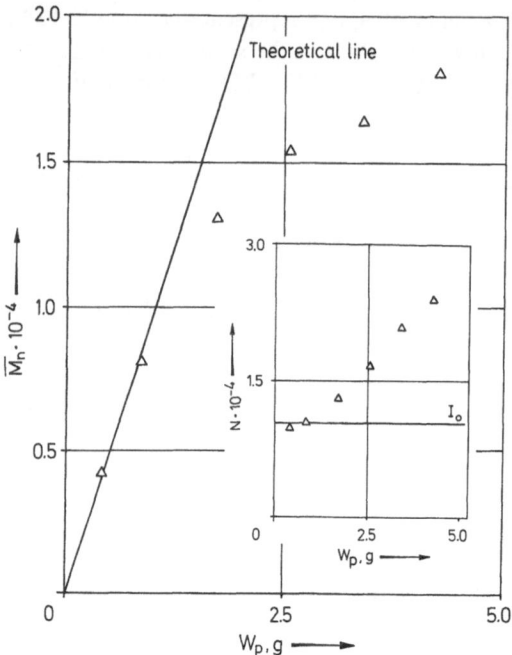

Fig. 2. Molecular weights ($\bar{M}_n$) and number of polymer chains (N) as a function of the weight of poly(β-PIN) obtained (W_p) with the TMPCl/EtAlCl$_2$/DtBP system ([M] = 0.25 mol/L, [EtAlCl$_2$] = [TMPCl] = 4×10^{-3} mol/L, MeCl/MCHx = 50/50, $-80\,^\circ$C, AMI conditions)

Evidently, with the DCE/TiCl$_4$ and TMPCl/TiCl$_4$ systems chain transfer is operational under the experimental conditions examined. For example in Exp. 4, which represents the first point in a series of AMI experiments, chain transfer is absent (I_{eff} = 95%), however, in Exp. 5. (the 6th point in the same AMI series) I_{eff} = 253%, which is due to significant chain transfer to monomer.

Expts. 18–20 carried out with the TMPCl/EtAlCl$_2$ system showed more promising results. For example, Exp. 20 yielded relatively high molecular weight poly(β-PIN), $\bar{M}_n$ = 8200, and I_{eff} = 105%, which is close to the theoretical value.

On the basis of these studies we decided to carry out a series of AMI and IMA experiments (2) with the TMPCl/EtAlCl$_2$/DtBP combination. Figures 1 and 2 show the results. The $\bar{M}_n$ versus W_p (g of poly(β-PIN) formed) plots and the N (number of moles of poly(β-PIN) formed) versus W_p plots (insets) indicate increasing deviation from the theoretical values (calculated for I_{eff} = 100%). According to these results chain transfer proceeds in these polymerizations, i.e., the systems are nonliving. Further experimentation would be necessary to develop satisfactory living conditions, in particular to investigate the effect of solvent polarity, temperature and electron donors on the mechanism.

4.2 Polymerizations with the "H$_2$O"/EtAlCl$_2$ System. The Synthesis of High Molecular Weight Poly(β-PIN)

In the course of the above preliminary experimentation we have discovered that the "H$_2$O"/EtAlCl$_2$ system (i.e. control experiments) readily yields unusually high molecular weight product, and that polymer molecular weights can be controlled by the polarity of solvent and the temperature. (The "H_2O" symbolism indicates adventitious moisture impurities). Thus a series of experiments were carried out to investigate the effect of solvent polarity on the molecular weight. Table 2 summarizes the results (Expts. 1–4). The molecular weight apparently reaches a maximum at MeCl/MCHx = 50/50. Contrary to these results, Kennedy et al. found that the molecular weights increase monotonically by increasing the polarity of the solvent [8]. Based on the above results we selected the MeCl/MCHx = 50/50 solvent composition for subsequent experiments.

Figure 3 shows the results of a series of experiments with the „H_2O/ EtAlCl$_2$/β-PIN system carried out in the -37 to -100 °C range. Evidently, in this temperature range molecular weights in the 14,000 to 27,600 range can be readily prepared. The Arrhenius plot over the temperature range examined is linear with a very low slope. This suggests that the contribution of individual activation enthalpy differences of elementary processes (i.e. propagation, chain transfer, termination) remain constant and the molecular weight determining mechanism does not change in this range. The slope of the log $\bar{M}_n$ vs. 1/T plot gives $\Delta H_{DP} = -0.83$ kcal/mol for the overall activation energy difference of β-PIN polymerization. Evidently the temperature has little effect on the molecular weight

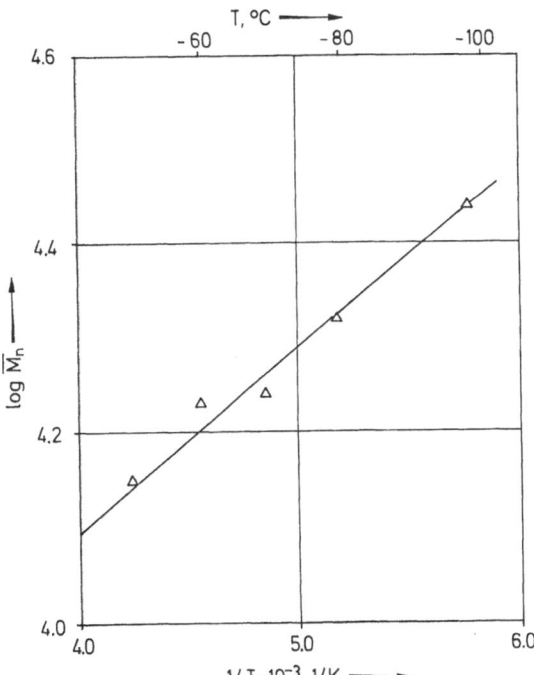

Fig. 3. The Effect of Temperature on Poly(β-PIN) Molecular Weight (EtAlCl$_2$ = 4×10^{-2} mol/L, β-PIN = 0.25 mol/L, MeCl/MCHx = 50/50, conversions ~100%)

Table 1. Effect of Conditions on the Polymerization of β-PIN*

Exp.	I_0 10^3 mol/L	Coin. 10^2 mol/L	β-PIN mol/L	ED 10^3 mol/L	DtBP 10^3 mol/L	t min	T °C	conv. %	$\bar{M}_n$ g/mol	$\bar{M}_w/\bar{M}_n$	I_{eff}^a %
1[b]	TMPCl, 10	TiCl₄, 16	0.50	DMA, 20	20	1	-50	89	2400	2.04	253
2	10	16	0.10	—	30	6	-50	~100	950	1.24	143
3	10	16	0.60	—	30	36	-50	~100	2900	1.71	238
4	6	9.6	0.20	—	6	15	-80	23	1100	1.28	95
5	6	9.6	1.20	—	6	90	-80	58	6250	1.53	253
6	DCE, 6	9.6	0.20	—	6	6	-80	42	1650	1.13	118
7	6	9.6	1.20	—	6	36	-80	30	4600	1.28	176
8[b]	—	10	0.25	—	—	6	-50	~100	6450	2.57	—
9	—	8	0.10	DMA, 6	6	15	-50	28	1750	2.13	—
10	—	6	0.25	—	—	15	-50	6	600	2.17	—
11	—	2.4	0.45	—	—	180	-80	~100	11200	1.99	—
12	—	9.6	0.25	—	—	15	-80	13	2200	2.41	—
13	—	9.6	0.25	—	12	15	-80	2	1700	2.38	—
14	DCE, 2.5	EtAlCl₂, 2.5	0.51	—	7.6	15	-35	7	9200	2.57	25
15	8	4	0.13	—	—	6	-80	~100	20050	2.61	11
16	4	4	0.50	—	—	6	-80	12	10000	3.51	12
17	CumOH, 8	4	0.13	—	—	6	-80	~100	23250	2.37	9
18	TMPCl, 4	4	0.13	—	—	6	-80	~100	3050	1.68	71
19	16	1.6	0.25	—	—	6	-80	~100	2600	1.65	83
20	4	4	0.25	—	4	6	-80	~100	8200	1.86	105
21	—	2.5	0.25	—	—	15	-35	90	20800	2.04	—
22	—	5	0.25	5	5	30	-80	~100	26850	2.56	—
23	—	5	0.25	—	—	1	-80	~100	23100	2.35	—

* Experiments 2 and 3; 4 and 5; and 6 and 7 are the first and the last n(sixth) experiments of AMI series. MeCl/MCHx = 50/50, ^{a} I_{eff} = 100 (g of poly(β-PIN)/$\bar{M}_n$)/I_0. ^{b} MeCl

Table 2. Polymerization of β-PIN with "H_2O"/$EtAlCl_2$ at $-80\,°C$*

Exp.	DMA 10^3 mol/L	DtBP 10^3 mol/l	MeCl/MCHx %	t min	conv. %	$\bar{M}_n$ g/mol	$\bar{M}_w/\bar{M}_n$
1	–	–	MeCl	20	~100	10400	3.40
2	–	–	75/25	20	~100	15850	3.17
3	–	–	50/50	20	~100	19450	2.55
4	–	–	25/75	20	~100	18950	2.27
5	–	2	50/50	6	~100	37350	2.22
6	–	4	50/50	6	~100	36250	2.27
7	–	20	50/50	6	~100	39900	2.12
8	DMA, 4	–	50/50	6	~100	24850	2.28
9	20	–	50/50	6	~100	28700	2.20
10	4	4	50/50	6	88	35550	2.06

* $EtAlCl_2 = 4 \times 10^{-2}$ mol/l, β-PIN = 0.13 mol/L, $V_0 = 25$ mL, T = $-80\,°C$

of poly(β-PIN). A further increase in the molecular weights can be achived using various additives, such as electron donors or a proton trap. The results are shown in Table 2. These products are certainly the highest molecular weight poly(β-PIN)s described to date.

5 Characterization of High Molecular Weight Poly(β-PIN)

We routinely estimated relative number average molecular weights $\bar{M}_n$'s of our poly(β-PIN)s by GPC calibrated with polyisobutylene standards. We have also determined the weight average molecular weights $\bar{M}_w$'s of select poly(β-PIN) samples by low angle laser light scattering analysis. We found that the $\bar{M}_n$'s obtained by calculation from the absolute LALLS method were substantially larger than those obtained by the relative GPC method. Table 3 shows representative molecular weights obtained by using two poly(β-PIN) samples. According to this information the $\bar{M}_n$ data shown in Tables 1 and 2 are in fact lower than the true values.

Figure 4 shows the ^{1}H-NMR spectrum of a poly(β-PIN) of $\bar{M}_n = 112000$ together with assignments. The doublet centered at 4.6 ppm is associated with the

Table 3. Molecular Weights of the Same Poly(β-PIN) Samples obtained by GPC and LALLS

Sample	GPC			LALLS		
	$\bar{M}_n$ g/mol	$\bar{M}_w$ g/mol	$\bar{M}_w/\bar{M}_n$	$\bar{M}_n$ g/mol	$\bar{M}_w$ g/mol	$\bar{M}_w/\bar{M}_n$
1	2050	3050	1.47	3100	4000	1.29
2	6800	10750	1.58	9600	11300	1.18

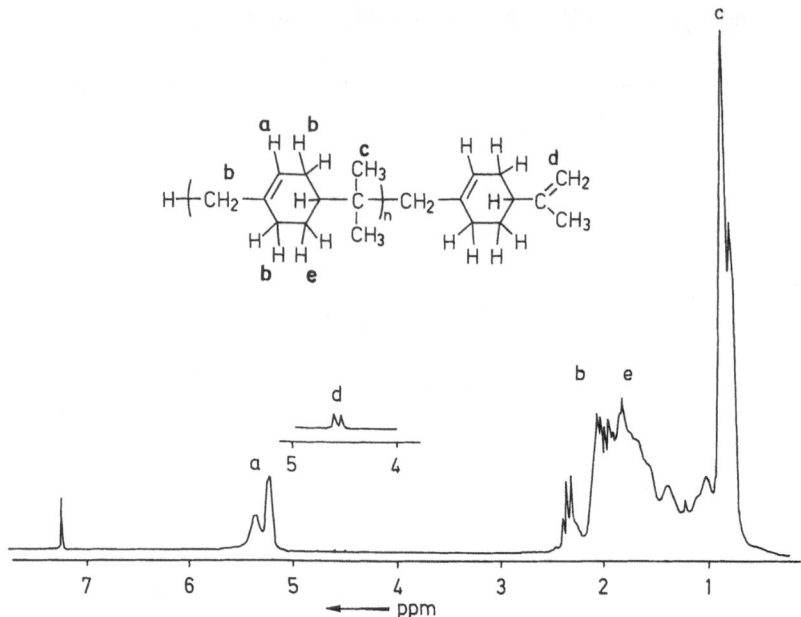

Fig. 4. ^{1}H-NMR Spectrum of Poly(β-PIN), $\bar{M}_n = 11\,200$

$\diagdown$C$=$C$\underline{H}_2$ protons of the terminal exo unsaturation that arises by chain transfer to monomer. The doublet centered at ~ 5.3 ppm is due to the endo unsaturation in the repeat unit. The resonances in the region from 0.5 to 2.5 ppm are assigned to the vartious protons at saturated positions (4).

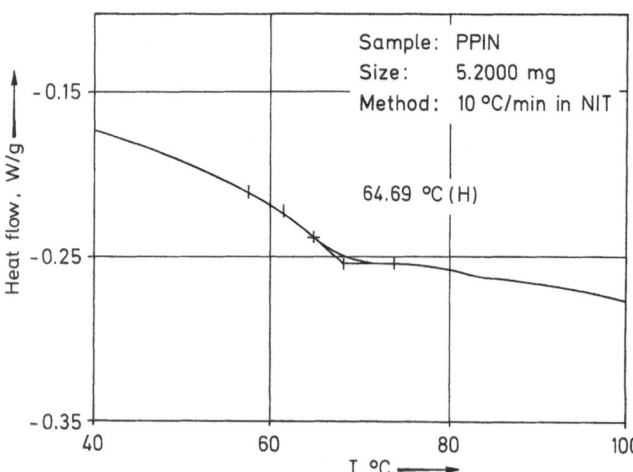

Fig. 5. DSC Trace of Poly(β-PIN), $\bar{M}_n = 39\,900$; instrument: General V2.2A DuPont 9900

Figure 5 shows the DSC trace of a poly(β-PIN) of $\bar{M}_n$ = 39,900. According to this evidence the T_g of poly(β-PIN) is 65 °C. This seems to be the first reported T_g of poly(β-PIN).

We have found that high molecular weight poly (β-PIN)s readily yields fibers by manual drawing from the melt. X-ray analysis of melt-drawn fibers showed a characteristic amorphous halo with a d spacing at ~6.2 Å.

Acknowledgement. This research was supported by the National Science Foundation under Grant DMR-89-20826.

6 References

1. Faust R, Kennedy JP (1986) Polym Bull, 15, 317
2. Faust R, Kennedy JP (1987) J Polym Sci, Polym Chem, A25, 1847
3. Kaszás G, Puskás J, Kennedy JP (1988) Macromol Chem, Macromol Symp, 13/14, 473
4. Kennedy JP, Chou T (1976) Adv Polym Sci, 21, 2
5. Roberts WJ, Day AR (1950) J Am Chem Soc, 72, 1226
6. Bates TH, Best JVF, Williams TF (1962) J Chem Soc, 1531
7. Snyder C, McWer W, Sheffer H (1977) J Appl Polym Sci, 21, 131
8. Kennedy JP, Liao T-P, Gunhaniyogi S (1982) J Polym Sci Polym Chem Ed, 20, 3219
9. Kaszás G, Györ M, Tüdös F, Kennedy JP (1982–83) J Macromol Sci, Chem Ed, A18, 1397
10. Mishra MK, Kennedy JP (1987) Polym Bull, 17, 7

Editor: J. P. Kennedy
Received April 25, 1991

Organic Polymer Hybrids with Silica Gel Formed by Means of the Sol-Gel Method

Yoshiki Chujo and Takeo Saegusa
Department of Synthetic Chemistry, Faculty of Engineering, Kyoto University,
Yoshida, Sakyo-ku, Kyoto 606, Japan

This article describes organic polymer hybrids with silica gels formed by means of the sol-gel method. The incorporation of organic polymers at the level of molecular dispersion into a metal oxide matrix was accomplished by the hydrolysis-condensation of tetraalkoxysilane (orthosilicate ester) in the presence of the appropriate organic polymer.

Especially, organic polymers consisting of amide groups such as poly(2-methyl-2-oxazoline) were found to form molecular hybrids with silica gel through strong hydrogen bonding. The obtained organic-inorganic polymer hybrids were homogeneous and transparent glassy composite materials in a wide range of the contents of organic polymers. The homogeneity of the hybrid may be due to the molecular dispersion of organic polymer in the silica gel matrix, which is ascribed to the formation of hydrogen bonds between the polymer amide group and silanol group of the silica gel. The occurrence of this hydrogen bond was actually supported by FT-IR spectra of hybrids. Organic-inorganic polymer hybrids having chemical bonds at the ends of organic polymer segments were also prepared starting from polyoxazoline silane coupling agents. The water-adsorption properties of these hybrids showed the hydrophilic or amphiphilic modification of the silica gel.

Another interesting material is porous silica gel with controlled-size pores which is prepared by the pyrolysis of the hybrid at a temperature far below the fusing point of silica gel.

Advances in Polymer Science, Vol. 100
© Springer-Verlag Berlin Heidelberg 1992

1 Introduction

Recently, composite materials by the combination of inorganic materials and organic polymers have been attracting attention for the purpose of creating high-performance or high-functional polymeric materials. Previously, the addition of inorganic materials as a filler into organic polymers was very popular to prepare composite materials having high mechanical properties and increased stability. In these cases, organic polymers and inorganic materials form independent phases, in which the interaction at interface plays an important role. On the other hand, organic polymers and inorganic materials have recently been combined at the molecular level as "polymer hybrids"; the word "hybrid" referring to combination blends of at molecular-level dispersion.

Generally, metal oxides may be regarded as three-dimensional network inorganic polymer consisting of metal-oxygen bonds. The molecular-level dispersion of silica gel is impossible. However, tetraalkoxysilane, a precursor of silica gel in the so-called sol-gel procedure, can be dissolved in organic solvents such as alcohol. Thus, the sol-gel reaction makes possible to incorporate the organic polymer segments in the network matrix of inorganic polymer.

This article describes the organic-inorganic polymers hybrids formed by means of the sol-gel procedure. In these hybrids, the organic polymer segments are dispersed in the silica gel matrix at the molecular level. As recent topics concerning organic-inorganic polymer hybrids, we also describe simulteneous filling, ceramers, and simultaneous interpenetrating hybrid networks in this report.

2 Simultaneous Curing or Filling of Elastomers

Mark and his co-workers reported the reinforcement of poly(dimethylsiloxane) networks by silica gel particles [1–6]. For example, bis(silanol)-terminated poly-(dimethylsiloxane) was reacted with tetraethoxysilane in the presence of acid-catalyst to produce the reinforced siloxane networks. The reaction proceeded homogeneously. The content of the silica filler can be controlled by the feed ratio of polysiloxane and tetraethoxysilane.

Some results of the properties of the silica-filled polysiloxane networks are shown in Table 1. These composite materials showed improved ultimate properties in comparison with those of polysiloxane networks without silica fillers. In other words, the modulus, the toughness, and the density were increased. The molecular weights of polysiloxanes have an influence on the stress-strain properties of the obtained composite gels. The effect of the number of junctions in the filler networks were also examined by using vinyltriethoxysilane or phenyltriethoxysilane, which possessed three junctions. As summarized in Table 2, low sol fraction and very large stress at high elongation were observed in these cases in comparison with the case of tetraethoxysilane having four junctions.

As a filler in the elastomeric matrix of polysiloxane, spherical particles of polystyrene were also used and provided considerable reinforcement of the

Table 1. Preparation and properties of the silica-filled networks

network	$10^{-3} M_n$	r^a	sol fraction	v_{2m}^b	ϱ, g cm^{-3}	wt% SiO$_2$ from Δw	wt% SiO$_2$ from ϱ	α_u^c	α_r	$(f/A^*)_r^d$ N mm^{-2}	$10^3 E_r^e$ J mm^{-3}
1	21.3	1.0	0.043	0.293	0.955	0.00	0.00		2.68	0.481	0.489
2	21.3	19.5	0.034	0.319	0.966	2.28	1.80	1.42	3.24	1.25	1.18
3	21.3	39.9	0.033	0.326	0.967	4.56	1.96	1.42	3.31	1.58	1.49
4	21.3	60.4	0.032	0.328	0.987	6.75	5.12	1.37	3.32	2.21	1.93
5	21.3	80.8	0.031	0.334	0.990	8.83	5.59	1.33	3.37	2.59	2.03
6	21.3	101.3	0.030	0.338	0.993	10.83	5.74	1.27	3.54	3.39	2.93
7	21.3	142.2	0.029	0.373	1.010	14.56	8.61	1.25	3.37	3.99	3.73
8	8.00	1.0	0.046	0.324	0.954	0.00	0.00		1.59	0.310	0.120
9	8.00	5.2	0.047	0.363	0.962	1.29	1.16	1.68	1.87	0.507	0.240
10	8.00	10.0	0.047	0.384	0.983	5.02	4.44	1.47	2.05	0.730	0.417
11	8.00	20.4	0.050	0.409	1.002	8.31	7.50	1.24	2.49	2.11	1.21

[a] Feed ratio of OC_2H_5 TEOS groups to OH chain ends. [b] Volume fraction of polymer present at swelling equilibrium in benzene at room temperature. [c] Elongation at initial upturn in modulus. [d] Ultimate strength as represented by the nominal stress at rupture. [e] Energy required for rupture

Table 2. Network characteristics and stress-strain results

end-linker molecule	ϕ[a]	short chains[b] mol%	wt%	sol fraction %	α_u[c]	α_r[d]	$(f/A^*)_r$[e] N mm^{-2}	10^3 E_r[f] J mm^{-3}
Si(OC$_2$H$_5$)$_4$	4	100.0	100.0	4.4		1.12	0.568	0.039
		99.9	96.7	4.7	1.15	1.20	0.856	0.082
		99.7	90.0	4.1	1.13	1.17	0.719	0.061
		99.4	83.3	4.1	1.17	1.21	0.695	0.068
		98.5	66.7	4.6	1.22	1.33	0.774	0.133
		97.0	50.0	4.9	1.23	1.66	1.45	0.397
		95.1	40.0	5.3	1.30	2.24	3.19	1.17
		94.2	33.3	5.4	1.41	2.27	2.45	0.963
		92.1	26.7	5.3	1.39	2.19	1.40	0.623
		90.8	23.3	5.2	1.55	2.42	1.64	0.769
ViSi(OC$_2$H$_5$)$_3$	3	100.0	100.0	3.9		1.17	0.557	0.052
		99.9	96.7	3.0	1.18	1.23	0.656	0.087
		99.7	90.0	3.2	1.24	1.27	0.700	0.105
		99.4	83.3	3.3	1.23	1.29	0.655	0.117
		98.5	66.7	3.6	1.29	1.58	0.834	0.256
		97.0	50.0	4.0	1.39	1.96	1.33	0.572
		94.2	33.3	4.1	1.38	2.19	1.29	0.640
		90.8	23.3	4.5	1.53	2.23	0.933	0.512
PhSi(OC$_2$H$_5$)$_3$	3	100.0	100.0	3.3		1.15	0.560	0.044
		99.9	96.7	3.5		1.25	0.658	0.082
		99.7	90.0	3.5	1.17	1.25	0.626	0.071
		99.4	83.3	3.7	1.23	1.40	0.935	0.195
		98.5	66.7	4.4	1.26	1.35	0.622	0.116
		97.0	50.0	4.7	1.31	1.78	1.08	0.397
		94.2	33.3	4.7	1.41	2.47	1.79	0.858
		90.8	23.3	4.6	1.45	2.40	1.30	0.710

[a] Junction functionality. [b] Having a number-average molecular weight of 660, in mixtures with long chains having 21.3×10^3. [c] Elongation at upturn in the modulus. [d] Elongation at rupture. [e] Ultimate strength, as represented by the nominal stress at rupture. [f] Energy required for rupture

networks [7]. In these anisotropic materials, the modulus in the direction parallel to the original stretching direction was found to be higher than that of the isotropic polysiloxane-polystyrene elastomer, whereas in the perpendicular direction it was lower. The particles themselves were characterized by using scanning and transmission electron microscopy.

Recently, Mark and co-workers also reported on organophilic silica formed by the combination of the sol-gel procedure and water-in-oil micro-emulsion method, in which methacryloyloxypropyltrimethoxysilane was used as one component of silica matrix [8]. The size of the silica particle was controlled by the content of water and emulsifier used. The surface of the particles was effectively covered with methacryloyl organic groups. This organophilic silica is expected to be used as a novel component of composite materials.

3 Ceramers

Recently, interesting composite materials incorporating polymeric materials into the sol-gel glasses have been reported by Wilkes and his co-workers [9]. These materials are named "ceramers". The properties of ceramers strongly depend on the reaction conditions, i.e., acidity, water content, reaction temperature, the amount of organic polymer, the molecular weight of polymer, solvent, and so on.

As an organic polymer segment, first, silanol-terminated telechelic poly(dimethylsiloxane) was used for the sol-gel reaction with tetraethoxysilane [10–12]. The idea of the incorporation of polysiloxane as an organic polymer segment was based on the similar structure of polysiloxane consisting of Si-O-Si bonds as that of silica gel. In addition, the polysiloxane segment has good thermal stability. The composite gels obtained were transparent and monolithic glasses and showed no cracking. The result of SEM micrographs of the hybrid prepared indicated no phase separation. The flexibility, gelation point, and the elongation at break were increased with an increase of acid content. An example of the effect of acid content on the mechanical properties of ceramers is shown in Fig. 1. The brittleness or flexibility of these hybrid materials also depended on the content of polysiloxane. For example, the modulus and the stiffness were increased when the content of tetraethoxysilane was increased. The low-molecular-weight polysiloxane produced more uniform hybrid materials, i.e., the problem of phase separation was decreased.

As an organic polymer, poly(tetramethylene oxide) was also used for the preparation of ceramers. The mechanical properties in these cases were much improved in comparison with those for hybrids from polysiloxanes. In these poly(tetramethylene oxide)-silica hybrids, the effect of the number of functional triethoxysilyl groups was examined [13]. As shown in Fig. 2, more multifunctional organic polymer produced more crosslinked hybrid networks. This means that the more rigid the structure in the hybrids is, the higher the modulus and the lower swelling property.

In the sol-gel procedure for the preparation of hybrids, polymeric acid catalysts such as poly(styrene sulfonic acid) were also used instead of hydrogen chloride [14]. The polymeric acid catalyst was effective for the preparation of hybrids at a similar level to that of hydrogen chloride catalyst. In some cases, the increased modulus was observed due to the higher extent of reaction. No difference was observed in morphologies between the hybrids prepared with polymeric and small molecule acid catalysts. The method using polymeric acid catalyst may depress the ion-conductive property, characteristic to the mobile acidic small molecules. Polymeric catalyst may also influence the rheology of the resulting hybrids.

By using a similar procedure for the preparation of hybrids of silica, hybrids materials consisting of other metal oxides were also prepared by the group of Wilkes [15]. For example, titania was incorporated into organic polymers by using the chemically controlled condensation (CCC) method for the preparation of poly(tetramethylene oxide)-silica or poly(dimethylsiloxane)-silica hybrids. Especially, in the case of the hybrid with poly(tetramethylene oxide), the modulus or ultimate strength of the hybrid increased in the presence of titania component, as shown in Table 3. This phenomenon was explained by the catalytic ability of

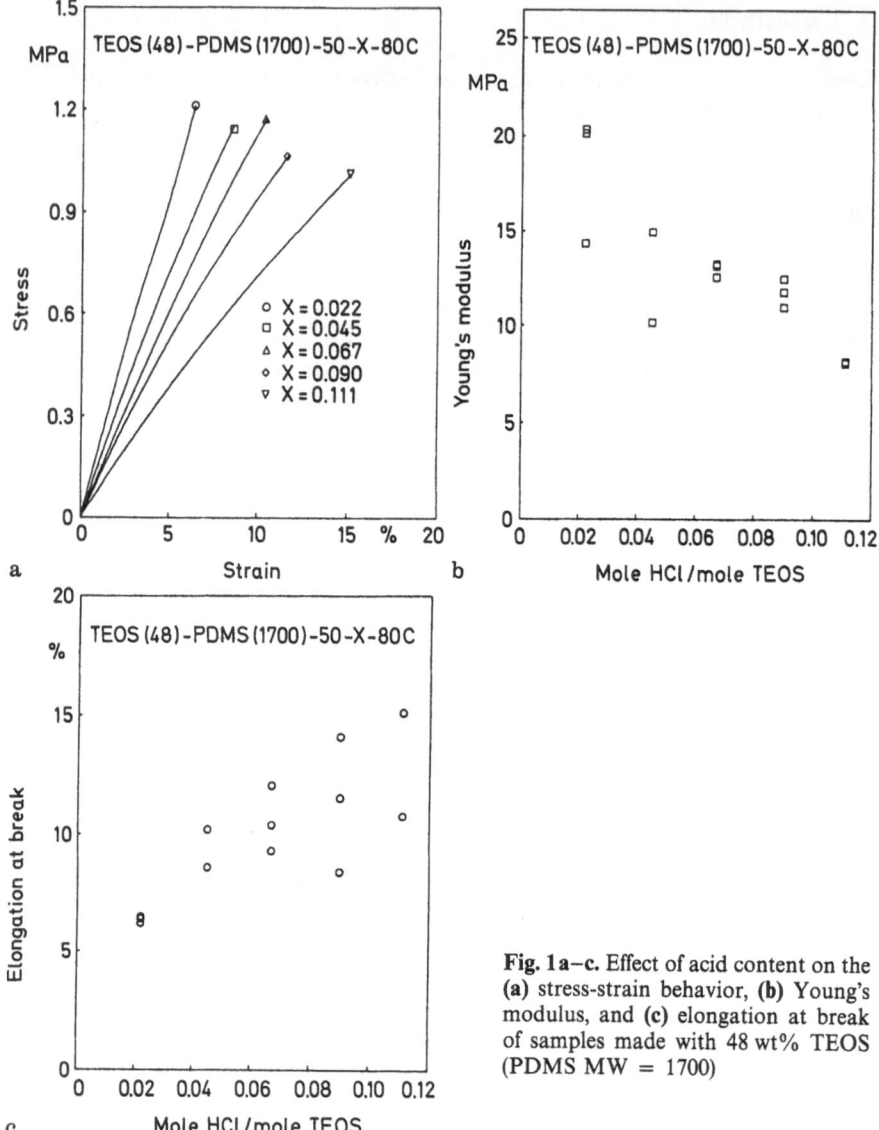

Fig. 1a–c. Effect of acid content on the (a) stress-strain behavior, (b) Young's modulus, and (c) elongation at break of samples made with 48 wt% TEOS (PDMS MW = 1700)

titanium for the sol-gel condensation reaction. As a result, a similar microphase separation was observed in this CCC method in comparison with the conventional sol-gel method. This was supported by results of small angle X-ray scattering measurements. Zirconia was also incorporated by using a similar procedure [16].

Alumina, zirconia, or zinc hybrids with poly(tetramethylene oxide)s were prepared by means of the conventional sol-gel method starting from metal

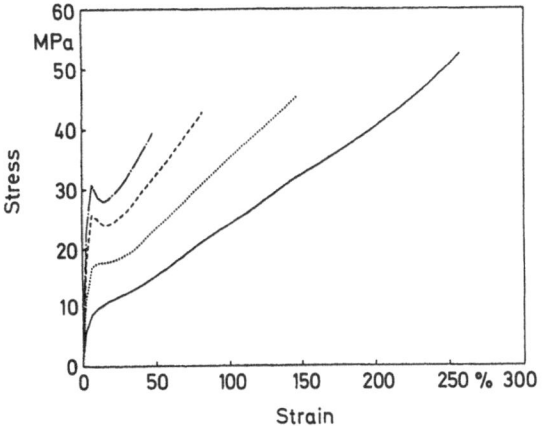

Fig. 2. Stress–strain behaviour of materials prepared with 50 wt% TEOS, 100% water content, and PTMO with various numbers of tri-EOS groups: TEOS(50)–PTMO(58-X)–100.   X =: ———, 2; ··········, 3; ———, 4; —·—·—, 5

Table 3. Tensile properties of Ti containing PTMO systems

wt% Ti-isop	Elongation at break (%)	Ultimate strength (MPa)	Young's Modulus (MPa)
0	104	1	7
15	136	11	22
30	57	11	67
30 [a]	61	12	71

[a] Opaque samples (different reaction batch)

acetylacetonate complexes [17, 18]. The rate of the gelation reactions strongly depended on the nature of the metals used. For example, the gelation of zirconium acetylacetonate was completed within a half hour, while the gelation of the corresponding zinc complex required several days. These gelation times had an influence on the shrinkage of the resulting hybrid. In the case of alumina hybrid, co-hydrolysis of aluminum secondary butoxide and aluminum ethyl acetylacetate was carried out without precipitation in the presence of poly(tetramethylene oxide). The modulus and glass transition temperature of the obtained hybrid depended on the composition of two kinds of aluminum ligands.

Poly(arylene ether ketone) and poly(arylene ether sulfone) were also tried to be incorporated into the hybrids with silica gel by means of the sol-gel procedure [19, 20]. For example, triethoxysilyl-terminated organic polymer was subjected to co-hydrolysis with tetraethoxysilane. A systematic change in mechanical and physical properties of the hybrid glass has been found with the content of organic polymer and the annealing temperatures.

4 Simultaneous Interpenetrating Networks

Generally, the glassy materials prepared by the sol-gel method show shrinkage (75–80%) after drying the solvents. Recently, Novak and his co-workers reported the new idea to prepare non-shrinking hybrids by so-called simultaneous interpenetrating polymer networks [21]. In this method, radical polymerization or ring-opening metathesis polymerization was carried out together with the hydrolysis-condensation of tetraethoxysilane (sol-gel method). Scheme 1 illustrates this idea in the case of radical polymerization.

Scheme 1

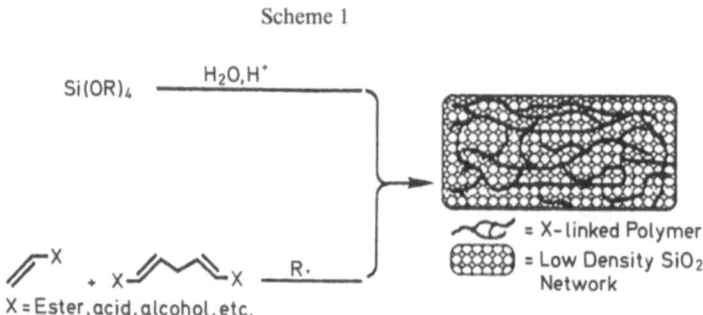

As an example, radical polymerization and hydrolysis-condensation of tetra-(acryloyloxyethoxy)silane were simultaneously carried out in the presence of the corresponding alcohol. The obtained hybrids showed to be clear and transparent with non-shrinking properties. It should be noted that preformed organic polymer was insoluble in normal sol-gel solutions. The formation of these organic-inorganic composites can also be accomplished using 5,6-dimethoxymethyl-7-oxanorbornene (1), which polymerizes by ring-opening metathesis.

(1)

5 Organic-Inorganic Polymer Hybrids Through Hydrogen Bonding

The basic reactions of the sol-gel procedure are shown in Eqs. (1–3), in which the species of Si−OH are the key intermediates. These Si−OH groups are known as Brønsted acids. On the other hand, the amide carbonyl groups are

$$-\overset{|}{\underset{|}{Si}}-OEt \quad + \quad H_2O \quad \xrightarrow[-EtOH]{} \quad -\overset{|}{\underset{|}{Si}}-OH \qquad (1)$$

$$-\overset{|}{\underset{|}{Si}}-OH \quad + \quad HO-\overset{|}{\underset{|}{Si}}- \quad \xrightarrow[-H_2O]{} \quad -\overset{|}{\underset{|}{Si}}-O-\overset{|}{\underset{|}{Si}}- \qquad (2)$$

$$-\overset{|}{\underset{|}{Si}}-OH \quad + \quad EtO-\overset{|}{\underset{|}{Si}}- \quad \xrightarrow[-EtOH]{} \quad -\overset{|}{\underset{|}{Si}}-O-\overset{|}{\underset{|}{Si}}- \qquad (3)$$

known as strong acceptors of acidic hydrogen to form hydrogen bonding. Accordingly, when the sol-gel procedure is carried out in the presence of an organic polymer consisting of amide groups, the organic polymer is to be incorporated into the three-dimensional network of silica gel [22].

Typical examples of organic polymers satisfying the above requirements are as follows: poly(2-methyl-2-oxazoline) (2), poly(N-vinylpyrrolidone) (3), or poly(N,N-

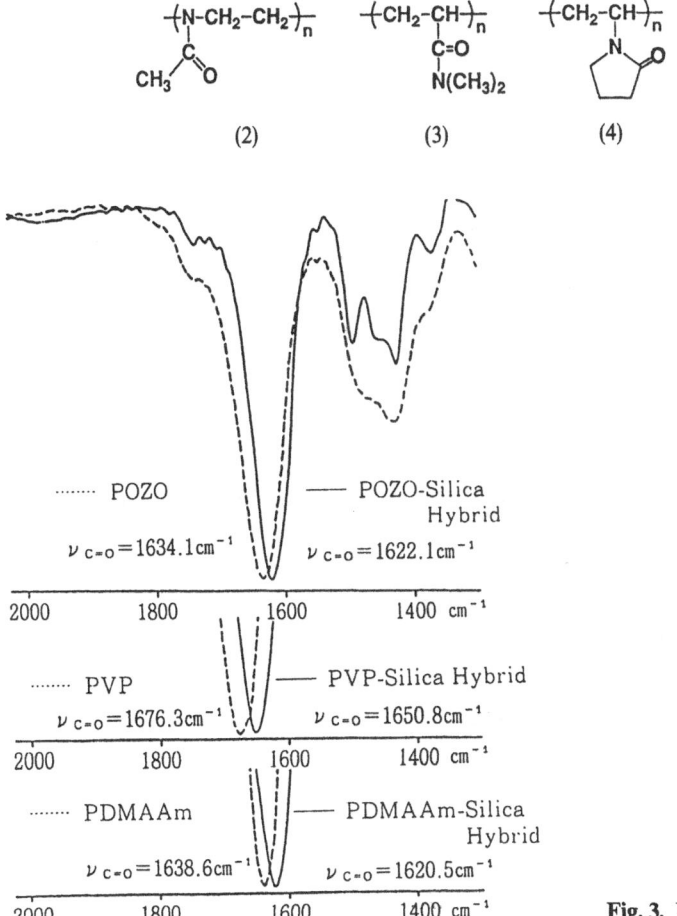

Fig. 3. IR spectra of hybrids

Typical examples of organic polymers satisfying the above requirements are as follows: poly(2-methyl-2-oxazoline (**2**), poly(N-vinylpryrrolidone) (**3**), or poly-(N,N-dimethylacrylamide) (**4**) [23]. One of these organic polymers was dissolved in ethanol with tetraethoxysilane. The resulting mixture was subjected to acid-catalyzed sol-gel reaction. After drying, a homogeneous and transparent glass material was produced. These colorless, transparent, and homogeneous hybrids were obtained in a wide range of compositions, which demonstrated the characteristic properties of these organic polymers.

The homogeneity of the present hybrid suggests that the organic polymer segments and inorganic ones were blended at the molecular level. This may be due to the strong hydrogen bonding as described above between amide and silanol groups. As illustrated in Fig. 3, the hydrogen bondings in these hybrids were supported by FT-IR spectra, in which the stretching bands due to amide carbonyl groups were shifted to the low wave-number region. The interaction between poly(2-methyl-2-oxazoline) and silica gel through hydrogen bonding is illustrated in Fig. 4. The molecular dispersion of organic polymers in the silica matrix also has been supported by the size of pores of porous silica prepared by the pyrolysis of hybrids (vide infra).

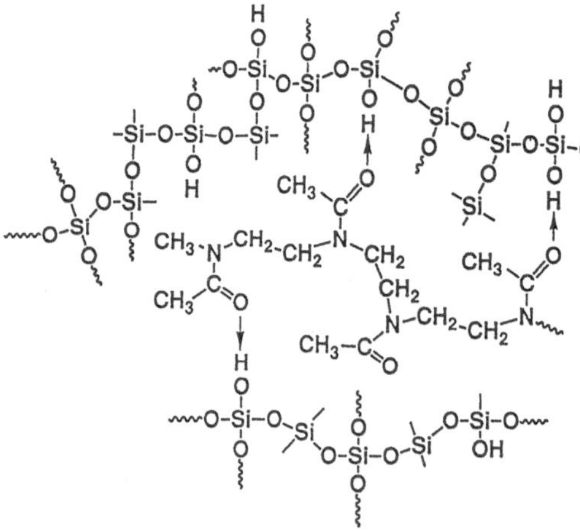

Fig. 4. Schematic representation of polyoxazoline-silica gel hybrid

6 Polyoxazoline Silane Coupling Agents

Scheme 2

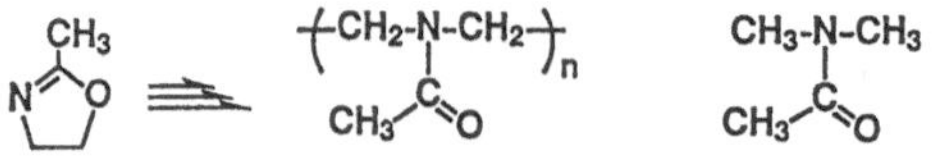

POZO   DMAc

Ring-opening polymerization of 2-methyl-2-oxazoline produces poly(*N*-acetyl-ethylenimine) (Scheme 2) [24–27]. The resulting polymer can be regarded as a polymer homolog of *N,N*-dimethylacetamide (DMAc). DMAc is known as a unique solvent to mix with water freely and to solubilize several organic polymers which are insoluble in common organic solvents. For example, poly-oxazoline (poly(*N*-acetylethylenimine)) has high hydrophilicity and good compatibility with commodity organic polymers such as poly(vinyl chloride) or polyamides.

The polymerization of 2-methyl-2-oxazoline is a clean reaction, which is not disturbed by chain transfer and termination. In this polymerization, the propagating species having the structure of an oxazolinium salt is not fragile, which is conveniently utilized for syntheses of block copolymers and end-reactive polymers [28].

Thus, the ring-opening polymerization of 2-methyl-2-oxazoline followed by the treatment of the resulting oxazolinium propagating end group with 3-aminopropyltriethoxysilane produced successfully triethoxysilyl-terminated poly-oxazoline as shown in Scheme 3 [29].

Scheme 3

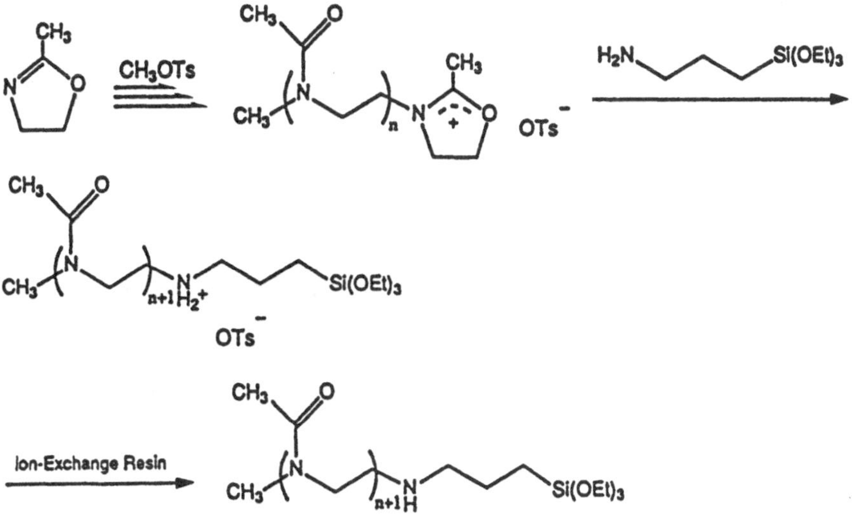

Scheme 4

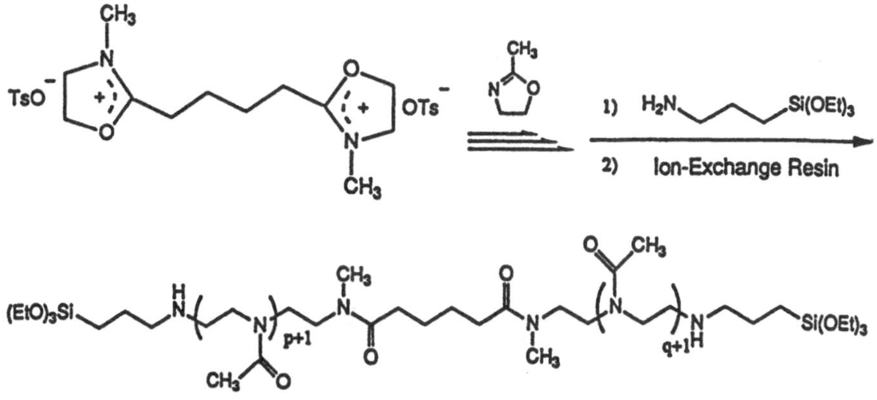

As shown in Scheme 4, using bifunctional initiator in the ring-opening poly-merization of 2-methyl-2-oxazoline, triethoxysilyl groups were introduced at the α- and ω-positions of polyoxazoline after treatment of telechelic oxazolinium end groups with 3-aminopropyltriethoxysilane. Table 4 summarizes the results of the preparation of polyoxazoline silane coupling agents according to Schemes 3 and 4, respectively. The conversions were almost quantitative and the molecular weights of the obtained silane coupling agents were easily controlled by the feed ratio of initiator and monomer in all cases of Table 4.

Table 4. Preparation of triethoxysilyl-terminated POZO

Run	Initiator	Time (h)	Yield (%)	$\overline{M}_n$ [a]
1	CH_3OTs	5	66	510
2	CH_3OTs	11	100	1040
3	CH_3OTs	10	100	1610
4	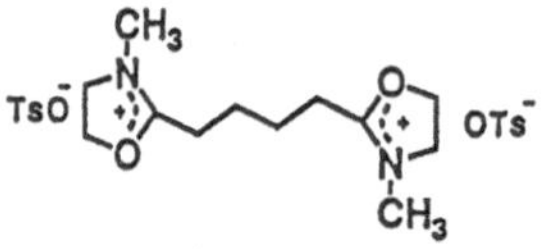	8.5	84	1960

in CH_3CN, 80 °C. [a] Calculated from ^{1}H-NMR

On the other hand, triethoxysilyl-terminated polyoxazolines were also prepared by the hydrosilation reactions of the $C=C$ bonds at the end of polyoxazolines with triethoxysilane [30]. As shown in Schemes 5 and 6, polyoxazoline silane coupling agents having two or three triethoxysilyl groups were obtained by using this procedure. As summarized in Table 5, a wide variety of polyoxazoline silane coupling agents was successfully prepared by termination with 3-aminopro-pyltriethoxysilane or the hydrosilation method.

Scheme 5

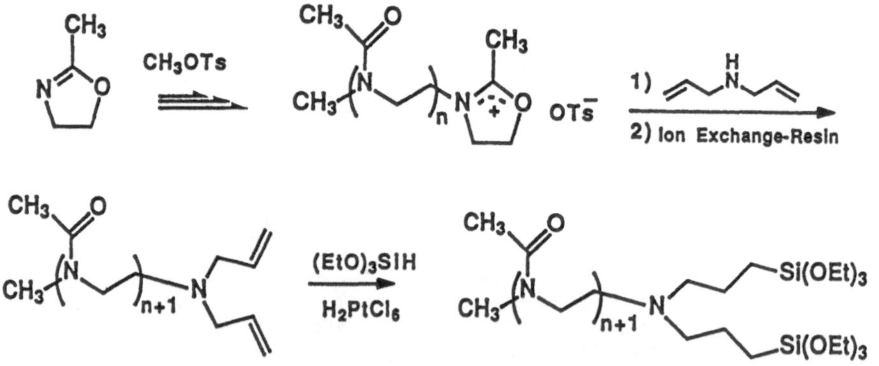

Scheme 6

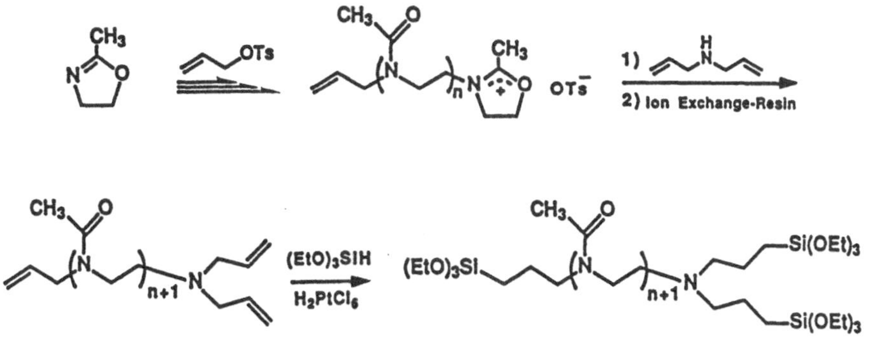

Table 5. Preparation of poyoxazoline silane coupling agents

	$HSi(OEt)_3$	$H_2N(CH_2)_3Si(OEt)_3$ (A)	N H and $HSi(OEt)_3$
~~POZO-N (+, O)	——	POZO-●	POZO<(●, ●)
POZO-N (+, O)	●-POZO	POZO-● (A) and $HSi(OEt)_3$ ●-POZO-●	●-POZO<(●, ●)
N-POZO-N (+, O) (+, O)	——	●-POZO-●	(●, ●)>POZO<(●, ●)

● = Triethoxysilyl group

As illustrated in Scheme 7, the reaction of trimethoxysilyl-terminated polyoxazoline and the silanol group of silica gel was examined in order to demonstrate the usefulness of the present novel silane coupling agent. For example, a mixture of polyoxazoline silane coupling agent and silica gel was heated in acetonitrile under nitrogen. From the results of the elemental analysis, polyoxazoline segments were found to be easily introduced to the surface of silica gel. From *N*% of the resulting modified silica gel, the content of polyoxazoline segments can be calculated.

Scheme 7

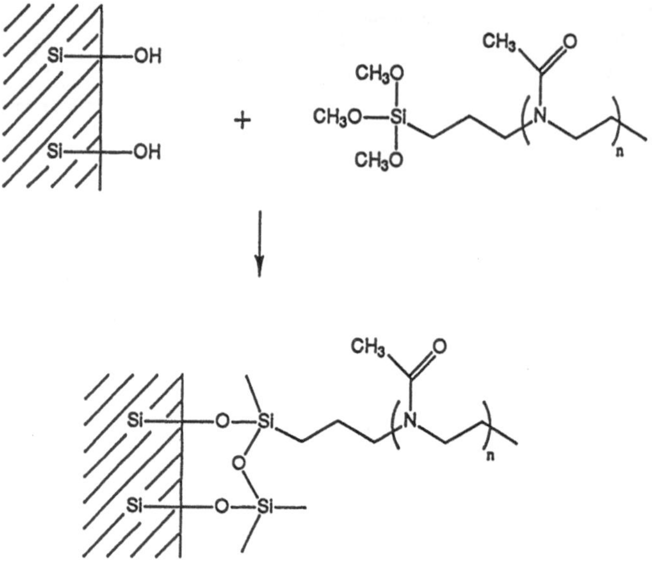

The dried polyoxazoline-modified silica gel was immersed into distilled water. The adsorption property of the resulting gel was estimated by the water content. The water uptake was calculated from an expression of $(W'-W)/W$, where W is the weight of dried gel and W' is the weight of water-absorbed gel. The modified gel showed a higher water-adsorption property than that of untreated silica gel, which absorbed 10.8 multiples of water. The water uptake of modified gel was up to 13.7 multiples of the weight of dried gel. Thus, silica gel has been made more hydrophilic by a polyoxazoline segment.

As illustrated in Scheme 8, acid-catalyzed co-hydrolysis of triethoxysilyl-terminated polyoxazoline and tetraethoxysilane produced the polyoxazoline-modified silica gel, which was homogeneous and transparent [31, 32]. The telechelic polyoxazoline also produced composite material in which polyoxazoline segments were connected through their α- and ω-positions to the silica matrix. The homogeneity of the obtained composite gel (hybrid) may be due to the interaction between amide groups in polyoxazoline and silica through hydrogen bonding as described before. In other words, the molecular level combination between organic polymer (polyoxazoline) and inorganic polymer (silica) has been accomplished.

The obtained gels were purified by Soxhlet extraction with chloroform to remove the unreacted polyoxazoline. Table 6 summarizes the results of the preparation of polymer hybrids together with their water adsorptions. In comparison with the silica gel without polyoxazoline segments, the modified silica with 50% polyoxazoline was found to show higher water adsorption.

Scheme 8

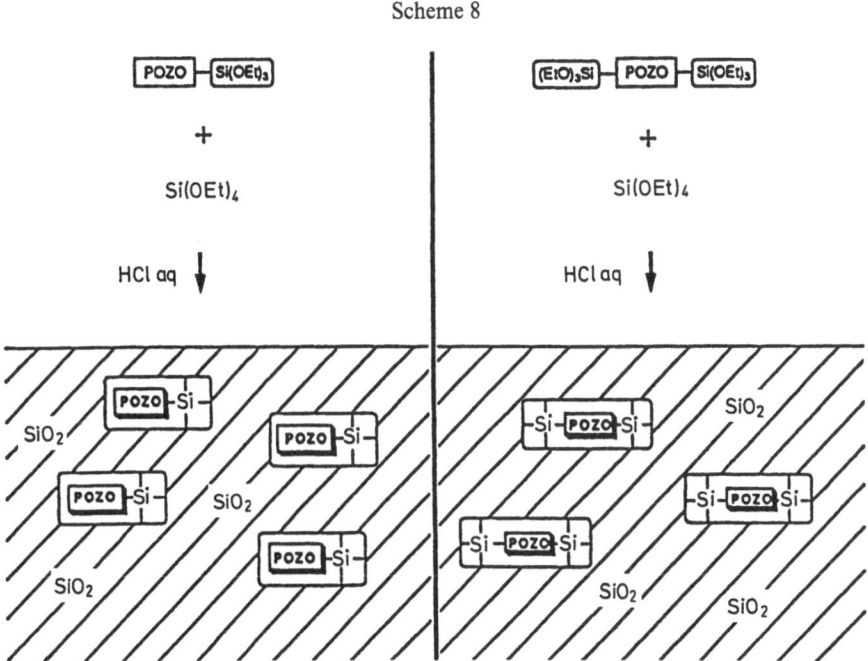

Table 6. Preparation and water adsorption property of POZO modified silica gel

Run	POZO (D.P.)[a]	POZO/ Si(OEt)$_4$	Weight loss (TGA) (%)	POZO (wt%)[b]	H$_2$O Content[c]
1		1/2	49.0	50.0	2.98
2	o———[d] (9.0)	1/10	29.4	15.6	1.87
3		1/2	47.5	47.2	2.26
4	o——— (14.4)	1/10	27.3	14.2	1.63
5		1/2	50.3	53.1	3.46
6	o———o (16.1)	1/10	28.5	17.9	1.92
7		0	18.2	0	1.53

[a] Calculated from feed ratio. [b] Calculated from elemental analysis. [c] g wet gel/g dried gel. [d] o = Si(OEt)$_3$

7 Amphiphilic Silica Gel

Generally, polyoxazolines cover a wide spectrum from hydrophilic to lipophilic depending on the N-acyl groups. By using a similar method for 2-methyl-2-oxazoline, various silane coupling agents were prepared by ring-opening polymerizations of 2-oxazolines (2-ethyl-, 2-n-butyl-, and 2-n-octyl-2-oxazoline) followed by the treatment of the propagating oxazolinium living end with 3-aminopropyl-triethoxysilane [33]. The molecular weights of the obtained silane coupling agents based on poly(2-alkyl-2-oxazoline)s were controlled by the feed ratios.

These poly(2-alkyl-2-oxazoline) silane coupling agents were copolycondensed with tetraethoxysilane by acid-catalyst to produce poly(2-alkyl-2-oxazoline)-modified silica gel. The composite gel from 2-ethyl-2-oxazoline was also homogeneous and transparent glass. Poly(2-alkyl-2-oxazoline)-modified silica gels, especially gels based on poly(2-ethyl-2-oxazoline) absorbed water and also organic solvents such as DMF or alcohols as shown in Table 7. This result means that the obtained composite gel shows the amphiphilic adsorption property.

Table 7. Adsorption property[a] of PROZO-modified silica gel

Run	POZO	$\overline{D.P.}$	P/S[b]	loss wt(%)	Solvent				
					H_2O	DMF	n-PrOH	Cl Cl	Toluene
1	Et	5.3	1/10	43	241	155	64	23	32
2	Et	11.7	1/2	40	143	36	16	16	20
3	Et	11.7	1/5	49	200	110	111	15	11
4	Et	11.7	1/10	40	214	120	52	59	54
5	nBu	7.7	1/5	42	146	49	22	14	19
6	nOct	12.5	1/5	18	54	11	4	0	0
7	–	–	–	–	119	16	8	12	15

[a] $(w' - w)/w \times 100$ w′ = weight of swollen gel. w = weight of dried gel. [b] POZO/Si(OEt)$_4$

8 Porous Silica by Pyrolysis of Hybrids

Blending at molecular-level dispersion between silica gel and the above organic polymers has been supported also by the number, size, and surface area of pores of porous silica gel prepared after burning the above hybrids at 600 °C [34]. The formation of porous silica by pyrolysis of hybrid is illustrated in Scheme 9. An example is the production of a porous silica with 800 m^2/g surface area and with 0.5 cc/g pore volume from polyoxazoline-silica hybrid by this procedure. From the results of pore-size distributions of the obtained porous silica, the average radius of the pores are around or even less than 10 Å. However, the molecular weight, the number of functional groups, and the content of organic polymer segments in the stage of hybrids had no noticeable influence on the size of pores of the resulting silica. In other words, almost the same size pores were formed in various cases of porous silica prepared by this procedure.

The control of the pore size of porous silica gel by the structure of the organic polymer has been accomplished by using the so-called starburst dendrimer as an

Scheme 9

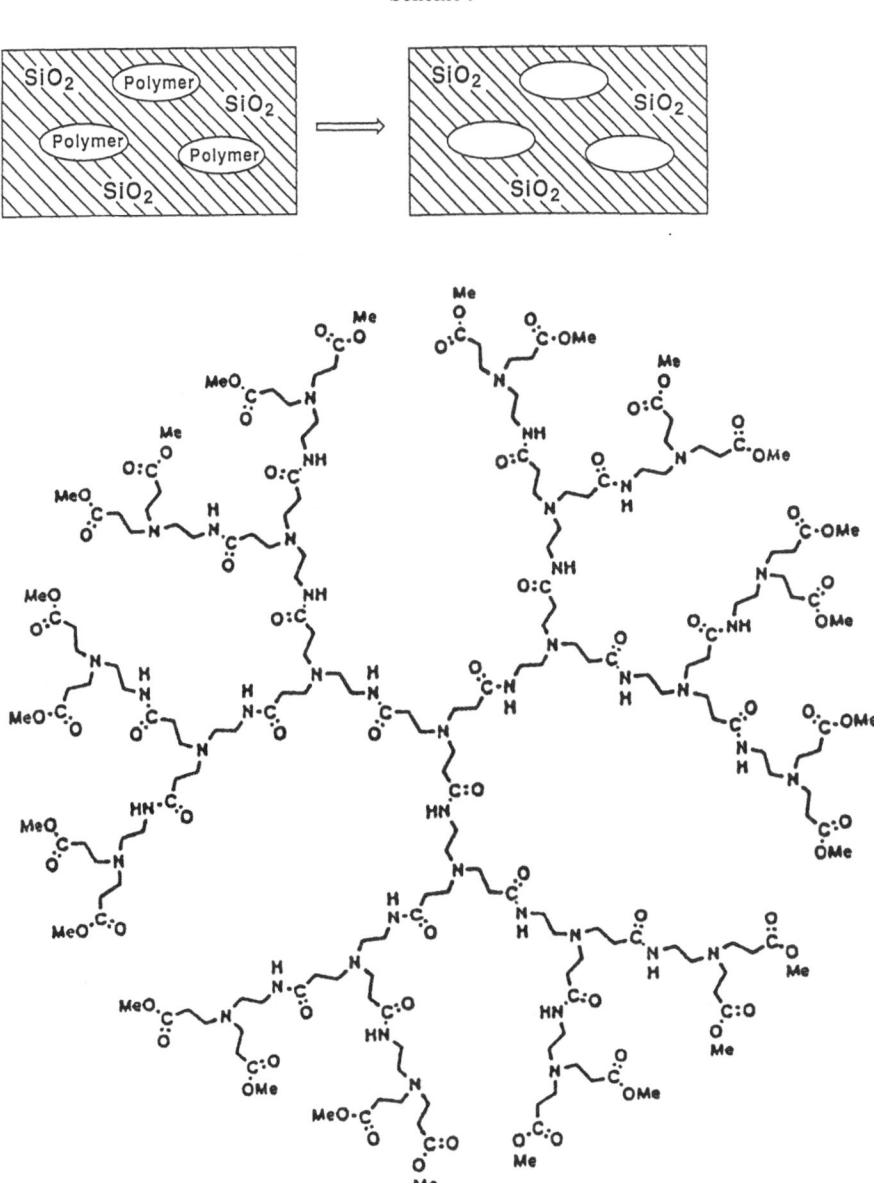

(5)

organic polymer segment. The starburst dendrimer (5) prepared by Tomalia's method [35] produced transparent hybrids with silica gel by means of the sol-gel procedure. The spherical shape of organic polymer in the hybrid formed the corresponding size pores after the pyrolysis.

9 Conclusions

Molecular hybrids between organic polymers and silica gel are expected to show many possibilities as new composite materials. First, the hybrids may show intermediate properties between plastics and glasses (ceramics). In addition, the composition of the hybrids can be widely varied. In other words, the hybrids can be used to modify the organic polymer materials or to modify the inorganic glassy materials. The hydrophilic modification as described before is a typical example.

Second, in the case of polyoxazoline hybrid, the characteristic property of high compatility of organic polymer with polar organic commodity polymers such as poly(vinyl chloride) or polyamide can be used as a type of compatibilizer. That is, it makes possible to incorporate the third organic polymer in the polyoxazoline-silica gel hybrid.

The porous silica having controlled size of pores is also important as a catalyst for size-selective organic reactions.

Such new materials should reveal various possibilities for application. Molecular hybrids between organic polymers and silica gel may become a novel type of composite material.

10 References

1. Mark JE, Pan SJ (1982) Makromol Chem Rapid Commun 3: 681
2. Ning YP, Tang MY, Jiang CY, Mark JE (1984) J Appl Polym Sci 29: 3209
3. Mark JE, Jiang CY, Tang MY (1984) Macromolecules 17: 2613
4. Tang MY, Mark JE (1984) Macromolecules 17: 2616
5. Mark JE (1988) Am Chem Soc, Polym Mat Sci & Eng, 58: 106
6. Clarson SJ, McCarthy DW, Mark JE (1989) Am Chem Soc, Polym Prep, 30(1): 298
7. Wang SH, Mark JE (1990) Macromolecules, 23: 4288
8. Espland P, Mark JE, Guyot A (1990) Polym Bull, 24: 173
9. Huang HH, Glaser RH, Wilkes GL (1988) Am Chem Soc, Symp Ser, 360: 354
10. Wilkes GL, Orler B, Huang HH (1985) Am Chem Soc, Polym Prep, 26(2): 300
11. Huang HH, Orler B, Wilkes GL (1985) Polym Bull, 14: 557
12. Huang HH, Orler B, Wilkes GL (1987) Macromolecules, 20: 1322
13. Huang HH, Wilkes GL, Carlson JG (1989) Polymer, 30: 2001
14. Brennan AB, Huang HH, Wilkes GL (1989) Am Chem Soc, Polym Prep, 30(2): 105
15. Glaser RH, Wilkes GL (1988) Polym Bull, 19: 51
16. Brennan AB, Wang B, Rodrigues DE, Wilkes GL (1990) Am Chem Soc, Polym Prep, 30(2): 146
17. Glaser RH, Wilkes GL (1989) Polym Bull, 22: 527
18. Wang B, Huang HH, Brennan AB, Wilkes GL (1989) Am Chem Soc, Polym Prep, 30(2): 146
19. Noell JLW, Wilkes GL, Mohanty DK, McGrath JE (1990) J Appl Polym Sci, 40: 1177
20. Wang B, Huang HH, Wilkes GL, Liptak S, McGrath JE (1990) Am Chem Soc, Polym Mat Sci & Eng, 62: 892
21. Novak BM, Ellsworth M, Wallow T, Davies C (1990) Am Chem Soc, Polym Prep, 31(2): 698
22. Saegusa T, Chujo Y (1989) Pacific Polym Prep, 1: 39
23. Kure S, Matsuki H, Jordan R, Chujo Y, Saegusa T (1990) Polym Prep Jpn, 39: 1684

24. Saegusa T, Kobayashi S (1976) Encyclopedia of polymer science and technology. Suppl vol 1: 220
25. Kobayashi S, Saegusa T (1984) Ivin K, Saegusa T (eds) in: Ring-opening polymerization. 2: 761. Elsevier, London
26. Kobayashi S, Saegusa T (1985) Makromol Chem, Suppl 12: 11
27. Saegusa T (1988) Makromol Chem, Macromol Symp, 13/14: 111
28. Saegusa T, Chujo Y (1989) in: Saegusa T, Higashimura T, Abe A (eds) Frontiers of macromolecular science (Proc 32nd IUPAC Int Symp on Macromolecules, 1988, Kyoto, Japan), Blackwell Scientific p 119
29. Chujo Y, Ihara E, Ihara H, Saegusa T (1989) Macromolecules, 22: 2040
30. Chujo Y, Ihara E, Ihara H, Saegusa T (1988) Polym Bull, 19: 435
31. Chujo Y, Ihara E, Kure S, Suzuki K, Saegusa T (1990) Am Chem Soc, Polym Prep, 31(1): 59
32. Chujo Y, Ihara E, Kure S, Suzuki K, Saegusa T (1991) Makromol Chem, Macromol Symp, 42/43: 303
33. Saegusa T, Chujo Y (1990) J Macromol Sci, Chem, A27: 1603
34. Kure S, Ihara E, Chujo Y, Saegusa T, Yazawa T, Eguchi K (1990) Polym Prep Jpn, 39: 1681
35. Tomalia DA (1990) Angew Chem, Int Ed Engl, 29: 138

Editor T. Saegusa
Received February 25, 1991

Tethered Chains in Polymer Microstructures

A. Halperin[1], M. Tirrell[2], T. P. Lodge[3]

Tethered polymer chains refers to macromolecular chains that are attached into micro-structures by their ends. Highly branched polymers, polymer micelles and end-grafted chains on surfaces are a few examples. This review brings out the common features of these seemingly widely disparate microstructures. Tethering can be reversible or irreversible and is frequently sufficiently dense that the chains are crowded. Densely tethered chains stretch to alleviate the interactions caused by crowding. They thus exhibit deformed configurations at equilibrium. These effects of tethering on the structure of the polymer chains are reflected in distinctive behavior and properties of microstructures containing tethered chains.

The topics we cover are: *structure*, in which the relationships of the free energy and characteristic dimensions of tethered chains to chain contour length, grafting density and geometry are developed; *aggregation*, by which we mean equilibrium microstructures formed by self-assembly such as micelles, adsorbed layers or ordered phases in block copolymers, where we can illustrate how the structural concepts come into play; *phase behavior*, which occurs, for example, in structures having more than one tethered component, such as mixed micelles; *block copolymer melts*, which are important systems which manifest the charac-teristics of tethered chains in both the aggregation and phase behavior, *interactions*, in which the behavior of two tethered layers in contact is explored, a subject of importance to applications involving colloidal stability; and *kinetics and dynamics*, where the chain deformation arising from tethering affects the kinetics of assembly and exchange processes and the entanglement behavior.

---

[1] Max-Planck-Institut für Polymerforschung, Postfach 3148, D-6500, Mainz, Germany.
[2] Department of Chemical Engineering and Materials Science, University of Minnesota, Minneapolis MN 55455, USA.
[3] Department of Chemistry, University of Minnesota, Minneapolis MN 55455, USA.

Advances in Polymer Science, Vol. 100
© Springer-Verlag Berlin Heidelberg 1992

1 Introduction

Highly branched polymers, polymer adsorption and the mesophases of block copolymers may seem weakly connected subjects. However, in this review we bring out some important common features related to the tethering experienced by the polymer chains in all of these structures. Tethered polymer chains, in our parlance, are chains attached to a point, a line, a surface or an interface by their ends. In this view, one may think of the arms of a star polymer as chains tethered to a point [1], or of polymerized macromonomers as chains tethered to a line [2–4]. Adsorption or grafting of end-functionalized polymers to a surface exemplifies a tethered surface layer [5] (a polymer "brush"), whereas block copolymers straddling phase boundaries give rise to chains tethered to an interface [6].

Focus on the idea of tethering is warranted if it is connected to common features or behaviors of the structures identified above, or other related microstructures. Figure 1 gives schematic illustrations of a range of important polymer micro-structures which comprise tethered chains. In addition to their prevalence in a wide range of systems, tethered chains, also known as grafted chains or terminally anchored chains, are of interest in their own right. Tethering introduces into the physics a new length scale, namely, the distance, d, between tethering or grafting points. When the tethering density is high and the d/R_g ratio is small, neighboring chains crowd one another. As a result, densely grafted chains tend to stretch out away from the grafting site. Under suitable conditions, strong deformations of the average dimensions may result. Remarkably, these stretched configurations are found under equilibrium conditions; neither a confining geometry nor an external field is required.

The significance of the unifying idea of tethered chains comes from a series of discoveries over the last five years which shows that the deformation of densely tethered chains affects many aspects of their behavior. The deformation of densely tethered chains is manifested in novel aggregation phenomena, phase transitions, interactions and dynamics. Development and use of the common concepts of tethered chains enables analysis of various features of many seemingly different structures to be pursued from a similar basis. This leads to the discovery of some analogies among these structures and brings out some distinctive common properties. Finally, the concept of tethered chains is of central importance in exploring the relationship between short chain amphiphiles and their ma-cromolecular counterparts, block copolymers and end-functionalized polymers. In turn, this allows a better understanding of typical technological uses of such polymeric surfactants, such as colloidal stabilization of suspensions of solid particles.

The distinctive properties of densely tethered chains were first noted by Alexander [7] in 1977. His theoretical analysis concerned the end-adsorption of terminally functionalized polymers on a flat surface. Further elaboration by de Gennes [8] and by Cantor [9] stressed the utility of tethered chains to the description of self-assembled block copolymers. The next important step was taken by Daoud and Cotton [10] in 1982 in a model for star polymers. This model generalizes the

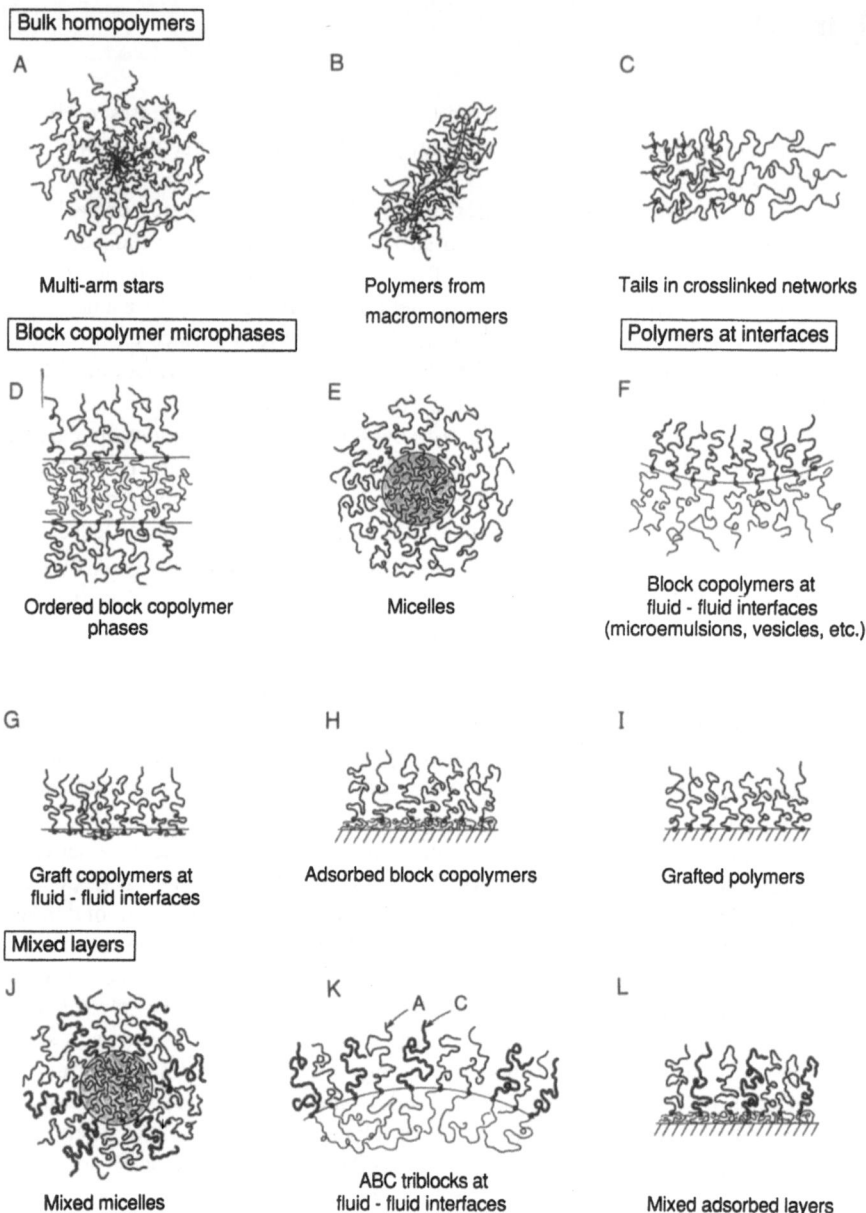

Fig. 1A–L. Examples of polymer microstructures comprising tethered chains

Alexander approach to spherical geometries, while making the connection between tethered chains and branched polymers. The internal structure of tethered layers was illuminated by numerical and analytical self-consistent field calculations, and by computer simulations.

Experimental activity in this field was slower to develop. An early attempt to test the Alexander model through study of the surface pressure vs area characteristics of block copolymers adsorbed at a liquid-air interface was made by Granick and Herz [6] in 1985. Star polymer configurations have been studied extensively [1], but without underlining the relationship to tethered chains. The present rapid growth in experimental activity in this field was initiated with the surface force measurements on layers of adsorbed block copolymers by Hadziioannou et al. [11], and Patel et al. [12], and on end-adsorbed polymers by Taunton et al. [13, 14]. Subsequent and current experimental work includes force balance measurements, neutron scattering and reflection, light reflection and ellipsometry, nmr, and dynamic investigations such as oscillatory squeezing and shearing in the surface forces apparatus. Experimental effort in this field is accelerating rapidly and new theoretical predictions are arising. This review will serve its purpose if it summarizes the current situation and focuses attention on important new directions.

Space limitations prevent a comprehensive review, so we have selected examples from diverse fields where the concept of tethered chains has led to insight and progress. We have also chosen to limit our discussion to emphasize the global structural features of tethered chains rather than the local ones. As we shall see, a rich variety of previously unanticipated behavior is uncovered at this level. Our aim is two-fold: to elucidate the common features of tethered chains and to explain the importance of the stretched configurations adopted by such chains. We will present the signature characteristics of tethered chains. Most of this can be done without analysis of detailed, local configurational features. We will point out where a more local description brings out interesting features [15]. We deal principally, but not exclusively, with tethered chains in good solvents since the chain stretching effects can be made strongest in that case. This is not to say, however, that all of the significant effects of tethering occur under good solvent conditions.

The topics we cover are: *structure*, in which the relationships of the free energy and characteristic dimensions of tethered chains to chain length, grafting density and geometry are developed; *aggregation*, by which we mean equilibrium microstructures formed by self-assembly such as micelles (Fig. 1E), adsorbed layers (Fig. 1H) or ordered phases in block copolymers (Fig. 1D), where we can illustrate how the structural concepts come into play; *phase behavior*, which occurs, for example, in structures having more than one tethered component, such as mixed micelles and related structures (Figs. 1J, 1K, 1L); *block copolymer melts* (Fig. 1D), which are important systems which manifest the characteristics of tethered chains in both the aggregation and phase behavior; *interactions*, in which the behavior of two tethered layers in contact is explored, a subject of importance to applications involving colloidal stability; and *kinetics and dynamics*, where the chain deformation arising from tethering affects the kinetics of assembly and exchange processes and the entanglement behavior.

2 Structure

The deformation of densely grafted chains reflects a balance between interaction and elastic free energies. In this discussion of structure, we assume that a dense tethering has been imposed on the chains by a mechanism that we do not yet specify. The grafting density should be imagined to be fixed and adjustable to arbitrary levels. Dense grafting enforces strong overlap among the undeformed coils. For tethered chains in a good solvent, this increases the number of monomer-monomer contacts and the corresponding interaction energy. This penalty is reduced by stretching the chains along the normal to the grafting sites, thereby lowering the monomer concentration in the layer and increasing the layer thickness, L. Stretching lowers the interaction energy per chain, F_{int}, at the price of a higher elastic free energy, F_{el}. The interplay of these two terms sets the equilibrium thickness of the layer.

2.1 Flat Interfaces

The Alexander model [7] for a flat tethered layer quantifies this argument in a simple way. It considers a flat, nonadsorbing surface bearing monodisperse tethered chains of N monomers of diameter, a. (We use the term "monomer" frequently as a shorthand; we actually mean statistical segment such that Na gives the full contour length of the chain). The average separation between chains at the tethering surface, d, is much smaller than the radius of a free, undeformed chain. As indicated above, the free energy per chain comprises two terms:

$$F = F_{int} + F_{el} . \tag{1}$$

The Alexander model is based on two assumptions that enable simple expressions for these two terms: (1) The concentration profile of the layer is step-like. That is, the monomer volume fraction within the layer, $\varphi \approx Na^3/d^2L$, is constant, independent of position; (2) The chains are uniformly stretched. That is, all chain ends are positioned on a single plane at a distance L from the surface. [In this paper, we use the symbol "$\approx$" to mean "approximately equal to" or "equal to within a numerical factor of order one"; we use "$\sim$" to mean "proportional to".] The first assumption simplifies the calculation of F_{int} while the second yields a simple expression for F_{el}.

Two schemes may be used to obtain explicit expressions for F. One is a so-called "Flory approximation" where the terms of Eq. 1 are estimated for ideal uncorrelated chains. The corresponding free energy per chain, in units of the thermal energy kT, is:

$$F/kT \approx v\varphi^2 d^2 L/a^3 + L^2/R_0^2 , \tag{2}$$

where v is a dimensionless excluded volume parameter ($\sim 1 - 2\chi$, with χ being the Flory-Huggins interaction parameter) and $R_0 \approx N^{1/2}a$ is the radius of an unperturbed, ideal coil. The first term accounts for binary interactions between monomers and the second for the elasticity of the Gaussian chains.

A second approach [7] allows for the effects of excluded volume correlations and self-avoidance by use of scaling arguments. In this picture, the layer is viewed as a semidilute solution characterized by a correlation length, $\xi \sim \varphi^{-3/4}$, and an interaction free energy density of kT/ξ^3. The excluded volume parameter v is considered to be near one in this rendition of the analysis, but this is not necessary. The layer is envisioned as a close-packed array of blobs of uniform size ξ (see Fig. 2) each assigned a free energy of kT. In this view, F_{int}/kT equals the number of blobs per chain, $\equiv N/g$, where $g \approx (\xi/a)^{5/3}$ is the number of monomers per blob. Therefore,

$$F_{int}/kT \approx N/g \approx N(\xi/a)^{-5/3} \approx N\varphi^{5/4} \approx \varphi^{9/4}d^2L/a^3 . \tag{3}$$

There is experimental support for this form of the interaction energy [16]. The elastic free energy must be modified as well to account for the nonideal character of the excluded volume chain. This is done in the spirit of the blob model by recognizing that in a semidilute solution the chain is Gaussian at large scales, but with the subunits being the excluded volume blobs. This Gaussian chain of blobs possesses an elastic energy of the form $F_{el} = L^2/R^2$, but here, $R \approx (N/g)^{1/2} \xi \approx N^{1/2}\varphi^{-1/8}$. This form of concentration-dependent coil contraction has been observed experimentally [17]. Consequently, both R and F_{el} are concentration dependent, in contrast to the Flory argument. Substituting these forms into Eq. 1 gives:

$$F/kT \approx (d^2L/a^3) \varphi^{9/4} + (L^2/Na^2) \varphi^{1/4} . \tag{4}$$

Notice that the result has been to multiply each term of Eq. 2 by a factor of $\varphi^{1/4}$, to account for the related effects of correlations on interactions between segments and on chain elasticity, respectively.

The equilibrium state of the layer is found by minimization of F with respect to L, bearing in mind that φ varies inversely with L. Both approaches (Equations 2 and 4) yield the correct equilibrium thickness:

$$L/a \approx N(a/d)^{2/3} . \tag{5}$$

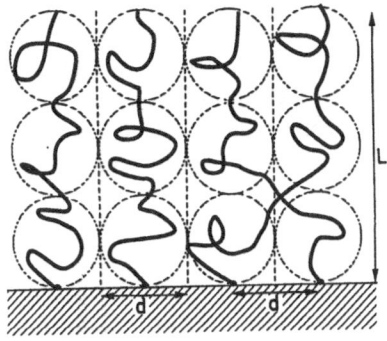

Fig. 2. Schematic of a flat, tethered layer. L is the average layer thickness while d is the average spacing between chain graft points on the surface

The Flory free energy is an overestimate:

$$F_{Flory}/kT \approx N(a/d)^{4/3} \tag{6}$$

owing to the neglect of the $\varphi^{1/4}$ corrections embodied in Eq. 4. The equilibrium blob size, $\xi \approx d$, and a more accurate free energy comes from Eq. 4:

$$F_{scaling}/kT \approx N(a/d)^{5/3} . \tag{7}$$

The characteristic result that the Flory argument gives accurate estimates of dimensions despite incorrect estimation of the individual terms in the free energy is well-understood [18, 19]. In the context of the study of tethered chains, it means that we can rely on the Flory argument for understanding certain features (eg., characteristic dimensions) but not others (e.g., force required to compress a tethered layer). Another more subtle pitfall of the Flory argument is in the consideration of the reversible, equilibrium assembly of structures (e.g., adsorption of block copolymers or end-functionalized polymers) where the two terms of Eq. 2 must be augmented by a third expressing the interfacial energy. Cancellation of errors in estimates of the individual terms can no longer be relied upon.

One of the signatures of densely tethered chains is expressed in Eq. 5, namely the linear variation of L with N. This stands in marked contrast with free chains where excluded volume interaction produces, at most, an $R \sim N^{3/5}$ distortion from the $R \sim N^{1/2}$ unperturbed dimensions. *Tethered layers are stretched and this is the origin of their interesting behavior.*

Equation 5 has been subjected to very limited experimental investigation. Relevant experiments are difficult since the behavior that these equations attempt to describe is only anticipated if $d/R_g \ll 1$. Small values of this ratio are difficult to achieve, in part because of the energetic barriers to adopting the stretched, tethered state. Patel et al. [12] studied a series of adsorbed block copolymers where one block adsorbed strongly, producing a small d, and the other block adsorbed negligibly. The average layer thickness was determined from the range of the onset of detectable repulsive forces exerted between the layers. For a series of polymers of nearly constant d and variable N (of the nonadsorbing block), the linearity of L with N of Eq. 5 was observed. Other experiments with adsorbed block copolymers and end-functionalized polymers [13, 14, 20, 21] have not seen this linearity (Fig. 3), but it has been recognized that in those studies d was not constant, but rather a decreasing function of N. These latter studies used more asymmetric block copolymers than those of Patel, et al. [12], and thus, the nonadsorbing block strongly influences the adsorbed amount. The physics of this will be discussed in the subsequent section on aggregation. The neutron reflection work of Satija, et al. [22] and of Cosgrove, et al. [23] and the nmr work of Blum, et al. [24], on selectively adsorbed block copolymers as above, are consistent with the idea of stretched chains but are too limited to test Eq. 5.

The strongest experimental support for Eq. 5 comes from the work of Auroy et al. [25], in which polydimethylsiloxane (PDMS) chains were chemically end-

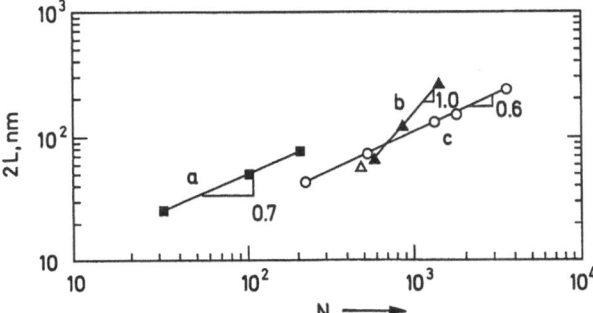

Fig. 3 a–c. Summary of data from different laboratories, obtained by surface force measurement, on the average layer thickness L as a function of tethered chain length for flat, tethered layers constructed by adsorption of amphiphilic polymers on mica. Adapted from Ref. 21. (**a**) Data of reference 20 on poly-tert-butylstyrene chains anchored by adsorbing blocks of poly-2-vinylpyridine. (**b**) Data of references 11 and 12 on polystyrene chains anchored by adsorbing blocks of poly-2-vinylpyridine. (**c**) Data of references 13 and 14 on polystyrene chains anchored by adsorbing zwitterionic groups [13] or by small adsorbing blocks of polyethyleneoxide [14]

grafted (inside a porous silica to obtain a large surface area) and the mean layer thickness measured by neutron scattering. It is important to note that the grafting density here does not correspond to equilibrium self-adjustment, but is irreversibly fixed by covalent bonding. Figure 4 illustrates their results in dichloromethane, an excellent solvent for PDMS. It shows not only the predicted linearity but also good agreement with the predicted inverse 2/3 power dependence on d, over more

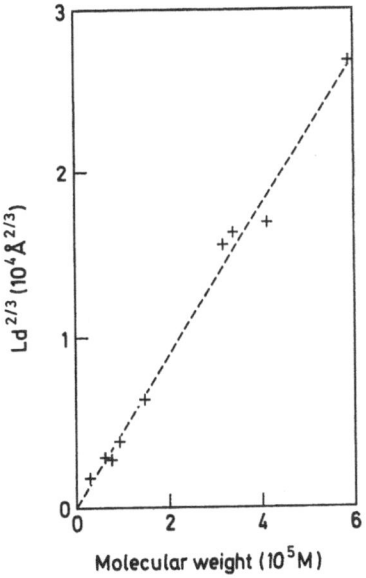

Fig. 4. Data of Auroy et al. [25] on polydimethylsiloxane chains covalently grafted in the interior of a silica porous medium. L is the measured layer thickness from neutron scattering; d is the average spacing per chain from the grafting density. The format of the plot is suggested by Eq. 5. The linearity of that relation is seen in these data

than a thirty-fold variation in N. In this work, the chain density on the surface exceeded the overlap threshold among the tethered chains by more than a factor of ten. In the block copolymer work, this ratio was at most about four. Further support for the linearity of L with N comes from the molecular dynamics simulation work of Murat and Grest [26]. Linearity is to be expected, in fact is unavoidable, if d ≈ a, since in that limit all the chains must close-pack (d ≈ a) and stand on end [27, 28]. An important point to recognize is that all of the results discussed so far pertain to $R_g > d \gg a$. The chains are not nearly close-packed, yet they are behaving in some ways as chains standing on end. Were they to approach the close-packed, fully extended limit, the estimate of stretching energy of a Gaussian chain would become inappropriate [27, 28].

The essential idea of the Alexander model, a global balance of interaction and stretching energies, can be applied to other situations involving tethered chains besides the good solvent case. In theta or poor solvents, the interaction term must be modified to account for poorer solvent quality. A simple limit is precisely at the theta point [29, 30] where binary interactions effectively vanish ($\chi = 1/2$ or $v = 0$). The leading term in F_{int} now accounts for three-body interactions:

$$F/kT = w\varphi^3 d^2 L/a^3 + L^2/Na^2 , \qquad (8)$$

where w is a dimensionless third virial coefficient. Minimizing F with respect to L gives:

$$L/a \approx N(a/d) . \qquad (9)$$

The linearity of L with N is maintained at the theta point. Relative to Eq. 5, the chains have shrunk by a factor of $(a/d)^{1/3}$ but the linear variation indicates that the chains are still distorted at the theta point and characteristic dimensions do not shrink through a series of decreasing power laws as do free chains [29–31]. Experimentally, Auroy [25] has produced evidence for this linearity even in poor solvents. Pincus [32] has recently applied this type of analysis to tethered polyelectrolyte chains, where the electrostatic interactions can produce even stronger stretching effects than those that have been discussed for good solvents. Tethered polyelectrolytes have also been studied by others [33–35].

2.2 Curved Interfaces, Spherical and Cylindrical Symmetry

The Alexander model is directly applicable only to flat grafted layers because of the assumed step-like concentration profile. In spherical and cylindrical geometries, the volume accessible to each chain grows with distance from the grafting site. Consequently, even a set of uniformly stretched chains, with all their chain ends on a shell at fixed distance from the grafting site (analogous to assumption (2) under Eq. 1), will result in monotonically decreasing concentration profiles. This differs from assumption (1) under Eq. 1. The Alexander model does provide guidance, however, in the construction of an analysis of the global, scaling features expected for curved tethered layers. In a flat layer, according to the Alexander model, each tethered chain is, in effect, confined to a cylindrical region of radius d and height L. The size of the cylinder sets the size of the blobs, so that, each chain appears equivalent to a linear stack of blobs of constant size, $\xi \approx d$. Thus, the layer consists of a stratified, multilayer array of blobs such that each chain contributes a single blob on average to each of the monolayers. This last view is directly translatable to curved tethered layers, as shown in Fig. 5.

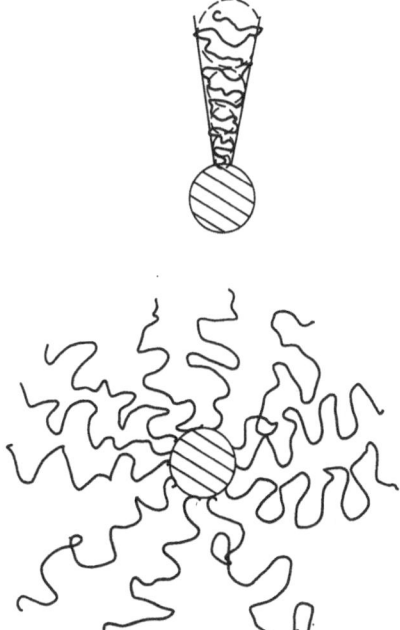

Fig. 5. Schematic for spherical tethered layers. The chains are confined to a conical region which expands with distance from the center of the structure

Daoud and Cotton [10] pioneered this geometrical analysis of tethered layers with spherical symmetry, which was later extended by Zhulina et al. [36] and Wang et al. [37] to cylindrical layers. The subsequent analysis is purely geometrical and requires no free energy minimization. The tethered layer consists of a stratified array of blobs such that all blobs *in a given sublayer* are of equal size, ξ, but blobs in different layers differ in size. This corresponds to the uniform stretching assumption of the Alexander model.

For a layer comprised of f grafted chains, the surface area, S, of the shell containing a given sublayer is given by $S \approx f\xi^2$. For a flat layer, all sublayers are of equal area and this translates to the Alexander result, $\xi \approx (S/f)^{1/2} \approx d$. However, for curved surfaces, S, and consequently ξ, depend on the distance, r, from the grafting site. Thus, a spherical layer is characterized by $S \sim r^2$, leading to $\xi \approx r/f^{1/2}$. For a cylindrical layer of length H, we have $S \approx rH$ and $\xi \approx (rH/f)^{1/2}$. Once $\xi(r)$

is known, the state of the system is completely specified within this model. Since $\xi \sim \varphi^{-3/4}$ in a good solvent, the segment concentration profile is given directly by:

$$\varphi(r) \approx (a/\xi)^{4/3} . \tag{10}$$

The average thickness of the spherical grafted layer is determined by the requirement that the integral over the segment concentration profile must account for all the monomers in the layer:

$$N \approx \int_{R_{in}}^{R_{in}+L} (\xi/a)^{5/3} \, \xi^{-1} \, dr , \tag{11}$$

where N is the number of monomers in one grafted chain, dr/ξ counts the number of blobs and $(\xi/a)^{5/3}$ gives the number of monomers per blob. R_{in} is the radial position of the grafting surface. The free energy per chain is given by the kT per blob ansatz:

$$F/kT \approx \int_{R_{in}}^{R_{in}+L} \xi^{-1} dr . \tag{12}$$

The predicted variations of F, L, and $\varphi(r)$ with tethering density and chain length for each of the geometries considered are summarized in Table 1.

Table 1. The scaling behavior of grafted layers of various geometries in a good solvent

Geometry	ξ	φ	L/a	F_{corona}/kT
flat	d	$(a/d)^{4/3}$	$N(a/d)^{2/3}$	$N(a/d)^{5/3}$
cylindrical	$(rH/f)^{1/2}$	$(fa^2/rH)^{2/3}$	$(fa/H)^{1/4} N^{3/4}$	$(fa/H)^{5/8} N^{3/8}$
spherical	$r/f^{1/2}$	$f^{2/3}(a/r)^{4/3}$	$f^{1/5}N^{3/5}$	$f^{1/2} \ln (R_{in} + L)/R_{in}$

It is interesting, for comparison, to apply the Flory approximation, parallel to Eq. 2, to a curved geometry. As an example, consider a star polymer with f arms and radius R. The segment volume fraction within the star can be written, $\varphi \approx fNa^3/R^3$, so that F_{int}/kT per arm can be written as $\approx v(fN)^2 \, a^3/fR^3$, giving rise to a total free energy per arm:

$$F/kT \approx v(fN)^2 \, a^3/fR^3 + R^2/Na^2 . \tag{13}$$

Minimization with respect to R gives:

$$R/a \approx f^{1/5}N^{3/5} . \tag{14}$$

This is identical to the result cited in Table 1, however the free energy of Eq. 13 is substantially higher than the more accurate Eq. 12 which includes correlations.

The results on characteristic dimensions display some interesting features, which have been subjected to very limited experimental investigation. Equation 14 says that the average size of an arm in a star polymer (or a chain densely grafted to a small spherical surface) varies with molecular weight in the same way as for a free linear chain in a good solvent. However, the average dimensions expand with tethering density embodied in the $f^{1/5}$ factor. The traditional parameter used to measure the dimensions of branched molecules relative to linear molecules is $G \equiv R_{g\,star}^2 / R_{g\,linear}^2$, where the ratio is taken at equal total molecular weight, that is, $N_{linear} = f N_{arm}$. Using Eq. 14 gives:

$$G \approx f^{-4/5} \tag{15}$$

for highly branched star polymers in good solvents. Zilliox [38] has studied a dozen pairs of linear and star polymers with 7 to 16 arms which obey the constancy of the product $Gf^{-4/5}$ predicted by Eq. 15 to within 10%. Until recently [1], the ability to test predictions such as Eqs. 14 and 15 has been limited by the difficulty of obtaining samples of both monodisperse arm-number distributions and high f [39].

Molecular dynamics simulations [40–42], as well as exact enumeration and Monte Carlo calculations on lattices [43, 44] have been used to study highly branched polymers. The work of Grest and coworkers [40–42] found excellent agreement with both the stretching of the average dimensions predicted by Eq. 14 and the density profile predicted by Eq. 10, over a range of f from 6 to 50. Furthermore, there is also good agreement with experimental data for R_g in good solvents of polystyrene and polyisoprene stars of 3 to 18 arms synthesized and studied by Fetters et al. [45]. There are complications (such as a highly dense interior region) [10] and modifications (to include concentration and temperature effects) [46] which are important to a more complete analysis of tethered chains in star polymers.

Synthesis of highly branched polymers via polymerization of macromonomers, in an effort to produce molecules with axisymmetric branching around a central backbone contour, has also been achieved [2, 3] (see Fig. 1B). Such molecules would correspond locally, along the central backbone, to the cylindrical symmetry of Table 1. Excluded volume interactions among the tethered side chains might be expected to effectively stiffen the central contour [47]. Though "bottlebrush" polymers such as this have been synthesized [2], we have seen no case to date where the backbone is substantially longer than the side chains. Tsukahara et al. [3] have determined G for a series of fourteen such polymers and found that the values were not very different from those of stars of comparable molecular weight. In the case where the backbone is comparable to the side chain in length, interactions of the side chains may still effectively stiffen the backbone, but the effect may be difficult to detect in the average dimensions. Furthermore, in contrast to stars where there is no reason to break the spherical symmetry, if a "bottlebrush" polymer is able to curve its backbone contour, the side chains could alleviate crowding by favoring the convex side of the contour, thus breaking the axial symmetry. One may speculate that if the tethering density along the backbone is not exceedingly high, the backbone may tend to curve to alleviate side chain interaction. For short backbones, this would produce a ring-like backbone contour that would give the polymacromonomer overall dimensions similar to a star; for longer backbones, a helical conformation of the backbone could result. Temperature and polymer concentration effects have been studied theoretically for this type of highly branched polymer, too [48]. This field is ripe for incisive experimental work. Further discussion of the effects of nonplanar geometries is given in the sections on aggregation and interaction.

2.3 Block Copolymer Microphases

The Alexander approach can also be applied to discover useful information in melts, such as the block copolymer microphases of Fig. 1D. In this situation the density of chains tethered to the interface is not arbitrary but is dictated by the equilibrium condition of the self-assembly process. In a melt, the chains must fill space at constant density within a single microphase and, in the case of block copolymers, minimize contacts between unlike monomers. A sharp interface results in this limit. The interaction energy per chain can then be related to the energy of this interface and written rather simply as $F_{int} = \gamma kT(N/L\varrho)$, where γkT is the interfacial energy per unit area, ϱ is the number density of chain segments and the term in parentheses is the reciprocal of the number of chains per unit area [49, 50]. The total energy per chain is then:

$$F/kT = \gamma(N/L\varrho) + L^2/Na^2 \tag{16}$$

which gives on minimization with respect to L:

$$L/a \approx N^{2/3}(\gamma/\varrho a)^{1/3} . \tag{17}$$

Here, the chains are expected to be stretched, as indicated by the 2/3-power dependence of L on N, but less strongly than in solvent. The experimental evidence available to examine this argument is discussed in the section on block copolymer melts.

2.4 Self-Consistent Field Calculations

The Alexander model and its descendants impose strong restrictions on the allowed chain configurations within the tethered assembly. The equilibrium state thus found is subject to constraints and may not attain the true minimum free energy of the constraint-free system. In particular, the Alexander model constrains the segment density to be uniform and all the chain ends to be at the same distance from the grafting surface. Related treatments of curved systems retain only the second constraint. A constraint-free, *ab initio* calculation of features such as the density profile and the distribution of chain end positions is clearly desirable.

This is possible within the framework of the self-consistent field (SCF) approach to polymer configurations, described more completely elsewhere [18, 19, 51, 52]. Implementation of this method in its full form invariably requires numerical computations which are done in one of two equivalent ways: (1) as solutions to diffusion- or Schrödinger-type equations for the polymer configuration subject to the SCF (in which solutions to the continuous-space formulation of the equations are obtained by discretization) or (2) as solutions to matrix equations resulting from a discrete-space formulation of the problem on a lattice.

The numerical implementation of this approach for flat grafted layers was done first by Hirz [53] and subsequently by Cosgrove et al. [54], and by Skvortsov et al. [55]. The concentration profiles found rose sharply to a maximum at short distances from the surface and thereafter were monotonically decreasing with increasing distance (z) from the grafting surface. At moderate grafting densities, the profiles appeared to fall from the maximum with a roughly parabolic (concentration falling like z^2) shape, except at the periphery where an exponential tail was evident. This parabolic form was explicitly recognized and explained by analytical SCF theories developed independently by Milner et al. [56] and by Zhulina et al. [57].

The key to these analytical SCF theories is founded on a key observation of Semenov [58]. The interactions among the chains produce stretching which in turn strongly reduces the number of configurations sampled by the polymer chain. The most probable configurations may then be viewed as fluctuations around a dominant, stretched trajectory. If the configurations of a weakly deformed chain are analogous to the possible trajectories of a quantum-mechanical particle (which can only be specified in a probabilistic sense), then the dominant configuration of a strongly stretched chain is reminiscent of the trajectory of a classical particle. This analogy was exploited by Milner et al. [56] to obtain the density profile in this limit directly analytically, avoiding numerical solution of the full SCF equations. The various tethered chain configurations correspond to the trajectories of classical particles, initially at rest, falling into a potential well. The initial rest positions correspond to the free ends, while the bottom of the well corresponds to the wall. In the SCF formulation, the segment index number plays the role of "time" in the equations of motion, in the sense that the path laid out by the chain could be constructed by laying down one segment per unit time. Accordingly, the trajectories corresponding to monodisperse chains in the brush must all reach the wall at equal "times", that is, after an equal number of segments, irrespective of their starting position. It is in a harmonic potential where a particle falls to the bottom in equal time from any starting position. Thus, if the assumed potential is directly proportional to the local concentration, the harmonic potential gives a parabolic profile of segment concentration. However, this parabolic profile is not universal. As the grafting density grows, the concentration profile approaches the step function assumed in the Alexander model.

Inherent in the analytical SCF theory are: (1) that the free ends of the chains can sample any position within the brush rather than be constrained to reside at the peripheral surface of the layer and, (2) that there is strong stretching, that is little fluctuation, around the dominant trajectory, at every point along the chain contour. As opposed to the Alexander model, this theory does not apply to weakly stretched chains. Given this, the analytical SCF theory includes the same physics as the Alexander model, the balancing of osmotic interactions with elastic resistance to stretching, **but** the key point is that it does so locally at every point along the chain rather than globally over the entire chain. *All of the scaling properties — for example, for average layer thickness — coming from the analytical SCF theory are identical with those from the Alexander model,* which is very useful support for the Alexander model as far as it goes; SCF theory is necessary for local properties.

3 Aggregation of Block Copolymer Surfactants

3.1 Tethering Processes

Tethering may be a reversible or an irreversible process. Irreversible grafting is typically accomplished by chemical bonding. The number of grafted chains is controlled by the number of grafting sites and their functionality, and then ultimately by the extent of the chemical reaction. The reaction kinetics may reflect the potential barrier confronting reactive chains which try to penetrate the tethered layer. Reversible grafting is accomplished via the self-assembly of polymeric surfactants and end-functionalized polymers [59]. In this case, the surface density and all other characteristic dimensions of the structure are controlled by thermodynamic equilibrium, albeit with possible kinetic effects. In this instance, the equilibrium condition involves the penalties due to the deformation of tethered chains.

The self-assembly of polymeric surfactants occurs in the presence of selective media, such as selective solvents and selective surfaces, giving rise to selective solvation and selective adsorption, respectively. The detailed structure of the aggregates formed depends on the selectivities of these media and on the nature of the polymeric surfactant involved: the size, rigidity and number of blocks, interactions between the blocks and the molecular architecture of the polymeric surfactant. The possible aggregates include adsorbed layers, micelles of various geometries and microemulsions (see Fig. 1 for some illustrations). For brevity, we consider just the aggregation of flexible AB diblock copolymers. An ideal selective solvent is a precipitant for the B blocks and a good solvent for the A blocks. A selective surface or interface is one which preferentially adsorbs one of the blocks. Selective interfaces can be classified as penetrable or impenetrable. Diblock copolymers straddling a liquid-liquid interface consitute an example of aggregation at a penetrable interface. In such a case, two back-to-back tethered layers of similar structure are created. At impenetrable interfaces, such as solid-liquid, selective adsorption creates an inner "anchor" layer of B blocks which are attracted to the wall and an outer "buoy" layer of A blocks which is tethered to the surface via the anchors. In contrast to self-assembly at penetrable interfaces, an impenetrable interface breaks the symmetry of the two blocks and produces structurally dissimilar anchor and buoy layers.

To simplify our discussion, we will consider two specific cases: spherical micelles in a selective solvent and selective adsorption on to a solid surface from a selective solvent.

3.2 Micelles

AB diblock copolymers in a selective solvent aggregate into spherical micelles consisting of two concentric regions: an inner core of the insoluble B blocks and an outer corona of solvent-swollen A blocks [60]. We assume that the blocks are extremely incompatible and the solvent is highly selective. These two assumptions lead to a sharp core-corona interface. The junctions between the two blocks rest at the interface, effectively tethering both blocks. We assume that the B blocks in the core are in a melt state, above T_g. Therefore, the volume of the core in a micelle comprising f copolymers is $fN_B a^3$ and the radius of the core is $R_{core} \approx f^{1/3} N_B^{1/3} a$. We focus on the assembly of a single aggregate, which *inter alia* ignores polydispersity in the size distribution of aggregates and the entropic effects of mixing an ensemble of aggregates into a system. These assumptions require deeper analysis for full justification, but can be partially justified *a priori* on the basis of experience showing that micelle size distributions are typically sharp [61]. This simpler analysis should be reasonable at concentrations of block copolymer in solution well above the critical micelle concentration.

Our aims in analyzing a reversible aggregation process involving tethering, such as micellization, differ from those in the analysis of the structure of a tethered layer of fixed density. We are after the self-adjusting, mean aggregation number, or tethering density, in the aggregate. In turn, this determines all the properties of the aggregate. The free energy per aggregated copolymer in an assembly such as this consists of three terms: an interfacial free energy, $F_{interface}$, associated with the core-corona interface, and two penalty terms due to the stretching of the grafted coronal and core blocks F_{core} and F_{corona}. $F_{interface}$ is proportional to R_{core}^2/f, the surface area per chain. Since it is a decreasing function of f, it favors micellar growth, whereas F_{core} and F_{corona} tend to arrest it. The energy of the corona, for a certain aggregation number, can be written from the previous section summarized in Table 1, $F_{corona}/kT \approx f^{1/2} \ln (R/R_{core})$, which recognizes that the corona chains are tethered into a spherical layer. This term reduces to $N(a/d)^{2/3}$ for flat surfaces and to $f^{1/2}$ for the present case of highly curved surfaces. F_{core} reflects the deformation of the core blocks which is necessary to maintain the constant melt density of the core, even if R_{core} increases beyond $R_0 \approx N_B^{1/2} a$. Since the volume of the interior of the core is rather small, it is sufficient to deform relatively few chains to fill it to constant density. This has been substantiated by recent Monte Carlo simulations of micellization [62]. Nevertheless, the uniform stretching assumption as in Eq. 2 yields the correct scaling form for the stretching penalty in the core, $F_{core}/kT \approx (R/R_0)^2$ [58].

One may now distinguish between two simple limiting cases illustrated in Fig. 6. When $N_A \gg N_B$, the micelle develops an extended corona reminiscent of star

Fig. 6. Possible structures of block copolymer micelles. Top: "Hairy" micelle; Bottom: "Crew-cut" micelle

polymers. The dominant penalty term in such star-like micelles is F_{corona}. Thus, balancing the corona and interface contributions gives:

$$F/kT \approx (\gamma a^2/kT) f^{-1/3} N_B^{2/3} + f^{1/2} , \tag{18}$$

where γ is the surface energy density of the core-corona interface. Minimization with respect to aggregation number f leads to:

$$f \approx (\gamma a^2/kT)^{6/5} N_B^{4/5} \tag{19}$$

and:

$$R_{core} \approx f^{1/3} N_B^{1/3} a \sim N_B^{3/5} a . \tag{20}$$

The overall radius of the micelle, R, is dominated by the corona whose thickness is, from Table 1, $L_{corona} \approx f^{1/5} N_A^{3/5} a$ and leads to:

$$R \approx N_B^{4/25} N_A^{3/5} a \quad \text{(star-like, or "hairy", micelles)} \tag{21}$$

In the opposite limit, $N_A \ll N_B$, the corona is thin and similar in structure to a flat grafted layer. The dominant penalty term in such "crew-cut" micelles is F_{core}. Accordingly,

$$F/kT \approx (\gamma a^2/kT) f^{-1/3} N_B^{2/3} + f^{2/3} N_B^{-1/3} \qquad (22)$$

which gives on minimization with respect to f:

$$f \approx (\gamma a^2/kT) N_B \qquad (23)$$

and

$$R_{core} \approx N_B^{2/3} a . \qquad (24)$$

Since in this case the coronal thickness is small compared to the core, the overall radius is dominated by the core and:

$$R \approx N_B^{2/3} a \qquad \text{("crew-cut" micelles)} \qquad (25)$$

Numerical SCF calculations give theoretical predictions for the chain length dependences of the characteristic dimensions in reasonable agreement with these economical scaling arguments [63].

Several physical ideas emerge from these results. Owing to the spherical geometry, in hairy micelles the coronal block molecular weight is anticipated to exert negligible influence on the aggregation number. Thus, a series of copolymers of constant core block molecular weight and increasing corona block size are predicted to have the same aggregation number. Yet, because of the unswollen core block, the physical dimension of the micelle is determined much more strongly by the corona block. The insensitivity of f to corona molecular weight exists also for crew-cut micelles, but for a different reason. The coronal stretching penalty in this case is not just insensitive to molecular weight but rather is essentially negligible. Stretching of the core block is anticipated in both limits as indicated by the predicted variation of core size with N_B despite the fact that the core blocks are in a solvent-free state. Finally, corona blocks are clearly expected to be stretched in hairy micelles, as L_{corona} can be increased by increasing N_B, thereby increasing f.

Block copolymer micelles have received a great deal of experimental attention, beginning with the pioneering studies of Krause [64] in 1964. The subject has been reviewed recently by Tuzar and Kratochvíl [65]. However, the pattern of scaling behaviors embodied in Eqs. 18 to 25 has not been established. Several factors have contributed to this. The first is the most obvious, which is that to establish firmly such behavior requires systematic preparation of a large number of well-characterized samples. Second, perfect incompatibility, and thus a sharp core-corona interface, may not obtain. Third, several commonly encountered systems (polystyrene, polymethylmethacrylate, poly-2-vinylpyridine) become glassy at high concentrations near room temperature which, as micelle cores, may inhibit equilibration. Fourth, free chains in solution can influence the aggregation number if they are present in sufficient concentration [66]. Fifth, experimental determination of both core and corona dimensions is difficult. A typical method [65] is to measure the overall radius, R, by some scattering or hydrodynamic mobility technique and to measure the core dimensions by scattering in a medium where the corona block is contrast-matched. This limits solvent selection and leads to substantial uncertainty in determination of corona dimensions, both from small contrast-mismatch and from the fact that subtraction of two experimental quantities is necessitated.

Nevertheless, measurements of coronal dimensions indicate significant and substantial stretching away from the interface, based on comparison with the dimension of free chains of equivalent chain length. This has been observed even for macromolecules with $N_A < N_B$ [67], (though the selectivity of the solvent will shrink the core to melt dimensions and expand the corona so that the hairy micelle limit may obtain), as well as for the expected situation where $N_A \geq N_B$ [68, 69]. Recent data of Winnik et al. on polystyrene-polyethyleneoxide diblock copolymers of three different molecular weights show reasonable agreement with Eq. 21 [70]. Similarly, in general, the core radius is considerably larger than the unperturbed melt dimensions of the core block, suggesting significant stretching in the radial direction [67, 69, 71–74].

3.3 Adsorbed Layers

AB diblock copolymers in the presence of a selective surface can form an adsorbed layer, which is a planar form of aggregation or self-assembly. This is very useful in the manipulation of the surface properties of solid surfaces, especially those that are employed in liquid media. Several situations have been studied both theoretically and experimentally, among them the case of a selective surface but a nonselective solvent [75] which results in swelling of both the anchor and the buoy layers. However, we concentrate on the situation most closely related to the micelle conditions just discussed, namely, adsorption from a selective solvent. Our theoretical discussion is adapted and abbreviated from that of Marques et al. [76], who considered many features not discussed here. They began their analysis from the grand canonical free energy of a block copolymer layer in equilibrium with a reservoir containing soluble block copolymer at chemical potential μ_{ex}. They also considered the possible effects of micellization in solution on the adsorption process [61]. We assume in this presentation that the anchor layer is in a solvent-free, melt state above T_g. The anchor layer is assumed to be thin and smooth, with a sharp interface between it and the solvent swollen buoy layer.

In the limit where μ_{ex} is small, the analysis of Marques et al. shows that the assembly of the layer from solution is governed by a balance of two energies. In this situation, the relevant energies are the attractive energy for the adsorbing block (favorable for adsorption) and the stretching energy of the brush (unfavorable for adsorption). The expression for the energy of the anchor layer takes the form of a van der Waals energy, varying like the inverse square of the thickness of the anchor layer, Δ, and containing an effective Hamaker constant, A, that is a composite parameter of the relevant material Hamaker constants. Balancing this against the stretching energy of a layer (Eq. 7, for the stretching energy per chain, multiplied by $1/d^2$, which gives the number of chains per unit area of the layer) gives, for the assembly of unit area of the layer:

$$F/kT = A/(12\,kT\pi\Delta^2) + N_A a^{-2}(a/d)^{11/3}\,. \tag{26}$$

The anchor layer is dense, with volume fraction, $\varphi = N_B a^3/d^2\Delta$, near unity, so $\Delta \approx N_B a^3/d^2$. Inserting this into Eq. 26, and minimizing with respect to d, gives the equilibrium surface density.

It is useful for the presentation of this result to define a new variable. One expects the equilibrium surface adsorption number density, or average spacing between tethered A chains (d), to be a function of the molecular weights of both blocks of the diblock copolymers. The anchoring blocks compete for space on the surface while the buoy blocks crowd one another immediately above the surface; thus, increasing either should increase d. It is possible to combine the size effects of both of the two blocks into a single parameter expressing the relative size of the two blocks or the asymmetry of the block copolymer. Several choices are conceivable *a priori*, such as N_B/N_A. Parsonage et al. [77] found empirically that a good asymmetry parameter to correlate data on adsorption of polymers with varying block molecular weights is $\beta \equiv N_A^{6/5}/N_B^{2/3}$. This is the ratio of the solvent-swollen radius squared (or projected area) of the buoy block, divided by the solvent-collapsed radius squared (or projected area) of the anchor block. If one casts the minimization of Eq. 26 in terms of β, the result of the theory of Marques et al. [76] for the equilibrium spacing between chains is:

$$d \approx aN_A^{3/5}(A/kT)^{-3/23}\,(N_A^{6/5}/N_B^{2/3})^{-9/23} \tag{27}$$

or

$$d/R_{g,A} \approx (A/kT)^{-3/23}\,\beta^{-9/23}\,. \tag{28}$$

Thus, the spacing of the chains relative to the neutral, free, swollen size of the buoy blocks is, for a given chemical system and temperature, a unique function of the solvent-enhanced size asymmetry of the diblock polymer and a weak function of the effective Hamaker constant for adsorption. The degree of crowding of the nonadsorbing blocks, measured by a decrease in the left-hand side of Eq. 28, increases with increasing asymmetry of the block copolymer.

Figure 7 shows the results of measurements of adsorption density by Parsonage, et al. [77] on a series of eighteen block copolymers, with poly(2-vinylpyridine) [PVP] anchors and polystyrene [PS] buoys, adsorbed from toluene (selective for PS) of variable molecular weight in each block. The results are presented as the reciprocal square of Eq. 28, that is, as a dimensionless number density of chains $\sigma^* \sim (d/R_{g,A})^{-2}$. For all but the copolymers of highest asymmetry, Eq. 28 is in good agreement with the data of Fig. 7. The high asymmetry copolymers are in the regime of the data of curves (a) and (c) of Fig. 3 where the large relative size

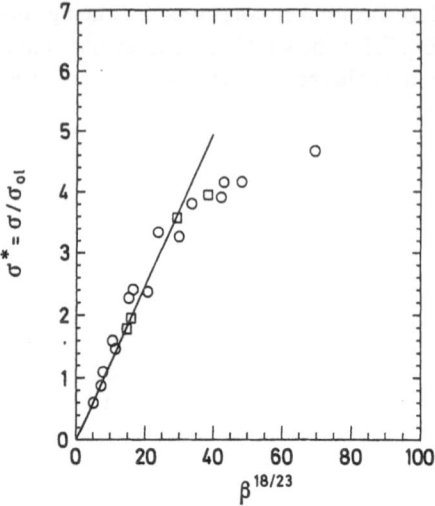

Fig. 7. Data of Parsonage et al. [77] on the adsorption of a series of block copolymers of polystyrene-poly-2-vinylpyridine. The ordinate is the measured surface density ($\sim d^{-2}$) reduced by the density required for the non-adsorbing chains to overlap; the abscissa is the solvent-enhanced size asymmetry of the block copolymer defined under Eq. 26. The form of this plot is that suggested by Eq. 28

of the nonadsorbing block is playing a significant role in reducing the surface density.

Clearly Fig. 7 *must* actually have a maximum at high asymmetry since this corresponds to negligible anchor block size and therefore to no adsorption ($\sigma^* = 0$). The lattice theory of Evers et al. predicts this quantitatively [78] and is, on preliminary examination, also able to explain some aspects of these data. From these data, the deviation from power law behavior occurs at a number density of chains where the number of segments in the PVP blocks are insufficient to cover the surface completely, making the idea of a continuous wetting anchor layer untenable. Discontinuous adsorbed layers and surface micelles have been studied theoretically but to date have not been directly observed experimentally [79].

Other interesting types of aggregates can be formed in polymer-solvent media including: micelles with rigid [80] or crystallizable [81] core blocks, cylindrical micelles, micelles with charged coronas, intramolecular micelles in multiblock copolymers [82], microemulsions (where the block copolymer is mixed with two immiscible solvents each of which is selective for one of the blocks) [83], and adsorbed layers where the solvent selectivity is different from the case discussed here. The point we hope to have made in this section is that the principles of tethered chains — namely aggregation energy balanced by stretching energy — provide a simple conceptual ansatz to explain or predict many interesting possibilities in the aggregation of polymeric surfactants.

4 Phase Behavior

One of the hallmarks of the special behavior of polymers is their tendency toward phase separation. Two phenomena come immediately to mind: the demixing of incompatible polymers and the phase separation *cum* collapse of polymers in a poor solvent. In this section we discuss how this behavior is modified for tethered chains, bringing to light certain features that have no counterpart for free polymers.

4.1 Transitions in Polymer Configuration

Like free chains, tethered chains shrink when immersed in a poor solvent. Comparison of Eqs. 5 and 9 shows the shrinkage expected on moving from a good to a theta solvent. Halperin has predicted [29], and Auroy et al. [25] have found experimentally, that the linearity of L with N is maintained even in poorer than theta solvents. This indicates that *dense tethering prevents the collapse* associated with the demixing phase transition and precipitation which occurs for free polymers in poor solvents. Zhulina and coworkers have recently studied these collapse phenomena in great detail via SCF calculations [84]. Some questions of current interest are whether the collapse of weakly overlapping tethered chains involves a phase transition and the nature of the collapse transition in layers of grafted polyelectrolytes [32], where the segmental excluded volume interactions are attractive but there are electrostatic repulsions among the segments.

4.2 Lateral Segregation

One can also envision a phase separation process between incompatible chains tethered into the same layer, such as mixed micelles or adsorbed layers, as illustrated in Fig. 1 J, K and L. Different scenarios result depending on the ability of the chains to exchange with the surrounding bulk and on whether the chains have lateral freedom to rearrange within the tethered layer. We focus our discussion on the situation where lateral equilibration within the layer is possible but exchange with the surroundings is negligible [85]. A concrete example might be a mixed monolayer at an immiscible liquid-liquid interface, constructed in the following way. The layer is comprised of two types of diblock copolymers, AC and BC, chosen so that the C blocks are miscible only in one liquid phase, while both the A and B blocks are miscible only in the second liquid phase. (This structure is similar to that of Fig. 1F but with two components.) The resulting monolayer is, in effect, a superposition of two tethered layers, a single component one and a mixed one. The single component layer plays no active role; its purpose is to provide an anchoring mechanism which allows lateral mobility but inhibits exchange with the bulk.

The simplest situation is the symmetrical one ($N_A = N_B$), with the solvent equally good for both blocks. We imagine that the excluded volume interactions of A and B are stronger than the A–B repulsive interactions so that the overall structure of the layer is like that of a single component; in other words, both components are equally stretched. The issue is whether or not they are homogeneously mixed with one another in the monolayer. This is essentially a two-dimensional random mixing process. In that spirit, we write the free energy per chain as:

$$F/kT \approx F_0/kT + \chi'_{AB}x(1 - x) + x \ln x + (1 - x) \ln (1 - x), \quad (29)$$

where x is the mole fraction of A chains in the layer and χ'_{AB} is the effective interaction parameter between tethered A and B chains. Note that this is different from the standard A–B segment-segment interaction parameter χ_{AB}, since here we are treating the entire tethered chains as the elementary components in the mixing process. F_0 is the free energy of a chain in a single component layer at the same grafting density (e.g. Eq. 7). The last two terms are the mixing entropy of a two-dimensional, ideal "fluid" of the tethered chains. (This is zero if the chains are irreversibly grafted in fixed positions.) Since we are interested in the demixing transition, variation in F_0, which is assumed to be independent of x, is immaterial. The distinctive features of tethered chains manifest themselves in the scaling behavior of the effective interaction parameter χ'_{AB} with respect to N and d.

In the Flory-Huggins picture, the inter*chain* interaction parameter should be related to the inter*segment* interaction parameter by

$$\chi'_{AB} = N\varphi\chi_{AB} \qquad (30)$$

so that

$$\chi'_{AB} = N(Na^3/d^2L) \chi_{AB} \approx N(a/d)^{4/3} \chi_{AB} \qquad (31)$$

using Eq. 5 for L. The Flory-Huggins approach allows all monomers to interact. However, some of the A–B contacts are shielded by excluded volume interactions. Since a semidilute solution is envisioned as a collection of space-filling blobs, χ'_{AB} is more accurately given by $\chi'_{AB} \approx (N/g) \chi^{blob}_{AB}$. A naive approximation is to use χ_{AB} for χ^{blob}_{AB}, which reduces the effective interaction parameter of Eq. 31 by an additional factor of $\varphi^{1/4}$ to account for the excluded volume correlations, leading to:

$$\chi'_{AB} \approx N(a/d)^{5/3} \chi_{AB}. \qquad (32)$$

A more accurate analysis of this problem incorporating renormalization results, is possible [86], but the essential result is the same, namely that stretched, tethered chains interact less strongly with one another than the same chains in bulk. The appropriate comparison is with a bulk-like system of chains in a brush confined by an impenetrable wall a distance R_F (the Flory radius of gyration) from the tethering surface. These confined chains, which are incapable of stretching, assume configurations similar to those of free chains. However, the volume fraction here is $\varphi = N(a/d)^2 R_F \approx N^{2/5}(a/d)^{5/3}$, as opposed to $\varphi = N(a/d)^2 L \approx (a/d)^{4/3}$ in the unconfined, tethered layer. Consequently, the chain-chain interaction parameter becomes $\chi'_{AB} \approx N^{3/2}(a/d)^{5/2} \chi_{AB}$. Thus, tethered chains tend to mix, or at least resist phase separation, more readily than their bulk counterparts because chain stretching lowers the effective concentration within the layer. The effective interaction parameters can be used in further analysis of phase separation processes within mixed monolayers or micelles [87]. An interesting feature, arising from the two-dimensional character of these mixtures, is that "line tension" between the demixing phases plays a role in these processes. Another intriguing possibility would be to use ABC triblock polymers of the type illustrated in Fig. 1K, where the A and B chains are covalently linked, so that they are somewhat analogous to 2D block copolymers which could not phase separate, but could conceivably undergo a microphase separation or disorder-order transition in the plane. Experimental work on these points thus far is essentially nonexistent.

4.3 Other Phase Transitions of Tethered Layers

Other phase transitions of tethered chains involve surface interactions. One interesting scenario is that of tethered chains incorporating weakly adsorbing monomers, i.e., weakly adsorbing homopolymers carrying strongly adsorbing terminal monomers. The adsorbed layer is expected to undergo a first-order phase transition in the vicinity of the overlap threshold [7]. For low coverages, the chains are essentially uniformly adsorbed. Once the chains begin to overlap, one expects a coexistence of weakly deformed, uniformly adsorbed chains and strongly stretched tethered chains. Some experimental evidence for this has been demonstrated [88].

Another scenario involves tethered chains in lamellae formed by rod-coil diblock copolymers, that is, diblock copolymers incorporating a rigid block. Such lamellae comprise an inner core of aligned rod blocks and a corona of flexible tethered chains swollen by solvent. These lamellae are expected to undergo a first-order phase transition between tilted and untilted phases [80, 89–91]. The tilt causes an increase in the exposed surface, and therefore the surface energy, of the rods. At the same time, tilting increases the area per tethered flexible chain, thereby relieving the stretching energy. The phase transitions result from the interplay of these effects.

5 Block Copolymer Melts

5.1 Bulk Block Copolymers

The distinguishing feature of block copolymer melts is their manifestation of disorder-order transitions. These involve self-assembly into a variety of microstructures such that the inter-block junctions are constrained to the interfacial region between microdomains. While the tethering of individual blocks produces stretching that is weaker than in the solvent case ($L \sim N^{0.5-0.67}$ vs. $L \sim N$) [92], block copolymer melts are nevertheless pertinent to this review. Block copolymers are arguably the most technologically important polymer system illustrated in Fig. 1 [93]. Chain stretching appears to be connected to the phase diagram near the order-disorder transition (ODT) [94]. The degree of stretching varies with distance from the transition. Furthermore, chain stretching reduces the penetrability of ordered microdomains to homopolymers or other block copolymers, thereby affecting the mixing of block copolymers with other polymers.

The enthalpic interaction between blocks in a copolymer melt is measured by the segment interaction parameter χ, while the entropy of mixing varies as N^{-1}; thus, the thermodynamic state of the melt is governed by χN, with the ODT predicted to occur for χN of order 10 [95, 96]. The vicinity of the transition is traditionally classified as the "weak-segregation limit", where, in principle, the chains are not appreciably distorted ($L \sim N^{0.5}$). For $\chi N \gg 10$, the "strong-segregation limit" (SSL) is obtained. The SSL is where Eq. 17 is valid, and is the region where the strongest tethering to the interface occurs and, therefore, where the consequences of tethering, such as chain stretching, are manifest. Four characteristic morphologies have been identified in the SSL, as illustrated in Fig. 8: a cubic array of spheres, a hexagonal array of cylinders, the ordered bicontinuous double diamond, and lamellae; other morphologies are certainly possible [50, and references therein]. The disposition of each block within its microdomain, the dimensions of the microdomains and the morphologies themselves can all be understood via a free energy balance (the simplest form of which is expressed in Eq. 1) in which the system strives to minimize the interfacial area while maintaining the appropriate bulk density in each microdomain. The stretching of the individual blocks away from the interface, plus the relatively unimportant confinement of the inter-block junction to the interfacial layer, provide the entropic contributions which balance the free energy associated with domain boundaries. As with chains in micelles or anchored to surfaces, block copolymer melts can order into, and thereby tether chains to, planar or curved surfaces; the distinctive feature of block copolymer melts is the constraint of maintaining the bulk density.

The first thorough treatment of the SSL, by Helfand and Wasserman [97, 98], employed a numerical SCF scheme to minimize the free energy, in which it was necessary to assume a priori the existence of a narrow interface, as well as the morphology itself. The interfacial layer thickness, t, was predicted to be $a/\chi^{1/2}$, in good agreement with several experiments, where typical valaues for t center around 20 Å [99–101]. The characteristic domain dimension, L, (i.e., lamellar period, cylinder radius and spacing, etc.) was predicted to scale asymptotically ($\chi N \to \infty$) as $aN^{0.643}\chi^{0.143}$, with a prefactor depending on the geometry. More recently, the

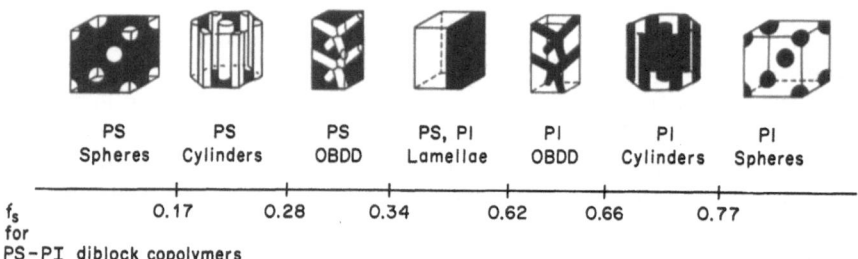

| PS Spheres | PS Cylinders | PS OBDD | PS, PI Lamellae | PI OBDD | PI Cylinders | PI Spheres |

f_s for PS-PI diblock copolymers

0.17   0.28   0.34   0.62   0.66   0.77

Fig. 8. Effect of varying composition on the ordered-phase symmetry in polystyrene-polyisoprene (PS-PI) diblock copolymers [92]

work of Semenov [58] and Ohta and Kawasaki [49] provide an analytical SCF approach which yields the asymptotic scaling $L \sim aN^{2/3}\chi^{1/6}$, in close agreement with Helfand and Wasserman, as well as with Eq. 17, if it is recognized that the interfacial tension $\gamma \sim t^{-1} \sim \chi^{1/2}$ [97, 98].

Experimentally, the stretching of block copolymer chains has been addressed in two ways: by measuring L as a function of N, and by measuring the components of R_g of the block chains both parallel and perpendicular to the interface. The domain dimensions have been studied most extensively for styrene-isoprene and styrene-butadiene block copolymers; X-ray and neutron scattering are the methods of choice. The predicted SSL scaling of $L \sim N^{2/3}$ has been reported for spheres, cylinders and lamellae [99, 102–106], but not in all cases. For example, Bates et al. found $N^{0.37}$ for styrene-butadiene spheres [100], and Hadziioannou and Skoulios observed $N^{0.79}$ for styrene-isoprene lamellae [107]. In the sphere case, kinetic limitations to equilibration were felt to be an important factor [100].

These can arise in the following way. Polystyrene is a glass below 100 °C, while the polydienes are chemically unstable above 100 °C. Consequently, the bulk samples are prepared by casting from a neutral solvent. Even in this process, movement in the polystyrene domains may become too sluggish to allow much rearrangement during the later stages of solvent evaporation. For spherical morphologies, this is particularly problematic since, unlike cylinders and lamellae, the domain dimensions cannot change appreciably without individual chains either entering or leaving a given sphere. It would be useful to re-analyze this issue with nonglassy block copolymers in the SSL.

Recently, Bates and coworkers found $L \sim N^{0.8}$ for nonglassy ethylenepropylene-ethylethylene lamellae [94], reminiscent of the Hadziioannou and Skoulios result [107]; however, in this case, the system was near the ODT, and thus probably not in the SSL at all. The fact that the exponent is larger than either the SSL prediction or the unperturbed result is thought to reflect a broad crossover regime but, interestingly, it persisted into the disordered state. The individual chains thus deviate more from Gaussian chain-like behavior in the crossover at the ODT than in the SSL [94]. That chain stretching exists in the disordered state calls into question the existence of a weakly segregated ordered state. This surprising result merits further investigation.

The individual chain dimensions are accessible by neutron scattering, and have been determined by several groups [106, 108–112]. This requires interspersing deuterium-labelled blocks into an unlabelled sample of identical chain length. A severe experimental difficulty is the presence of strong domain scattering, the signal that is used to determine L. This may be minimized by contrast matching, in which the proportion of deuterium labelling in one domain is adjusted to make the average scattering length of that domain equal to that of the other. It is a delicate balance to achieve such contrast matching to a degree adequate to measure the single chain scattering accurately. Nevertheless, in several cases it has been possible to determine the components of R_g parallel and perpendicular to the interface, in lamellar systems. The results indicate a compression to approximately 70% of the unperturbed dimension in the parallel direction, and stretching by a factor of about 1.6 normal to the interface [107, 110–112]. In a recent, related study, Matsushita et al. have examined the chain stretching as a function of position in the block, by labelling at different positions along the chain [113]. A subsection of the block at the free end was found to be unperturbed, whereas a subsection near the interface exhibited lateral contraction comparable to that of the block as a whole.

The chain stretching, predicted and observed, for blocks in copolymer melts has implications in the behavior of these microstructures that are analogous to solvent-swollen tethered chains. Stretched lamellae will resist mutual interpenetration with other stretched lamellae, thereby affecting the mechanical and dynamic properties of these materials. Stretched lamellae will also resist interpenetration with free melt chains, a feature of polymer brushes recognized early by de Gennes [8]. Leibler has analyzed this situation in more detail [114]. The relevant parameter, besides N and d of the brush chains, is P, the degree of polymerization of the melt chains. The simplest system to analyze is that where $P < N$, where one can show that the P chains penetrate into the N brush only if $d > aP^{1/4}$ [8]. Leibler has coined the terms "wet brush" for the penetration regime and "dry brush" for the impenetrable regime [114]. Data on the degree of miscibility of free chains with block copolymer lamellae are beginning to appear along with information on the configurations of the individual chains, which will be useful in determining where the free chains reside in the mixtures. Hashimoto et al. [115] and Winey [116] have shown that small free melt chains freely penetrate into block copolymer lamellae while larger homopolymers are excluded.

5.2 Block Copolymers at Melt-Melt Interfaces

Favorable enthalpic interactions between the tethered chains and the free chains can alter this picture in important ways, as has been explored by Brown and coworkers [117]. They have studied the profile of tethered segments, and thereby obtained information on chain configurations of block copolymer chains emanating from the interface between two immiscible homopolymer melts. They found, inter alia, that when the block was extending into a melt of chemically identical chains ($\chi = 0$) there was *less* chain stretching than when the block was extending into a melt of miscible chains ($\chi < 0$). An argument similar to that of Eqs. 1 and 2 can be used to understand this qualitatively, if the interaction term of Eq. 2 is replaced by a Flory-Huggins type of enthalpy, leading to:

$$F/kT \approx \chi N\varphi_P + L^2/Na^2 ,\tag{33}$$

where $\varphi_P \approx 1 - \varphi_N = 1 - Na^2/d^2L$ is the volume fraction of free chains that have intermingled with the brush. Minimization with respect to L then yields:

$$L \approx (-\chi)^{1/3} aN(a/d)^{2/3}\tag{34}$$

which is essentially Eq. 5 with a different interaction parameter. This equation underscores the point that negative enthalpy of mixing favors chain stretching.

This additional chain stretching leads to the practical conclusion that it may be desirable to join polymers with enthalpically favorable interfacial agents, rather than using block copolymers of the two homopolymers, at least in applications such as mechanical adhesion where the *depth* of interpenetration is of some significance [118]. Block copolymers straddling interfaces between immiscible homopolymers are of general interest as possible agents for emulsification. Several analyses have focused on how tethered chains affect the bending energy of such interfaces, a key determinant of microemulsification behavior [9, 47, 114, 119–122]. Another intriguing issue for future investigation concerns mixtures of block copolymers of different molecular weights. In a binary mixture, the two molecular weights would either be able to form a single lamellar size, with stronger stretching of the shorter chains, or separate into coexisting phases [123].

6 Interactions

6.1 Flat Layers

Comprehension of the interactions among microstructures composed of tethered chains is central to the understanding of many of their important properties. Their ability to impart stability against flocculation to suspensions of colloidal particles [52, 124, 125] or to induce repulsions that lead to colloidal crystallization [126] are examples of practical properties arising from interactions among tethered chains; many more are conceivable but not yet realized, such as effects on adhesion, entanglement or on the assembly of new block copolymer microstructures. We will be rather brief in our treatment of interactions between tethered chains since a comprehensive review has been published recently of direct force measurements on interacting layers of tethered chains [127].

The simplest situation to consider is that of the direct interaction of two flat, tethered layers in a good solvent, at a distance $D < L$ between the tethering surfaces. In fact, Eq. 2 can readily be adapted to this purpose; the basic ingredients of the balance between interaction and elastic energy remain. An additional ansatz, originally suggested by de Gennes [128], is needed: the recognition of the reluctance of stretched, tethered chains to interpenetrate one another. With this, one can convert Eq. 2 into a formula for two interacting layers through the realization that, at any interaction distance, the layer thickness L must be half the separation D. Substitution of D/2 for L in Eq. 2, followed by subtraction of the free energy of the two noninteracting layers (F, from Eq. 6 or 7) leads [128, 129] to a formula for the free energy of two interacting polymer brushes, compressed to D:

$$F_{comp} \approx (2F/3)\,(1/u + u^2/2)\,, \tag{35}$$

where $u = D/2L$. Formulae for the interactions between parabolic brushes have been calculated, as have corrections for polydispersity [130]. The parabolic brush results in a third term in Eq. 35, varying as u^5. These features, arising from a more complete local description of the brush, are important, in principle, for an accurate calculation of the weak compression regime. Inclusion of the u^5 term leads to a longer range onset and milder increase in forces with initial compression than predicted by Eq. 35. It becomes unimportant as the layers are compressed to the point that the stretching energy part of the formula is diminished, concomitantly reducing the importance of local variation in stretching along the chain, and the repulsion results mainly from the osmotic pressure of the segments in the gap (the first term in Eq. 35). Detailed comparisons of surface forces data with Eq. 35, and improvements which require independent estimation of parameters d, a and v, show that Eq. 35 underestimates the weak compression part of the data [13, 129, 131]. Quantitative comparisons of this sort depend strongly on accurate determinations of d, which have not always been available.

These comparisons give prima facie support to the supposition of noninterpenetration between the symmetrical interacting layers, which is discussed further in section 6.3. Molecular dynamics simulations give more explicit, quantitative support and further information [26]. Polydispersity corrections are surprisingly important to the accord with experiment, even for polymers of quite uniform molecular weight [130, 131]. The high compression data are in good agreement, as anticipated, since in this region any local differences in the profiles have been squeezed out.

6.2 Tailoring Interfaces

The importance of polydispersity is an interesting clue that it may be possible to tailor the weak interactions between polymer brushes by controlled polydispersity, that is, designed mixtures of molecular weight. A mixture of two chain lengths in a flat tethered layer can be analyzed via the Alexander model since the extra chain length in the longer chains, like free chains, will not penetrate the denser, shorter brush. This is one aspect of the vertical segregation phenomenon discussed in the next section.

In one of several conceivable combinations of molecular weight and surface density, the binary brush forms a mixed, shorter layer and the extra length in and number of the longer chains is sufficient that these extra segments crowd one another in the outer layer and form an outer, stretched tethered layer. The Alexander analysis applied to the two-layer situation gives for the outer layer thickness of the mixture [132]:

$$L_m/a = N_S(a/d)^{2/3} \left[1 + (\alpha - 1) \varrho^{2/3} \right], \tag{36}$$

where $\alpha = N_L/N_S$, the ratio of long to short chain lengths, and $\varrho = d_T/d_L$, the ratio of the chain spacing in the mixed, inner layer to that of the long chains alone. (This equation assumes that the number of segments of long and short chains in the inner brush is equal. This constraint can be relaxed). The effect of mixing this on the interaction profile can be calculated as in Eq. 35. The essential result is that, for example for a 50% mixture, the longest-range part of the force is close to that of a pure layer of long chains at d_T, because L is much more sensitive to N than to tethering density, while at higher compressions the force becomes controlled by the osmotic pressure, and therefore by the total number of segments, irrespective of the chain lengths.

Chakrabarti and Toral have studied mixed layers by Monte Carlo simulation, displaying features consistent with the view above [133]. Tirrell et al. have studied [132] experimentally the self-assembly of, and interactions between, mixed layers constructed from the block copolymers of Fig. 7, and found reasonable agreement between the range of the forces and Eq. 36. There are many more possibilities in this regard. Mixed layers provide the means to force some chain ends toward the periphery of a brush, as discussed briefly in the next section. This might be useful if these chain ends bear a chemical functionality designed to interact or react on direct contact with its surroundings. Another aspect of tailoring interactions comes in hetero- or asymmetric interactions. The first experiments in this vein indicate that interpenetration may be a more important phenomenon in interaction between layers of different density [134–136]. Very few studies have been made on interactions between tethered layers immersed in solvents of less than very good quality.

6.3 Vertical Segregation, Interpenetration and Fluctuations

The ansatz of impenetrability leading to Eq. 35 is directly related to some of the interesting characteristics of the tailored interfaces just discussed. Just as a dense monodisperse brush is nearly impenetrable in a symmetric interaction with another identical brush, so, too, is a dense inner brush impenetrable to the extra segments in the longer chains in a bimodal layer. The extra length of longer chains is segregated to the outer brush. Homogenous, vertical mixing of the "extra" length of longer chains into the inner, shorter layer would require the inner layer to accommodate more segments; being stretched already, the inner layer resists this penetration. This is one aspect of the phenomenon of vertical segregation. Similarly, even in a monodisperse brush, as grafting density increases, chain segments more distant along the chain contour from the tethered end will reside farther from the grafting surface [137].

This tendency to vertical segregation has further implications for the mixing of chains in multicomponent tethered layers. Tethering of chains promotes homogeneous lateral mixing of multicomponent layers. Mixtures of tethered chains of different length will tend to mix homogeneously laterally within a given layer since the longer chains discover the advantage of less crowding, and therefore less stretching energy, when they are surrounded by shorter chains. Witten and Milner [138] have studied the case of strongly incompatible, chemically different, irreversibly grafted chains. In this case, one anticipates vertical demixing based on chemical composition but with the stretching energy of the chains playing a role.

As described in Sect. 2.4, SCF calculations are useful in determining local details of density profiles. A more local examination of profiles is indeed necessary to study the question of interpenetration in more detail. The analytical SCF theory [56, 57] shares with the adapted Alexander model embodied in Eq. 35 the characteristic of impenetrability. The full numerical SCF theory is necessary to

62   A. Halperin, M. Tirrell, and T. P. Lodge

study the extent of possible interpenetration. The profiles calculated from the analytical SCF theory agree with those from the full SCF theory reasonably well, except for the exponential tail on the latter. Figure 9 shows a comparison of the two calculations for profiles at equal N and d. The deviations from the parabolic profile that occur at the far edge of the grafted layers are due to fluctuations around the stretched trajectory which must become important at some distance from the grafting surface as the segment density drops off to a level where the interactions no longer induce stretching [139, 140].

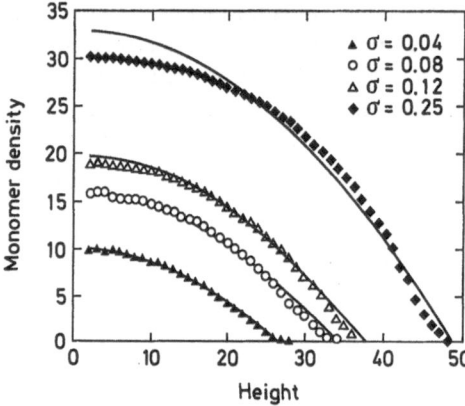

Fig. 9. Comparison of the analytical SCF model [56] with the full numerical SCF calculation [53] for the segment density profile in flat, grafted layers at various surface densities (σ is the fraction of the maximum possible surface coverage of grafted ends). The analytical profile is parabolic to its tip, while the numerical calculation shows that the density at the periphery of the layer drops off exponentially

This region, which can only be modelled accurately by a full SCF treatment including correlations, or simulated by Monte Carlo or molecular dynamics methods, plays an important role in determining the interactions of grafted layers. Since the parabolic region (or the step-function region in the simpler Alexander model) is stretched in response to the high density of interchain interactions, this portion of the brush, when juxtaposed with another brush of the same density, will contract to relieve stretching energy rather than take on the additional stretching necessary to allow outside chains to penetrate. The relatively unperturbed, nonparabolic region is penetrable, however, thus giving it control over the degree of intermingling and entanglement.

No informative experimental data have been obtained on the precise shape of segment profiles of tethered chains. The only independent tests have come from computer simulations [26], which agree very well with the predictions of SCF theory. Analytical SCF theory has proven difficult to apply to non-flat geometries [141], and full SCF theory in non-Cartesian geometry has been applied only to relatively short chains [142], so that more detailed profile information on these important, nonplanar situations awaits further developments.

The impenetrability of assemblies of tethered chains has broad implications. Some of these are discussed in more detail in the sections on interactions, on dynamics and on block copolymer melts. Examples of phenomena that have been studied or anticipated related to the impenetrability of tethered assemblies of chains include the (difficulty of) mixing of grafted layers [8, 143] or highly branched polymers [144] with free linear polymers, the diminishment of entanglement between juxtaposed, tethered lamellae [137] and the intermolecular correlation contribution to the structure factor in the scattering from nondilute solutions of highly branched polymers [39, 125, 145, 146].

6.4 Spherical Surfaces

Interactions among tethered layers are the key determinants of the organization of some higher level structures in polymer systems. Important examples of this involve the progressive increase of density of spherical objects bearing tethered chains, such as stars (Fig. 1A), micelles (Fig. 1E) or chains tethered to solid spheres. As the density of these objects is increased, they interact with one another repulsively which produces, at some density, a transition to an ordered array of spheres. This is manifest macroscopically as pronounced intermolecular correlations in concentrated solutions of highly branched polymers [39, 125, 145, 146], as a disorder to bcc order transition in concentrated solutions or melts of asymmetric block copolymers [50, 92, 147], or as colloidal crystallization in suspensions of sterically stabilized particles [126, 148]. The details of the interactions between curved layers are understood in less detail than for those between flat layers, which is related to some of the features of curved layers discussed in the section on structure. One would expect that curved layers would be more penetrable, and therefore less repulsive on initial overlap, than flat layers at equal tethering density. Some direct evidence for this has been adduced in block copolymer melts of bcc symmetry [149].

7 Kinetics and Dynamics

We use the term kinetics here to refer to those time-dependent processes which occur mainly in reversibly tethered systems, and involve the rates of equilibration and exchange, as well as the time evolution of the assembly or dissolution of tethered microstructures. By dynamics, we intend to describe another category of rate processes which occurs for all tethered layers, such as the dynamics of shape fluctuations and the response of the layers to shear or compression. The kinetic processes of tethered chains are particularly distinct from those of uniformly adsorbed chains. Tethered chains, by definition, have few, most often one (though tethered loops are also possible [150]), anchoring points to the tethering interface. Uniformly adsorbed chains have many. Tethered chains are typically anchored at much higher density as well, giving rise to a much higher and wider potential barrier to insertion and expulsion of chains.

7.1 Kinetics of Expulsion

To illustrate some of these points, we analyze one situation: the expulsion of a tethered chain from a flat, dense layer immersed in a good solvent [151, 152]. We imagine that the expulsion process has the stages schematically represented in Fig. 10. A tethered layer has its interfacial anchor point severed at some instant of time (Fig. 10, middle) and begins to move out of the layer. At some intermediate stage of the expulsion process (Fig. 10, right), part of the chain has emerged from the layer and relaxed its stretching energy, while the other part remains stretched within the layer. The process represents the final stage of a desorption or exchange from an adsorbed layer or micelle. The expulsion process is characterized by a time constant τ_{exp}; the objective of our discussion is the dependence of τ_{exp} on N and d.

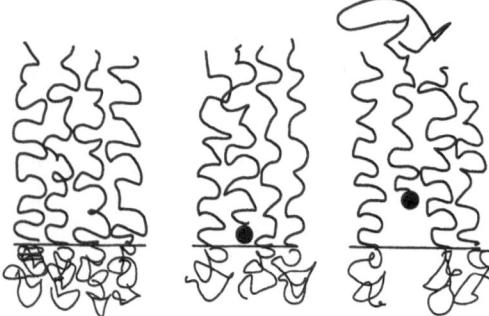

Fig. 10. Various stages in the expulsion of a chain that has been tethered to an interface. Left: Block copolymer straddling interface. Center: Initiation of expulsion process. Right: Chain partially expelled, and therefore partially relaxed from the deformation induced by tethering

The Alexander model allows a simple approach to this problem. Within this model, each tethered chain is, in effect, confined within a cylindrical capillary of diameter d. Combining Eq. 5 and 7, we can express the stretching energy as:

$$F/kT \approx L/d .\tag{37}$$

In the intermediate stages of the expulsion, the chain comprises a segment which is embedded in the layer to a depth of $a \leq x \leq L$, and an expelled segment. The configuration of the expelled chain segment is similar to that of a free coil of comparable size. The excess free energy of the partly expelled chain is thus primarily due to the embedded segment:

$$F/kT \approx x/d .\tag{38}$$

The outward motion of the expelled chain is reminiscent of reptation since the chain of blobs moves, in effect, along its own contour. However, this case is markedly different from the familiar reptation, as encountered in polymer melts and solutions, on two counts. One, the virtual tube in our case does not arise from entanglements. Rather, the origin is in thermodynamics. Two, in reptation the chain is always confined to a tube but the tube changes with time; a chain diffuses through an infinite series of tubes. In the present case, the tubes are finite and short. Accordingly, the distinctive features of the expulsion process may be viewed as end effects. While bulk reptation is typically a purely diffusive process, the expulsion motion is also driven by a non-random force, f, in the outward direction. f is given by the gradient of the free energy of Eq. 38, $f \approx -\partial F/\partial x$, leading to:

$$f \approx -kT/d . \tag{39}$$

One can readily show [153] that this is more important than random diffusion in driving the outward flux of chains (though this is not equally true for curved tethered layers). Assuming that the force acts only on the embedded segment, and the resulting motion is not influenced by the expelled segment, we obtain the expulsion velocity, $-dx/dt$, by balancing the driving force and the frictional force. The friction coefficient is expected to be proportional to the embedded length: $\approx 6\pi\eta x$, with η being the medium viscosity, giving rise to the differential equation for the the embedded length:

$$dx/dt \approx -(6\pi\eta x)^{-1} kT/d . \tag{40}$$

Integrating x from L to 0 over the time period from 0 to τ_{exp} gives:

$$\tau_{exp} \sim (\eta/kT) N^2 (a/d)^{1/3} . \tag{41}$$

This result indicates a strong molecular weight dependence of the expulsion time constant. Direct experimental observation of this expulsion process is difficult since another process affects the rate of exit and exchange with a tethered layer. This is the breakage of the tethering anchor. If the layers are diblock copolymers at an interface, as suggested by Fig. 10, severing of the tethering anchor actually involves extraction of another block from its favored phase. This early stage corresponds to escape out of a potential well. This aspect has been treated via an adapted Kramers rate theory [153], which yields the relevant rate constant. In turn, this allows the calculation of the residence time of polymeric surfactants within an aggregate, as well as the characteristic times for relaxation experiments [154]. Residence times may be obtained by the use of fluorescence techniques [155]. Relaxation experiments involve changing the prevailing solvent or temperature conditions. The development of the appropriate experimental tools to observe such a process in polymeric systems is in its early stages. Bednár et al. [156] have developed a stopped-flow light scattering apparatus for the observation of the time-dependent change in micelle size in response to an abrupt change in solvent conditions. They find that a multiexponential relaxation function is necessary to describe their data. The curved geometry of the micelle causes it to differ in detail from the effects expected for flat layers.

7.2 Kinetics of Adsorption

The assembly and dissolution of a tethered brush, spanning the range from a bare surface to a saturated layer has been studied theoretically by Ligoure and Leibler for the adsorption of end, grafted polymers on a flat surface [157]. Johner and Joanny [61] studied the kinetics of adsorption of diblock copolymers from a selective solvent on a flat selective surface from a micellar solution, where they also account for the influence of the rate of expulsion of adsorbable free chains by the micelle. Stretching energy in the assembling layer becomes an important kinetic barrier and a slow approach to final equilibrium is anticipated. Experimentally, the kinetics of diblock copolymer adsorption has been studied by surface plasmon spectroscopy [158], ellipsometry [159], internal reflection interferometry [160] and by radiolabelling [77, 161]. A very slow terminal time for the final assembly of the layer has been seen [161, 162]. The influence of concentration and micellization of the block copolymers in solution has also been observed [158, 160]. Further experimental studies in this area would be very valuable.

7.3 Dynamics

Studies of dynamic processes involving the unique features of tethering are very few, but the potential for progress in this area is large. Hydrodynamic measurements of flow in pores bearing layers of selectively adsorbed block copolymers have shown that the layer thickness deduced in this way is in good agreement with that determined from direct force measurement [163]. Fredrickson and Pincus [164] have studied the permeability of semidilute polymer layers to solvent with a view to calculating the rate of squeezing of solvent from dense layers under time-dependent compression. Rabin and Alexander [165] studied how stretching of a dense grafted brush by application of shearing stress affects the swelling of the brush. The analysis of Witten et al. [166] has focused on the connection between the impenetrability of tethered layers and the characteristic time for stress relaxation. Since flat layers interpenetrate only sparingly, it follows that chains in these layers, when they interact with another dense layer, must retract, on average, by about that distance to become disengaged from any interlamellar entanglements. In this respect, they resemble the dynamic interactions among highly branched polymers. The possible influence of other relaxation mechanisms in dense polymer fluids on the dynamics of interacting polymer brushes remains to be investigated. Shearing experiments in the surface forces apparatus are an attractive experimental avenue to explore the response of polymer brushes to imposed forces.

Measurements of diffusion of tracer polymers in ordered block copolymer fluids is another potentially informative activity, since molecular diffusion is one of the most basic dynamic characteristics of a molecule. Balsara, et al. have measured the retardation of diffusion due to ordering in the diffusion of polystyrene tracer homopolymers in polystyrene-polyisoprene matrices of various domain sizes [167]. Measurement of the tracer diffusion of block copolymer molecules will also be important. Several interesting issues are directly addressable via measurements of the components of the diffusion tensor in samples with "single-crystal", macroscopic orientation of lamellar or cylindrical domains. Mobility parallel to the interface would provide information on the lateral diffusion, the N-dependence of which could be informative on how chain entanglement may be modified in fluids of tethered chains. Mobility perpendicular to the interface would determine the height of the potential barrier presented by the microdomains and the degree of confinement of the junctions [168].

8 Conclusions

Development of novel polymeric materials involves new arrangements of monomers more than it does development of new monomers. Many new avenues can be reached by permutations and rearrangements of existing monomers. Highly branched polymers, block copolymers and modification of macromolecules and surfaces by grafting reactions are examples of useful new arrangements that differ significantly from linear homopolymers. All of these structures contain chains "tethered" by their ends. Tethered chains become particularly interesting and distinct in their behavior when the lateral number density of chains is high. Stretched equilibrium configurations are the signature and unifying characteristic of tethered chains.

In this work we have examined the unique structural features of tethered chains, and how these result in particular features in the aggregation phenomena, phase behavior, interactions, kinetics and dynamics of materials containing tethered chains. Since an important category of tethered chains is polymeric amphiphiles, such as block copolymers and end-functionalized polymers, one can recognize specific traits emanating from the features that distinguish polymeric surfactants from their small molecule counterparts, namely their large size and configurational flexibility. There are significant new opportunities for research on tethered chains, both in their own right, and for the general insight they can provide into the behavior of soft, microstructured materials.

Acknowledgement: The authors would like to acknowledge with thanks support and assistance from several sources during the preparation of this article: AH, the hospitality of Professor E. W. Fischer and the Max-Planck-Institut für Polymerforschung; AH and MT, the Institute for Theoretical Physics at the University of California, Santa Barbara, where considerable work on this article was done; MT and TPL, the Center for Interfacial Engineering, an NSF Engineering Research Center at the University of Minnesota (NSF-CDR-8721551); MT, the Shell Companies Foundation; and, especially, the Department of Chemical Engineering at the University of Wisconsin, which supported MT during this writing via the stimulating environment of the Olaf A. Hougen Visiting Professorship. We also acknowledge with gratitude discussions and collaboration with many whose work is cited here, especially Shlomo Alexander, Nitash Balsara, Frank Bates and Jean-François Joanny.

9 References

1. Burchard W (1983) Adv Polymer Sci 48: 1
2. Meijs GF, Rizzardo E (1990) J Macromol Sci Rev Macromol Chem C30: 305
3. Tsukahara Y, Mizuno K, Segawa A, Yamashita Y (1989) Macromolecules 22: 1546
4. Jalal N, Duplessix R (1988) J Phys (Paris) 49: 1775
5. Clarke J, Vincent B (1981) J Colloid Interface Sci 82: 208
6. Granick S, Herz J (1985) Macromolecules 18: 460
7. Alexander S (1977) J Phys (Paris) 38: 977
8. de Gennes P-G (1976) J Phys (Paris) 37: 1443; (1980) Macromolecules 13: 1069
9. Cantor R (1981) Macromolecules 14: 1186
10. Daoud M, Cotton JP (1982) J Phys (Paris) 43: 531
11. Hadziioannou G, Patel S, Granick S, Tirrell M (1986) J Amer Chem Soc 108: 2869
12. Patel S, Hadziioannou G, Tirrell M (1987) Proc Nat Acad Sci 84: 4725
13. Taunton HJ, Toprakcioglu C, Fetters LJ, Klein J (1988) Nature 332: 712; (1990) Macromolecules 23: 571
14. Taunton HJ, Toprakcioglu C, Klein J (1988) Macromolecules 21: 3333
15. Milner ST (1991) Science 251: 905
16. Ferry JD (1980) Macromolecules 13: 1719; Noda I, Kato N, Kitano T, Nagasawa M (1981) Macromolecules 14: 668
17. Daoud M, Cotton JP, Farnoux B, Jannink G, Sarma G, Benoit H, Duplessix R, Picot C, de Gennes P-G (1975) Macromolecules 8: 804
18. de Gennes P-G (1979) Scaling concepts in polymer physics. Cornell University, Ithaca
19. Freed KF (1987) Renormalization group theory of macromolecules. John Wiley, New York; des Cloizeaux J, Jonnink G (1990) Polymers in solution Clarendon Oxford
20. Ansarifar MA, Luckham PF (1988) Polymer 29: 329
21. Gast AP, Munch MR (1988) Polymer Communications 30: 324
22. Satija SK, Majkrzak CF, Russell TP, Sinha SK, Sirota EB, Hughes GJ (1990) Macromolecules 23: 3860
23. Cosgrove T, Heath TG, Phipps JS, Richardson RM (1991) Macromolecules 24: 94; Cosgrove T (1990) J Chem Soc Faraday Trans. 86: 1323
24. Blum FD, Sinha BR, Schwab FC (1990) Macromolecules 23: 3592
25. Auroy P, Auvray L, Léger L (1991) Phys Rev Lett 66: 719
26. Murat M, Grest GS (1989) Phys Rev Lett 63: 1074; Macromolecules 22: 4054; (1991) Macromolecules 24: 704
27. Shim DFK, Cates ME (1989) J Phys (Paris) 50: 3535
28. Lai P-Y, Halperin A (1991) Macromolecules (in press)
29. Halperin A (1988) J Phys (Paris) 49: 547
30. Auroy P (1990) Thèse, Université de Paris-Sud
31. Auroy P, Auvray L, Léger L (1991) Macromolecules 24: 2523
32. Pincus P (1991) Macromolecules 24: 2912
33. Miklavic SJ, Marčelja S (1988) J Phys Chem 92: 6718
34. Granfeldt MK, Miklavic SJ, Marčelja S, Woodward CE (1990) Macromolecules 23: 4760
35. Misra S, Varanasi S, Varanasi PP (1989) Macromolecules 22: 4173; Misra S, Varanasi S (1991) Macromolecules 24: 322
36. Birshtein TM, Zhulina EB (1984) Polymer 25: 1453; Zhulina EB, Birshtein TM (1985) Polymer Sci USSR 27: 570; Birshtein TM, Borisov OV, Zhulina EB, Khokhlov AR, Yurasova TA (1987) Polymer Sci USSR 29: 1293
37. Wang Z-G, Safran SA (1988) J Chem Phys 89: 5323
38. Zilliox GJ (1972) Makromol Chem 156: 121
39. Dozier WD, Huang JS, Fetters LJ (1991) Macromolecules (in press)
40. Grest GS, Kremer K, Witten TA (1987) Macromolecules 20: 1376
41. Grest GS, Kremer K (1986) Phys Rev A 33: 3628
42. Grest GS, Kremer K, Milner ST, Witten TA (1989) Macromolecules 22: 1904

43. Lipson JEG, Gaunt DS, Wilkinson MK, Whittington SG (1987) Macromolecules 20: 186
44. Batoulis J, Kremer K (1989) Macromolecules 22: 4277
45. Khasat N, Pennisi RW, Hadjichristidis N, Fetters LJ (1988) 21: 1100
46. Borisov OV, Birshtein TM, Zhulina EB (1987) Vysokomol Soedin 29 A: 1413; Birshtein TM, Zhulina EB (1989) Polymer 30: 170
47. Wang Z-G, Safran SA (1991) J Chem Phys 94: 679
48. Zhulina EB, Birshtein TM (1985) Vysokomol Soedin 27 A: 511
49. Ohta T, Kawasaki K (1986) Macromolecules 19: 2621
50. Bates FS, Fredrickson GH (1990) Ann Rev Phys Chem 41: 525
51. Dolan AK, Edwards SF (1975) Proc Roy Soc London A343: 427
52. Ploehn HJ, Russel WB (1990) Adv Chem Eng 15: 137
53. Hirz SJ (1986) MS Thesis, University of Minnesota
54. Cosgrove T, Heath T, van Lent B, Leermakers F, Scheutjens JHMS (1987) Macromolecules 20: 1692
55. Skvortsov AM, Pavlushkov IV, Gorbunov AA (1988) Polymer Sci USSR 30: 487; Skvortsov AM, Pavlushkov IV, Gorbunov AA, Zhulina EB, Borisov OV, Pryamitsin VA (1988) Polymer Sci USSR 30: 1706
56. Milner ST, Witten TA, Cates ME (1988) Europhys Lett 5: 413; Macromolecules 21: 2610
57. Zhulina EB, Borisov OV, Pryamitsin VA (189) Vysokomol Soedin 31 A: 185; Zhulina EB, Borisov OV, Pryamitsin VA (1990) J Colloid Interface Sci 137: 495
58. Semenov AN (1985) Sov Phys JETP 61: 733
59. Bug ALR, Cates ME, Safran SA, Witten TA (1987) J Chem Phys 87: 1824
60. Halperin A (1987) Macromolecules 20: 2943
61. Johner A, Joanny J-F (1990) Macromolecules 23: 5299
62. Rodrigues K, Mattice WL (1991) J Chem Phys 94: 761
63. Bluhm TL, Whitmore MD (1985) Can J Chem 63: 249
64. Krause S (1964) J Phys Chem 68: 1948
65. Tuzar Z, Kratochvíl P (1991) Colloids and Surfaces (in press)
66. Halperin A (1989) Macromolecules 22: 3806
67. Tuzar Z, Pleštil J, Koňak C, Hlavata D, Sikora A (1983) Makromol Chem 164: 2111; Brown DS, Dawkin JS, Farnell AS, Taylor G (1987) Europ Polym J 23: 463
68. Pleštil J, Baldrin J (1975) Makromol Chem 176: 1009
69. Kotaka T, Tanaka T, Hattori M, Inagaki H (1978) Macromolecules 11: 138
70. Xu R, Winnik MA, Hallett FR, Riess G, Croucher MD (1991) Macromolecules 24: 87
71. Rigby D, Roe RJ (1984) Macromolecules 17: 1778
72. Higgins JS, Dawkins JV, Maghami GG, Shakir SA (1986) Polymer 27: 931
73. Pleštil J, Baldrin J (1973) Makromol Chem 174: 183
74. Utiyama H, Takenaka K, Mizumori M, Fukuda M, Tsunashima Y, Kurata M (1974) Macromolecules 7: 515
75. Marra J, Hair ML (1989) Colloids and Surfaces 34: 215; Guzonas D, Boils D, Hair ML (1991) Macromolecules 24: 3383
76. Marques CM, Leibler L, Joanny J-F (1988) Macromolecules 21: 1051
77. Parsonage EE, Tirrell M, Watanabe H, Nuzzo RG (1991) Macromolecules (in press)
78. Evers OA, Scheutjens JMHM, Fleer GJ (1990) Macromolecules 23: 5221
79. Ligoure C (1991) Macromolecules 24: 2968
80. Halperin A (1989) Europhys Lett 10: 549; (1990) Macromolecules 23: 2724
81. Vilgis T, Halperin A (1991) Macromolecules 24: 2090
82. Halperin A (1991) Macromolecules 24: 1418
83. Cogan KA, Gast AP (1990) Macromolecules 23: 745
84. Zhulina EB, Borisov OV, Pryamitsin VA, Birshtein TM (1991) Macromolecules 24: 140
85. Halperin A (1987) Europhys Lett 4: 439
86. Broseta D, Leibler L, Joanny J-F (1987) Macromolecules 20: 1935
87. Halperin A (1988) J Phys (Paris) 49: 131
88. Ou-Yang HD, Gao Z (1991) Bull Am Phys Soc 36: 580

89. Halperin A, Alexander S, Schechter I (1987) J Chem Phys 86: 6550; (1989) J Chem Phys 91: 1383
90. Chen Z-Y, Talbot J, Gelbart WM, Ben-Shaul A (1988) Phys Rev Lett 61: 1376
91. Douy A, Gallot B (1987) Polymer 28: 147
92. Bates FS (1991) Science 251: 898
93. Goodman I (ed) (1985) Developments in block copolymers — 2. Applied Science, New York
94. Almdal K, Rosedale JH, Bates FS, Wignall GD, Fredrickson GH (1990) Phys Rev Lett 65: 1112
95. Leibler L (1980) Macromolecules 13: 1602
96. Fredrickson GH, Helfand E (1987) J Chem Phys 87: 697
97. Helfand E, Wasserman ZR (1976) Macromolecules 9: 879
98. Helfand E, Wasserman ZR (1980) Macromolecules 13: 994
99. Hashimoto T, Shibayama M, Kawai H (1980) Macromolecules 13: 1237
100. Bates FS, Berney CV, Cohen RE (1983) Macromolecules 16: 1101
101. Roe RJ, Fishkis M, Chang JC (1981) Macromolecules 14: 1091
102. Hashimoto T (1982) Macromolecules 15: 1548
103. Hashimoto T, Shibayama M, Kawai H (1983) Macromolecules 16: 1093
104. Richards RW, Thomason JL (1983) Macromolecules 16: 982
105. Hashimoto T, Tanaka H, Hasegawa H (1985) Macromolecules 18: 1864
106. Matsushita Y, Mori K, Mogi Y, Saguchi R, Nakao Y, Noda I, Nagasawa M (1990) Macromolecules 23: 4313
107. Hadziioannou G, Skoulios A (1982) Macromolecules 15: 258
108. Hadziioannou G, Picot C, Skoulios A, Ionescu M-L, Mathis A, Duplessix R, Gallot Y, Lingelser J-P (1982) Macromolecules 15: 263
109. Bates FS, Berney CV, Cohen RE, Wignall GD (1983) Polymer 24: 519
110. Hasegawa H, Hashimoto T, Kawai H, Lodge TP, Amis EJ, Glinka CJ, Jan CC (1985) Macromolecules 18: 67
111. Hasegawa H, Tanaka T, Hashimoto T, Han CC (1987) Macromolecules 20: 2120
112. Matsushita Y, Mori K, Mogi Y, Saguchi R, Noda I, Nagasawa M, Chang T, Glinka CJ, Han CC (1990) Macromolecules 23: 4317
113. Matsushita Y, Mori K, Saguchi R, Noda I, Nagasawa M, Chang T, Glinka CJ, Han CC (1990) Macromolecules 23: 4387
114. Leibler L (1988) Makromol Chem Macromol Symp 16: 1
115. Tanaka H, Hasegawa H, Hashimoto T (1991) Macromolecules 24: 240
116. Winey KI (1990) Ph D Dissertation, University of Massachusetts
117. Brown HR, Char K, Deline VR (1990) Macromolecules 23: 3383
118. Kausch HH, Tirrell M (1989) Ann Rev Mat Sci 19: 341
119. Szleifer I, Ben-Shaul A, Roux D, Gelbart WM (1986) Phys Rev Lett 60: 1966
120. Milner ST, Witten TA (1988) J Phys (Paris) 49: 1951
121. Wang Z-G, Safran SA (1990) J Phys (Paris) 51: 185
122. Wang Z-G, Safran SA (1990) Europhys Lett 11: 425
123. Broseta D, Fredrickson GH (1991) J Chem Phys 93: 2927
124. Napper DH (1983) Polymeric stabilization of colloidal dispersions. Academic, New York
125. Witten TA, Pincus PA (1986) Macromolecules 19: 2509
126. Oxtoby DW (1990) Nature 347: 725
127. Patel SS, Tirrell M (1989) Ann Rev Phys Chem 40: 597
128. de Gennes P-G (1985) C R Acad Sci (Paris) 300: 839
129. Patel S, Tirrell M, Hadziioannou G (1988) Colloids and Surfaces 31: 157
130. Milner ST, Witten TA, Cates ME (1989) Macromolecules 22: 853
131. Milner ST (1988) Europhys Lett 7: 695
132. Tirrell M, Parsonage EE, Watanabe H, Dhoot S (1991) Polymer Journal 23: 641
133. Chakrabarti A, Toral R (1990) Macromolecules 23: 2016
134. Watanabe H, Tirrell M (1991) Macromolecules (submitted)
135. Patel SS (1991) Macromolecules, in press

136. Shim DFK, Cates ME (1990) J Phys (Paris) 51: 701
137. Milner ST (1990) J Chem Soc Faraday Trans 86: 1349
138. Witten TA, Milner ST (1990) Mat Res Soc Symp Proc 177: 37
139. Milner ST, Wang Z-G, Witten TA (1989) Macromolecules 22: 489
140. Muthukumar M, Ho JS (1989) Macromolecules 22: 965
141. Ball RC, Marko JF, Milner ST, Witten TA (1991) Macromolecules 24: 693
142. Leermakers FAM, Scheutjens JMHM, Lyklema J (1983) Biophys Chem 19: 353; Dill KA, Naghizadeh J, Marqusee J (1988) Ann Rev Phys Chem 39: 425
143. Gast AP, Leibler L (1986) Macromolecules 19: 686
144. Roby F, Joanny J-F (1991) Macromolecules 24: 2069
145. Huber K, Bantle S, Burchard W, Fetters LJ (1986) Macromolecules 19: 1404
146. Witten TA, Pincus PA, Cates ME (1986) Europhys. Lett. 2: 137
147. Hashimoto T, Shibayama M, Kawai H, Watanabe H, Kotaka T (1983) Macromolecules 16: 361
148. Smits C (1990) Phase Transitions 21: 157
149. Thomas EL, Kinning DJ, Alward DB, Henkee CS (1987) Macromolecules 20: 2934
150. Balazs AC, Gempe M, Lantman CW (1991) Macromolecules 24: 168
151. Halperin A, Alexander S (1988) Europhys Lett 6: 329; (1989) Macromolecules 22. 2403
152. Halperin A, Alexander S (1989) Macromolecules 22: 2403
153. Halperin A (1989) Europhys Lett 8: 351
154. Aniansson EAG, Wall SN (1975) J Phys Chem 78: 857
155. Chu D-Y, Thomas JK (1987) Macromolecules 20: 2133
156. Bednár B, Edwards K, Almgren M, Tormod S, Tuzar Z (1988) Makromol Chem Rapid Comm 9: 785
157. Ligoure C, Leibler L (1990) J Phys (Paris) 51: 1313
158. Tassin JF, Siemens RL, Tang W-T, Hadziioannou G, Swalen JD, Smith BA (1989) J Phys Chem 93: 2106
159. Motschmann H, Stamm M, Toprakcioglu C (1991) Macromolecules 24: 3681
160. Munch MR, Gast AP (1990) J Chem Soc Faraday Trans 86: 1341
161. Huguenard C, Varoqui R, Pfefferkorn E (1991) Macromolecules 24: 2226
162. Leermakers FAM, Gast AP (1991) Macromolecules 24: 718
163. Webber RM, Anderson JL, Jhon MS (1990) Macromolecules 23: 1026
164. Fredrickson GH, Pincus PA (1991) Longmuir 7: 786
165. Rabin Y, Alexander S (1990) Europhys Lett 13: 49
166. Witten TA, Leibler L, Pincus PA (1990) Macromolecules 23: 824
167. Balsara NP, Eastman CE, Foster MD, Lodge TP, Tirrell M (1991) Makromol Chem, Macromol Symp (in press)
168. Frederickson GH, Milner ST (1990) Mat Res Soc Symp Proc 177: 169

Editor: John L. Schrag
Received April 25, 1991

Mechanochemical Degradation in Transient Elongational Flow

Tuan Q. Nguyen, Hans-Henning Kausch
Polymer Laboratory, Department of Materials Science, Swiss Federal Institute of Technology, CH-1015 Lausanne, Switzerland

Polymer molecules respond to the application of stress by disentanglement, chain orientation and bond rupture. Being the last step in this series of events, the efficiency of flow-induced mechanochemical degradation depends markedly on the details of the pervading flow field. From the results of fluid mechanics, it is usual to distinguish degradation in simple shear flow from degradation in elongational flow. Polymer dynamics require a further subdivision between transient elongational flow, characterized by a short residence time (less than the chain relaxation time τ_1) in the high strain rate region, from quasi-steady-state elongational flow which possesses a stagnation point and a long residence time ($\gg \tau_1$). This chapter reviews the main factors which influence the degradation kinetics in transient elongational flow, exemplified by flow through a narrow contraction and ultrasonic irradiation. The experimental results reveal a fundamental difference in the kinetics of chain scission in transient and stagnant conditions. Differences in the characteristic residence time prevalent in each type of flow permits one to probe the different stages of the chain unravelling process up to the point of bond rupture. It is suggested that the intricate degradation behavior observed in transient elongational flow stems from the compact nature of the weakly deformed molecular coil. This favors intramolecular interactions during the dynamics of deformation.

Advances in Polymer Science, Vol. 100
© Springer-Verlag Berlin Heidelberg 1992

List of Symbols

A   = rate constant pre-exponential factor
b   = excluded volume integral
c^* = molecular coil overlap concentration
C_∞ = characteristic ratio
D   = bond dissociation energy
D   = diffusion constant
E_A = Arrhenius activation energy
E_s = excess stress energy
ΔE_r = potential barrier for bond rotation
E_{e1} = molecular elastic energy
F   = mean force potential
f   = average force on the chain
f_b = bond breaking force
H_o = Hookean spring constant
k_B = Boltzmann constant
k   = elastic constant for valence bond deformation
k_c = rate constant for bond dissociation
K   = global scission rate constant
l   = separation distance between two consecutive bonds
l_p = separation distance between two consecutive links
l_b = bond length at break
l_{max} = maximum contour length
m   = bead mass in the bead-spring model
M   = polymer molecular weight
n   = number of backbone bonds in a chain
N   = number of beads in the bead-spring model
N_A = Avogadro number
p_{eq} = equilibrium probability distribution function for the chain end-to-end distance
P_{ent} = entrance pressure drop
R   = molar gas constant
r   = end-to-end vector
R   = root-mean-square end-to-end distance
R   = molar gas constant
R_g = radius of gyration
R_h = hydrodynamic radius
R_e = hydrodynamic equivalent radius
Re   = Reynolds number
S   = entropy
T   = absolute temperature
T_θ = theta temperature
U_{nm} = intramolecular segmental energy
U_0 = thermal activation energy for bond rupture
v_0 = volumetric average velocity

$v(r)$ = velocity field
x = ordinate of flow direction
η_s = solvent viscosity
η_0 = zero-shear viscosity
$[\eta]$ = intrinsic viscosity
τ_1 = chain ends relaxation time
τ_r = residence time
$\boldsymbol{\Omega}$ = Oseen tensor
ε = fluid strain
$\dot{\varepsilon}$ = fluid strain rate
λ = molecular extension ratio
σ_E = elongational stress
σ = macroscopic stress
σ = standard deviation of a Gaussian distribution
ω = vorticity tensor
$\dot{\gamma}$ = simple shear rate
$\dot{\boldsymbol{\gamma}}$ = rate of strain tensor
ψ = molecular stress
ξ = bead-solvent friction coefficient

1 Introduction

Mechanochemical degradation is ubiquitous in macromolecular systems and can be encountered practically in any field involving high molecular weight polymers. The formation of free radicals by mechanical stress has been detected during polymer processing, analysis, weathering and gel swelling; chain scission was also observed in applications as diverse as drag reduction, rubber mastication and mechanochemical synthesis. It was known centuries ago that the application of stress under certain conditions, like the combing of wool, the mastication of rubber or the malaxing of dough, can improve the physical properties of a few materials which were later recognized as being polymeric. Mechanochemical degradation is a classical field of research in polymer chemistry: the first scientific account of flow-induced degradation was reported by Staudinger practically at the same time as the concept of macromolecules [1]. Early experiments were performed in simple shear flow, across a capillary tube, in concentric rotating cylinders and in cone-plate geometry to improve flow field uniformity. In those days, it was generally believed that the presence of a velocity gradient is sufficient to orient and to degrade polymer chains [2, 3]. It seems now evident that a simple shear field, with a velocity gradient transverse to the flow direction, is ineffective for distorting and imposing a stress on isolated flexible polymer molecules. In the early 1960s, it was shown theoretically that only elongational flow with a velocity gradient parallel to the direction of flow, is capable of achieving a large degree of molecular coil extension [4–6]. No experiments were performed in this direction, however, until 1975. Following the impetus given by de Gennes' prediction of the coil-to-stretch transition [7], a number of experimentalists devised varying types of elongational flow with the purpose of verifying the anticipated coil stretching [8–11], while the observation of mechanochemical degradation was generally neglected or only secondary in importance. For this reason, much effort has been concentrated on the so-called "stagnant" elongational flow which possesses a center or axis of symmetry where the dwelling time of a fluid element is large as compared to the relaxation time of the macromolecule. Elongational flow which does not meet the above symmetry requirement has a short residence time and is called "transient elongational" or "fast transient" flow following the current terminology [12]. This type of flow is commonly found in convergent channels and is mainly investigated in conjunction with the viscoelastic problems of die extrusion. A few researchers in the mid 1970s attempted to determine molecular conformation in convergent flow by light scattering [13, 14] and their works can still be considered as pioneering up to the present time. As clarified during this review, most of the reported degradation of flexible polymer chains in solution had occurred under situations where the macromolecule retains its coiled conformation. In only rare circumstances has it been possible to study chain degradation in a highly stretched state. From the mechanistic point of view, degradation of fully aligned polymers is much simpler to comprehend than for partly uncoiled chains. Due to this simplicity, the theory is also better developed [15]. In the discussion, we will make frequent references to the stretched state as a starting point for further refinement into the coiled state.

The process of degradation is the final result of a succession of events which began with the input of mechanical force on the macroscopic level. Following the application of stress, the embedded macromolecules will attempt to follow the bulk deformation in order to minimize the total free energy of the system. A change in chain conformation generates strain at the molecular level, leading to a redistribution of electronic densities on the different molecular orbitals which results ultimately in the chemical effects globally known as mechanochemical degradation. This series of complex events embraces several fields of research, from rheology and polymer dynamics, to chemical kinetics and quantum mechanics. This paper aims at presenting an overall perspective for the process of mechanochemical degradation in transient elongational flow, and its relation to other types of flow-induced degradation like that in stagnant elongational flow and in simple shear flow. Such an analysis is necessarily incomplete since progress is being made steadily in each of the designated fields, particularly in the domain of chain dynamics simulation.

Flow-induced degradation is intimately related to the nonequilibrium conformation of polymer coils and any attempt to interpret the process beyond the phenomenological stage would be incomplete without a sound understanding of chain dynamics. To make the paper self-contained and to provide a theoretical basis for the discussion, we have included some fundamental models of polymer dynamics in the next section which may also serve as a guideline for future work in the field of polymer degradation in flow. For the first-time reader, however, this section is not absolutely necessary. Further, any reader familiar with molecular rheology or interested only in experimental results can skip this section, only to go back whenever a reference is needed.

2 Chain Dynamics in Flows

2.1 Flexible Chains in the Quiescent State

The static properties of an isolated chain constitute a good starting point to study polymer dynamics: many of the features of the chain in a quiescent fluid could be extrapolated to the kinetics theories of molecular coil deformation. As a matter of fact, it has been pointed out that the equations of chain statistics and chain dynamics are intimately related through the simplest notions of graph theory [16].

Even in the absence of flow, a polymer molecule in solution is in a state of continual motion set forth by the thermal energy of the system. Rotation around any single bond of the backbone in a flexible polymer chain will induce a change in conformation. For a polyethylene molecule having $(n + 1)$ methylene groups connected by n C$-$C links, the total number of available conformations increases as $\sim 3^n$. With the number n encompassing the range of 10^5 and beyond, the number of accessible conformations becomes enormous and the shape of the polymers can only be usefully described statistically.

In a real polymer chain, rotation around backbone bonds is likely to be hindered by a potential energy barrier of height ΔE_r. If $\Delta E_r < RT$, the population of the *trans* and *gauche* rotamers, as given by the Stefan-Boltzmann rule, are nearly equiprobable and the limit of high flexibility is obtained. For higher values of $\Delta E_r/RT$, the *trans* state will be definitely preferred and the chain is locally rigid. Early in the history of polymer science, it was recognized [17] that details of bond angles and rotational potentials which control the arrangement of nearest-neighbor atoms in the chain do not significantly affect the molecular shape at a distance of many atoms. More specifically, if details smaller than a certain characteristic length $l_p \approx l \exp (\Delta E_r/RT)$ are ignored (l is the separation distance between two consecutive bonds on the polymer backbone), the chain could be envisioned as a succession of N rigid links of length l_p connected by highly flexible joints and able to point in any direction in a completly uncorrelated manner (Fig. 1). This model, called the freely-jointed or the random flight chain [17], constitutes the simplest mechanical description of flexible macromolecules. For an ideally flexible polymer chain, the length of each link (l_p) coincides with the bond distance *l*. The ratio $(l_p/l)^2$, known as the characteristic ratio (C_∞), is tabulated in a number of books [19, 20] and can be used as an indication of chain flexibility. Some water-soluble polymers, like sodium polystyrene sulfonate [21] or xanthan [22] constitute an interesting class of polymers since their characteristic ratio could be changed over a broad range by simply adjusting the ionic strength or pH of the solution. Depending on the salt content, it is possible to obtain different classes of polymers from the same material ranging from rigid rods in pure water, to worm-like chains,

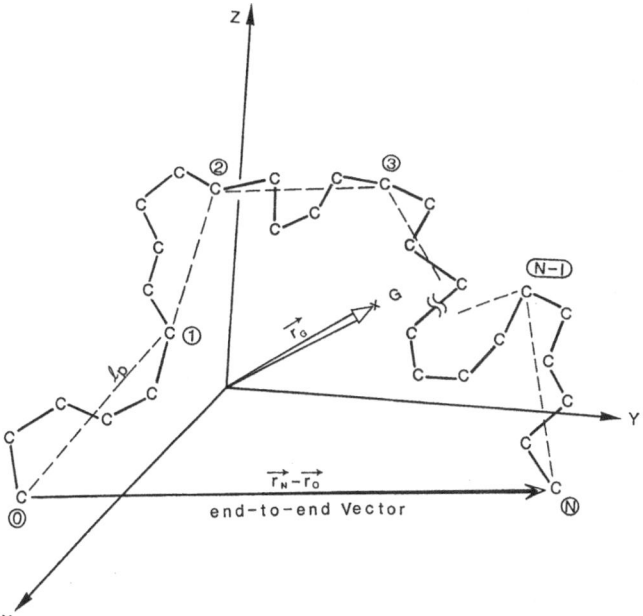

Fig. 1. Freely jointed bead-rod model of a chain formed by (N + 1) "beads" and N rigid links of length l_p

then to random-coils and finally to collapsed globules at increasing salt concentration. A different approach used the complexation effects with a surfactant (sodium dodecyl sulfate) to stabilize the collapsed state in poly(N-isopropyl acrylamide) [23].

The results of the three-dimensional random walk, based on the freely-jointed chain, has permitted the derivation of the equilibrium statistical distribution function of the end-to-end vector of the chain (the underscript "eq" denotes the equilibrium configuration) [24]:

$$p_{eq}(r) \approx (2/3 \cdot \pi N l_p^2)^{-3/2} \exp\left[-3r^2/(2Nl_p^2)\right]. \tag{1}$$

In the limit of large N, $p_{eq}(r)$ is a Gaussian function from which other average geometric properties of the chain can be obtained like the mean square end-to-end distance R^2:

$$R^2 = \langle (\vec{r}_N - \vec{r}_0)^2 \rangle = \int_0^\infty p_{eq}(r)\, r^2\, 4\pi r^2\, dr = N \cdot l_p^2. \tag{2}$$

Since the contour length of a fully extended C−C chain is $l_{max} = Nl_p$, a polymer molecule can be stretched in the extreme by a factor of l_{max}/R or $N^{1/2}$ from its equilibrium value.

Although R^2 is the easiest quantity to be obtained theoretically, there is no straigthforward experimental method for its determination. For this reason, two other quantities are widely in use to characterize the dimensions of a randomly coiled polymer molecule:

− the squared radius of gyration (R_g^2), which is readily determined from light scattering experiments and is defined as:

$$R_g^2 = 1/(N+1) \left\langle \sum_{j=0}^N (\vec{r}_j - \vec{r}_G)^2 \right\rangle. \tag{3}$$

$\vec{r}_G$ is the vector defining the center of gravity of the chain (Fig. 1):

$$\vec{r}_G = 1/(N+1) \sum_{j=0}^N \vec{r}_j. \tag{4}$$

Substituting Eq. (4) into Eq. (3) gives:

$$R_g^2 = 1/(N+1)^2 \left\langle \sum_{i=0}^N \sum_{j=0}^N (\vec{r}_i - \vec{r}_j)^2 \right\rangle = l_p^2/(N+1)^2 \sum_{i=0}^N \sum_{j=0}^N (j-i)^2$$

$$= \frac{1}{6} \cdot \frac{(N+2)}{(N+1)} \cdot Nl_p^2 \approx 1/6R^2 \tag{5}$$

— the hydrodynamic radius (R_h), which is defined as the radius of a friction-equivalent impermeable sphere:

$$1/R_h = 1/(N + 1)^2 \left\langle \sum_{i=0}^{N} \sum_{j=0}^{N} |\vec{r}_i - \vec{r}_j|^{-1} \right\rangle. \qquad (6)$$

The hydrodynamic radius reflects the effect of coil size on polymer transport properties and can be determined from the sedimentation or diffusion coefficients at infinite dilution from the relation $R_h \equiv k_B T/6\pi\eta_s D$ (D = translational diffusion coefficient extrapolated to zero concentration, k_B = Boltzmann constant, T = absolute temperature and η_s = solvent viscosity).

2.1.1 Instantaneous Shape of a Random Coil

The mean density of segments of a flexible long chain molecule about the molecular center of mass is spherically symmetric in space only in the long-time average. The instantaneous shape of a random flight chain, however, is higly asymmetric as can be assessed from any random walk simulation: this fact was recognized long ago by Kuhn [25]. Diverse analytical and Monte Carlo calculations indeed show that random flight coils at any given moment are on the average prolate ellipsoids with an aspect ratio of ≈ 4 having the long axis parallel to the end-to-end vector. To cite Solc and Stockmayer [26], these coils look "more like a cake of soap than like a tennis ball". This instantaneous anisotropy implies that a molecular coil, even in the absence of significant distortion, will be able to rotate along the flow line in presence of a velocity gradient. This tumbling motion may influence the rate of flow-induced degradation by bringing the molecular coil into the right configuration for chain scission. For linear flexible polymers, this should occur simultaneously as chain stretching since the rotational relaxation time is identical to the end-to-end relaxation time (τ_1) (Cf. Sect. 2.3).

2.1.2 Excluded Volume Effect

The interaction between immediate neighbors in a macromolecule is considered expressly by the Rotational Isomeric State model [19]. In the freely-jointed model, this short-range effect has been implicitly taken into account through the characteristic ratio C_∞. For a real polymer, the chain can coil upon itself and segments distant along the chain can come close to each other in space and interact. The nature of this interaction is complex and results from a short range steric repulsion between the polymer segments ("hard-core potential"), and a longer range ("soft") attractive potential due to the Van der Waals forces and other specific interactions mediated by the solvent (Fig. 2). It is usual to express the interaction between the polymer segments n and m by a potential energy $u_{nm}(r)$ which depends only on the separation distance r and to define a potential of mean force F as the total free energy of intramolecular segmental interaction [16]:

$$F = \sum_{n=0}^{N-1} \sum_{m=n+1}^{N} u_{nm}(r). \qquad (7)$$

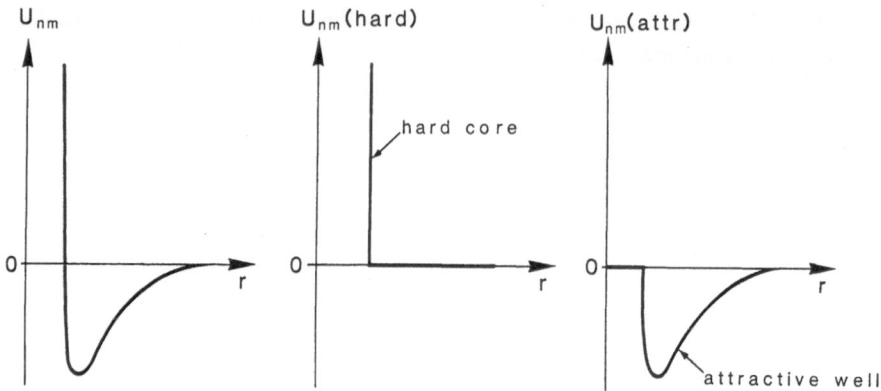

Fig. 2. Sketch of the interaction potential between segments m and n. The potential can be decomposed into a hard core repulsive potential U_{nm} (hard) and a weak attractive potential U_{nm} (attr)

In analogy to the virial theory of gases, the following integral was introduced:

$$b = \int_0^\infty [1 - \exp(-u_{nm}(r)/k_B T)] \, 4\pi r^2 \, dr. \qquad (8)$$

The quantity b has the dimension of a volume and is known as the "excluded volume" or the "binary cluster" integral. The mean force potential is a function of temperature (principally as a result of the soft interactions). For a given solvent or mixture of solvents, there exists a temperature (called the θ-temperature or T_θ) where the solvent is just poor enough so that the polymer feels an effective repulsion toward the solvent molecules and yet, good enough to balance the expansion of the coil caused by the excluded volume of the polymer chain. Under this condition of perfect balance, all the binary cluster integrals are equal to zero and the chain behaves like an ideal chain.

In good solvents, the mean force is of the repulsive type when the two polymer segments come to a close distance and the excluded volume is positive: this tends to swell the polymer coil which deviates from the "ideal chain" behavior described previously by Eq. (1). Once the excluded volume effect is introduced into the model of a real polymer chain, an exact calculation becomes impossible and various schemes of simplification have been proposed. The excluded volume effect, first discussed by Kuhn [25], was calculated by Flory [24] and further refined by many different authors over the years [27]. The rigorous treatment, however, was only recently achieved, with the application of renormalization group theory. The renormalization group techniques have been developed to solve many-body problems in physics and chemistry. De Gennes was the first to point out that the same approach could be used to calculate the MW dependence of global properties of polymers with excluded volume interactions [18]. The application of this method to the radius of gyration of a polymer chain in a good solvent yielded the following relation [29]:

$$R_g^2 = 0.406N^{2\nu} \quad \text{with} \quad \nu = 0.589 .\tag{9}$$

In his original calculations, Flory found a value of $\nu = 3/5$ which seems to be quite close to the value given by Eq. (9). This accord, however, was proved to be fortuitous as a result of the mutual cancellation of two terms neglected in Flory's treatment [30].

In a poor solvents below the θ-temperature, the mean force becomes attractive. The theoretical treatment of polymer solutions near and below θ-temperature is a complex problem and can be handled only with the renormalization group theory. The simple theory of Flory [19], when applied to $T < T_\theta$, predicts a rapid collapse of the polymer coil into a globule. This is in complete contradiction with the light scattering results which show a rather gradual decrease of the gyration and hydrodynamic radii with the reduced temperature $(T - T_\theta)/T_\theta$. According to the experimental data [31], the collapsed regime could be reached only at temperatures well below the θ-temperature. At the beginning of the collapsed regime, the radius of gyration of the coil contracts by only 15% but follows a quite distinct scaling law as compared to an ideal Gaussian chain, i.e. $R_g \sim N^{1/3}$ instead of the $N^{1/2}$ dependence (Eq. 5).

2.2 Free-Energy Storage in a Single Chain

The freely-jointed chain considered previously has no internal restraint, and hence, its internal energy is zero regardless of its present configuration. The entropy (S) is not constant, however, since the number of available configurations decreases with the chain end separation distance. The variation which follows from chain length change by a small amount (dr) at constant temperature (T) is given by the Boltzmann rule of statistical thermodynamics:

$$dS = k \, d \ln p(r) .\tag{10}$$

Application of the first law of thermodynamics permits us to relate the change in entropy with the work output from the chain:

$$dq = T \, dS = -f(r) \, dr .\tag{11}$$

The average force $f(r)$ in the chain when the ends are held a distance r apart could then be obtained from Eq. (10) providing the appropriate configuration distribution function $p(r)$ is known. In the limit of a small extension ratio, $p(r)$ is approximately proportional to $p_{eq}(r)$:

$$p(r) = C \cdot p_{eq}(r)\tag{12}$$

with $C = $ constant.

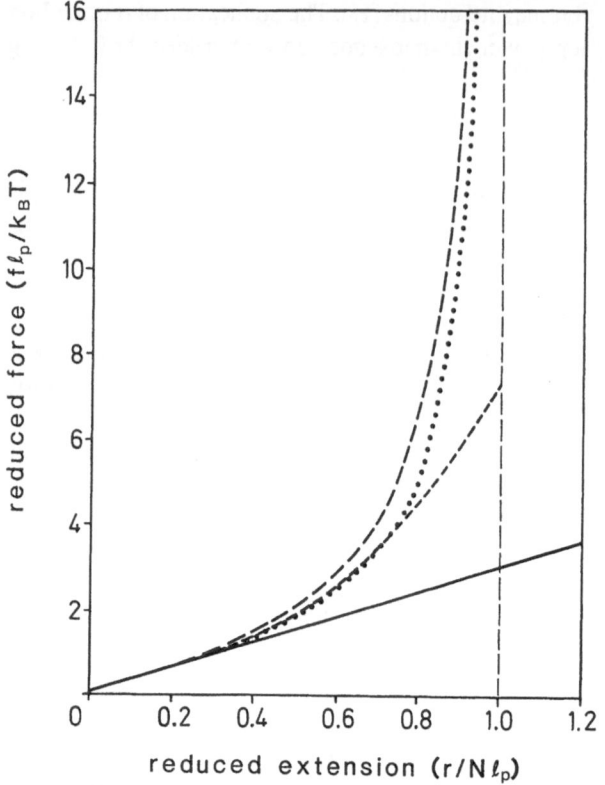

Fig. 3. Plots of approximate spring laws for the ensemble average tension f(r) in a freely jointed chain with end-to-end distance r. (———): Hookean; (–––––): anharmonic oscillator; (··········): inverse Langevin; (– – –): Warner

Substituting Eq. (12) into Eq. (11) permits us to derive the Hookean spring force law, well-known in the classical theory of rubber elasticity:

$$f(r) = H_o \cdot r,\tag{13}$$

where $H_o = (3k_BT/Nl_p^2)$ is the Hookean spring constant.

Equation (13) is valid for $r/Nl_p < 0.25$ (Fig. 3). At much higher extension ratios, the force must increase indefinitely since the molecule is almost straightened out. The thermodynamic approach to the problem of coil stretching for a freely-jointed chain was considered by Treloar [32], who obtained the following expression for the stress-strain relationship when the two chain ends are kept a distance r apart:

$$f(r) = (k_BT/l_p) \cdot L^{-1}(r/Nl_p),\tag{14}$$

where L^{-1} is the function inverse to the Langevin function defined by $L(x) = (\coth x) - 1/x$.

Derived from molecular arguments, Eq. (14) is correct for any extension ratio of the freely-jointed chain. In spite of its generality, the use of Eq. (14) is limited due to mathematical complexity. To account for the finite extensibility of the chain, the approximate *finitely extensible nonlinear elastic* (FENE) law proposed by Warner has gained popularity due to its ease of computation [33]:

$$f(r) = (H_o \cdot r)/(1 - (r/Nl_p)^2) \,. \tag{15}$$

The Warner function has all the desired asymptotical characteristics, i.e. a linear dependence of $f(r)$ on r at small deformation and a finite length Nl_p in the limit of infinite force (Fig. 3). In a non-deterministic flow such as a turbulent flow, it was found useful to model $f(r)$ with an anharmonic oscillator law which permits us to account for the deviation of $f(r)$ from linearity in the intermediate range of chain deformation [34]:

$$f(r) = H_o \cdot (r + \chi r^3) \,, \tag{16}$$

where χ is a parameter which determines the nonlinearity.

The work done by an external force to stretch the polymer chain raises the free energy of the molecule by a quantity which is usually referred to as the energy storage. Such a free-energy storage in deformed polymers plays an essential role in viscoelasticity, drag reduction and also in mechanochemical degradation, as suggested from the results obtained under transient elongational flow.

It is worth recalling that any of the molecular force laws given by Eqs. (13–16) are derived within the framework of the freely-jointed model which considers the polymer chain as completely limp except for the spring force which resists stretching; thus $f(r)$ is purely entropic in nature and comes from the flexibility of the joints which permits the existence of a large number of conformations. With rodlike polymers, the statistical number of conformations is reduced to one and $f(r)$ actually vanishes when the chain is in a fully extended state.

In a real flexible polymer, rotation around the backbone bonds is hindered and the chain consists of a statistical distribution of the *gauche* (g^+, g^-) and *trans* (t) conformations of the trimers. Calculation of the free-energy storage using the Rotational Isomeric State model [35] shows that these non-entropic effects account for a significant part in the free-energy storage of a real system (Fig. 4). The shape of the curves reported in Fig. 4 was interpreted as follows: if the interaction between polymer segment and the flow field is less than the potential barrier ($\Delta E_g \approx 11 \, kJ \cdot mol^{-1}$ for polyethylene), the relative ratio of trans to gauche conformation is unlikely to be altered significantly. A deformation of the molecular coil is still possible by a redistribution of the *trans* and *gauche* conformations in order to increase the projected chain length along the external field direction. At this stage, the total internal energy of the molecule is unchanged and the free-energy storage is entirely of entropic origin like in the Gaussian model. At stronger flow interaction, some gauche conformation start to change into the trans state by bond rotation according to the Boltzmann statistics with a con-

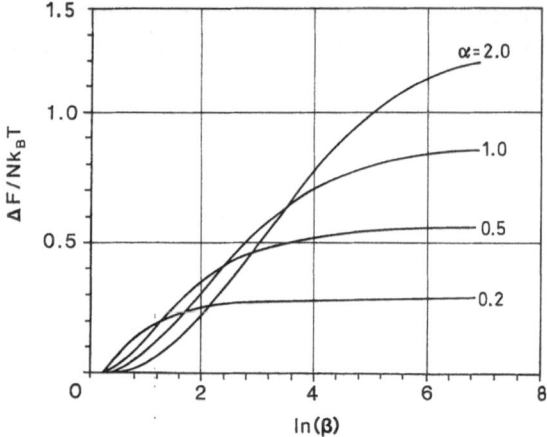

Fig. 4. Free-energy storage per monomer unit against monomer-flow field interaction energy $\alpha = \exp\left(-\Delta E_g/k_B T\right)$ with ΔE_g = relative energy of the *gauche* with respect to the *trans* conformation; $\beta = \exp\left(-\Delta U_{frict}/k_B T\right)$ = Boltzmann factor for the monomer-flow field interaction energy (ΔU_{frict})

comitant increase in free-energy. Ultimately, at a very high strain rates, the chain will be fully extended into an all-*trans* state and the free-energy reaches a plateau value.

Up to the present, a polymer chain has been considered as reversibly deformed at an infinitely small strain-rate. Intuitively, one may expect that the resistance to chain deformation should increase with the rate of deformation. Early in the history of polymer dynamics, it was recognized by Kuhn that the same internal barriers which account for the conformation energy difference could stiffen the coil during fast deformation; this phenomenon is distinguished by the time for a conformational change in the flow field and was designated as "kinetic rigidity", to differentiate from "static rigidity" which characterizes the average conformation of the macromolecule at rest. This additional resistance to chain stretching is known as "internal viscosity" with an associated force (f^{IV}) increasing with the stretching rate but decreasing with the polymer molecular weight [36]:

$$f^{IV}(\text{Kuhn}) = \xi \cdot \dot{r}/M, \tag{17}$$

where is a factor independent of strain rate.

In order to reconciliate high-frequency dynamic rheological data with theory, Cerf proposed an additional internal viscosity term which should be mainly molecular weight independent [37]:

$$f^{IV}(\text{Cerf}) = \xi \cdot \dot{r}. \tag{18}$$

The use of internal viscosity forces permit us to take into account kinetic effects associated with deformation rates which were beyond the scope of most polymer

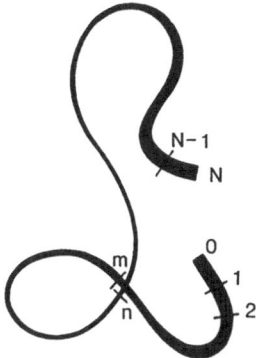

Fig. 5. Large loops model of internal viscosity

dynamics models. Internal viscosity seems to be, at present, the only available explanation for the minimal coil distortion at high shear rates in nonviscous solvents and for the experimental dependence of shear thinning on both molecular weight and solvent viscosity. Despite these important successes, the physical meaning of internal viscosity is still a subject of controversy because its description is often vague and arbitrary. Recently, de Gennes [38] proposed an interesting model of long-range contact points which was later refined by Rabin and Öttinger [39], to explain the Cerf's internal viscosity. In this model, the molecular origins of internal viscosity come from the direct monomer-monomer friction from long-range contact points during the stretching process (Fig. 5).

The inclusion of internal viscosity raises considerably the free-energy storage capacity of a rapidly deforming macromolecule as compared to the idealized Hookean spring model and could play a decisive role in mechanochemical reactivity in transient elongational flow.

2.3 Polymer Dynamics

The presence of large size macromolecules in a flowing polymer solution disturbs the streaming pattern of the Newtonian solvent; at the same time, viscous friction between the solvent and polymer molecules will induce changes in conformation of the latter. This dynamic coupling which occurs at a molecular level, is easy to observe macroscopically and forms the basis of two complementary fields of investigation: rheology which deals with the flowing property of the fluid in presence of polymer molecules, and polymer dynamics which seeks the response of polymer molecules when the solvent is set into motion.

The dynamics of any molecular system could in principle be calculated by considering all the acting intermolecular and intramolecular forces by means of interatomic potentials as a function of distance. Such precise modelling is, obviously, beyond our reach at present and different levels of approximation were imagined to mimic the behavior of a real macromolecule while restricting the number of degrees of freedom to within computational limits. One successful

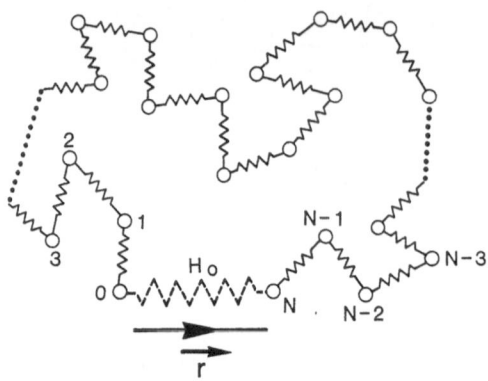

Fig. 6. Bead-spring and elastic dumb-bell representation of a polymer chain in solution

approach, illustrated by the bead-spring models, was to lump the viscous resistance and the molecular mass of the polymer into a low number of discrete beads equidistantly positioned along the chain and connected with frictionless springs (Fig. 6). The total number of beads (N + 1) is in a large measure arbitrary, providing the following criteria are met

— the "submolecule" between two beads should be sufficiently long to follow the random flight statistics
— the relaxation time within each submolecule should be fast enough to be in equilibrium with the overall chain deformation.

To study processes which affect the end-to-end vector r, it is sometimes informative to consider only the two beads localized at each chain end and connected by a single spring. This model, known as the elastic dumbbell, was originally proposed by Kuhn over half a century ago [40] and constitutes the simplest model of chain dynamics in flow.

The behavior of a bead-spring chain immersed in a flowing solvent could be envisioned as the following: under the influence of hydrodynamic drag forces ($\vec{f}^H$), each bead tends to move differently and to distort the equilibrium distance. It is pulled back, however, by the "entropic need" of the molecule to retain its coiled shape, represented by the restoring forces ($\vec{f}^S$) and materialized by the spring in the model. The random bombardment of the solvent molecules on the polymer beads is taken into account by "time smoothed" Brownian forces ($\vec{f}^B$). Finally inertial forces ($\vec{f}^I$) are introduced into the forces balance equation by the bead mass (m) times the acceleration ($\ddot{\vec{r}}$) of one bead relative to the others:

$$\vec{f}_i^H + \vec{f}_i^S + \vec{f}_i^B + \vec{f}_i^I = \vec{0} \tag{19}$$

with the subscript "i" denoting the ith-bead of the chain.

The hydrodynamic forces are usually described by a linear relation between drag resistance and relative fluid velocity:

$$\vec{f}_i^H = \xi(\vec{v} - \dot{\vec{r}}_i), \quad i = 0 \text{ to } N. \tag{20}$$

In this equation, ξ is a proportionality factor known as the "bead-solvent friction coefficient" which purports to account in some kind of average way for the complex molecular interactions as the polymer segments (schematized by the bead) move about in the solvent. Following Stokes' law of drag resistance, this friction coefficient is usually given as $\xi = 6\pi\eta_s a$, with a equal to the bead radius.

The term $(\vec{v} - \dot{\vec{r}}_i)$ is the difference between the statistically averaged velocity of the bead $(\dot{\vec{r}}_i)$ and the local velocity $(\vec{v})$ of the continuum background at the position of the bead. The drag force is zero when $(\vec{v} - \dot{\vec{r}}_i) = \vec{0}$, in which case the beads deform affinely with the surrounding medium.

The entropic force $\vec{f}^s$, given as the negative of the gradient of the spring potential energy, is represented by one of the various kinds of springs used in molecular models (Fig. 3).

Einstein's theory of diffusion states that the velocity of a particle fluctuates much more rapidly than its position under the influence of Brownian forces. On a time-scale longer than the time required for the bead to equilibrate $(t > m/6\pi\eta_s a \approx 10^{-12} \text{ s})$ its velocity distribution, a steady-state velocity distribution is reached and it is usual to omit the inertia term on the grounds of "the small mass and the sluggish motion of the beads" [41]. It was pointed out, nevertheless, that inertial effects may be consequential in flow with a high velocity gradient and its omission can introduce serious errors into the calculations [42]. Equation (19), which combines deterministic forces with random forces from the microscopic world (Brownian motion), is called a Langevin equation. Instead of considering explicitly the random forces, the other alternative would be to look at the effect of Brownian motion as a diffusion process and to derive the corresponding Smoluchowski equation. Both the Langevin equation and the Smoluchowski equation represent the same motion and can be used interchangeably.

Due to the presence of the stochastic term, a Langevin equation is not solvable as an ordinary differential equation but only in terms of a probability $p(r, t)$ for the bead-spring chain to have some value of the end-to-end vector $\vec{r}$ at a given time t. Solutions of the diffusion equation in flow sufficiently "strong" to deform the molecular coil show that the distribution function $p(r, t)$ is always highly localized [43]. The high localization of the beads means that the diffusive Brownian forces are weak in comparison to the other forces on the beads. The physical interpretation of this fact is fairly simple: since the elastic and drag forces are rapidly increasing functions of the chain extension, the Brownian forces which do not change so much will become eclipsed and eventually negligible at sufficiently high degrees of coil distortion.

For most systems in thermal equilibrium, it is sufficient to regard $\vec{f}^B$ as random forces which follows a Gaussian distribution function with mean value $\langle \vec{f}_i^B(t) \rangle = 0$ and standard deviation $\langle \vec{f}_i^B(t)\, \vec{f}_j^B(t') \rangle = 2k_B T \xi \delta(i - j)\, \delta(t - t')$ [44].

2.3.1 Free-Draining and Non-Free Draining Chains

In the Rouse model [45], the drag velocity is considered to be uniform on any bead and equal to the relative velocity of the center of mass of the bead-spring with respect to the surrounding fluid; moreover, a Hookean spring law was assumed for $\vec{f}^S$. These assumptions permit us to transform the average Langevin equation into a system of $(N + 1)$-linear equations:

$$\overset{\circ}{\vec{r}}_i = (-3k_BT/l_p^2) \cdot [(\vec{r}_i - \vec{r}_{i+1}) + (\vec{r}_i - \vec{r}_{i-1})] + \vec{f}_i^B, \qquad 1 < i < N$$

$$\text{(21a)}$$

and for the end beads:

$$\overset{\circ}{\vec{r}}_0 = (-3k_BT/l_p^2) \cdot (\vec{r}_0 - \vec{r}_1) + \vec{f}_0^B, \tag{21b}$$

$$\overset{\circ}{\vec{r}}_N = (-3k_BT/l_p^2) \cdot (\vec{r}_N - \vec{r}_{N-1}) + \vec{f}_N^B. \tag{21c}$$

In the limit where i can be regarded as a continuous variable, the system of Eqs. (21) represents the Brownian motion of N coupled oscillators in which each bead attached to a given spring feels forces from the adjacent springs. Rouse devised a "normal-mode" transformation which permits one to diagonalize the matrix equation into a system of $(N + 1)$ uncoupled equations. Solving each equation gives the relaxation time (τ_i) for the corresponding normal mode with the longest one (τ_1) defining the relaxation of the chain ends (Fig. 7):

$$\tau_p \, (\text{Rouse}) = N^2 l_p^2/(3\pi^2 k_B T \cdot p^2), \qquad p = 1 \text{ to } N. \tag{22}$$

When rearranged into a form which shows no dependence on the arbitrary value N, the following relation was obtained for τ_1:

$$\tau^R \equiv \tau_1 \, (\text{Rouse}) = \xi_0 P^2 l^2/(6\pi^2 k_B T). \tag{23}$$

Here, ξ_0 is the friction coefficient per monomer unit and P is the degree of polymerization.

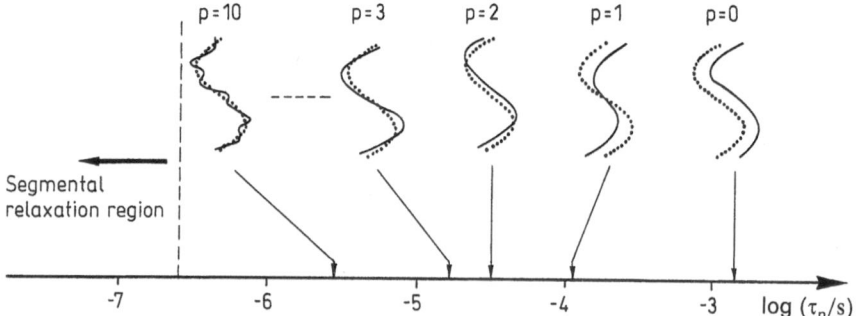

Fig. 7. Significance and time scale of the normal modes (the values given refer to a 10^6 MW PS in decalin; $p = 0$ corresponds to a chain translation)

For practical purposes, it is convenient to define the relaxation time in terms of macroscopic quantities which can be readily determined. Within the validity limit of Hookean connectors (Eq. 13), the low-shear viscosity of a polymer solution is given by the relation:

$$\eta_0 = Nk_BT \sum_{p=1}^{N} \tau_p + \eta_s, \tag{24}$$

where N is the number density of macromolecules, related to the mass concentration (c) by $N = c \cdot N_A/M$ (N_A: Avogadro's number). The intrinsic viscosity, on the other hand, is defined as:

$$[\eta] = \lim_{c \to 0} (\eta - \eta_s)/(\eta_s \cdot c). \tag{25}$$

Combining Eqs. (24) and (25) gives the longest relaxation time for the Rouse model as:

$$\tau^R = (6/\pi^2) ([\eta] \eta_s M/RT). \tag{26}$$

Equation (23) predicts a dependence of τ^R on M^2. Experimentally, it was found that the relaxation time for flexible polymer chains in dilute solutions obeys a different scaling law, i.e. $\tau_1 \approx M^{3/2}$. The Rouse model does not consider excluded volume effects or polymer-solvent interactions, it assumes a Gaussian behavior for the chain conformation even when distorted by the flow. Its domain of validity is therefore limited to modest deformations under θ-conditions. The weakest point, however, was neglecting hydrodynamic interaction which will now be discussed.

"Hydrodynamic interaction" is a long-range interaction mediated by the solvent medium and constitutes a cornerstone in any theory of polymer fluids. Although the mathematical formulation needs somewhat elaborate methods, the idea of hydrodynamic interaction is easy to understand: suppose that a force is somehow exerted on a Newtonian solvent at the origin. This force sets the surrounding solvent in motion; away from the origin, a velocity field is created which decreases as:

$$v(r) \sim 1/(\eta_s r) \tag{27}$$

according to the Oseen formula.

In a concentrated solution, characterized by an "effective" medium viscosity $\eta_e \gg \eta_s$, the hydrodynamic field decays much faster due to the shielding effect of the encountered polymer segments:

$$v(r) \sim 1/(\eta_e r). \tag{28}$$

It is usual to define a "screening length" (ξ_H) as the distance at which hydrodynamic interaction becomes negligible (Fig. 8) [44].

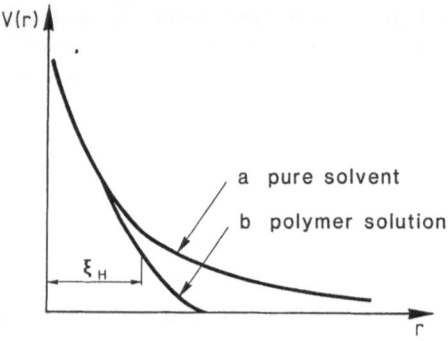

Fig. 8. Velocity field caused by a point force in a pure solvent and in a concentrated polymer solution

The same effect happens inside a random flight chain where the close proximity of the polymer segments offers mutual screening from the bulk flow field. The idea of a chain being "non-drained" was first considered by Debye & Bueche who introduced the concept of a "shielding length" defined as [46]:

$$l = (\eta_s/n \cdot \xi)^{1/2} \tag{29}$$

with n = bead density and ξ = friction coefficient of the bead. For most flexible polymer chains, the "shielding ratio" R_g/l is much larger than unity and the molecular coils behave much like impenetrable spheres with an equivalent radius equal to [47]:

$$R_e \approx \langle 5/3 \rangle^{1/2} R_g . \tag{30}$$

The "non-free draining" character of flexible polymer chains was considered in the Zimm model [48]. In this model, the effect of hydrodynamic interaction at the location of bead "i" is taken into account by an additional fluid velocity term $\vec{v}_i'$:

$$\vec{v}_i = \vec{v}_{i0} + \vec{v}_i' . \tag{31}$$

Due to the gradual decay of the hydrodynamic interaction (Eq. 27), the extra velocity component at bead i results from the motion of all the remaining beads and it is presumed that $\vec{v}_i'$ depends linearly on the hydrodynamic forces acting on these beads:

$$\vec{v}_i' = -\sum_{j \neq i} \mathbf{\Omega}_{ij} \vec{f}_j^h . \tag{32}$$

The $\mathbf{\Omega}_{ij}$ are coefficients of the Oseen tensor given by:

$$\mathbf{\Omega}_{ij} = (1/8\pi\eta_s|\vec{r}_{ij}|) \cdot (\hat{r}_{ij}\hat{r}_{ij} + \mathbf{I}), \qquad i \neq j \tag{33}$$

with $\vec{r}_{ij} = \vec{r}_i - \vec{r}_j$, $\hat{r}_{ij}$ = the unit vector in the direction of $\vec{r}_{ij}$ and $\mathbf{I}$ = the unit tensor defined by $\mathbf{I}_{\alpha\beta} = \delta_{\alpha\beta}$.

The Rouse model, as given by the system of Eq. (21), describes the dynamics of a connected body displaying local interactions. In the Zimm model, on the other hand, the interactions among the segments are delocalized due to the inclusion of long range hydrodynamic effects. For this reason, the solution of the system of coupled equations and its transformation into normal mode coordinates are much more laborious than with the Rouse model. In order to uncouple the system of matrix equations, Zimm replaced Ω_{ij} by its average over the equilibrium distribution function:

$$\Omega_{ij} \approx \langle \Omega_{ij} \rangle_0 = (I/6\pi\eta_s) \langle 1/|\vec{r}_i - \vec{r}_j| \rangle_0$$
$$= I/(6\pi^3|i - j|)^{1/2}\,\eta_s l_p . \tag{34}$$

In analogy with the Rouse model, the longest relaxation time (τ_1) according to the Zimm model can again be put into a form which does not depend on N [44]:

$$\tau^z \equiv \tau_1 \,(\text{Zimm}) = 0.398 \cdot \eta_s R^3/k_B T . \tag{35}$$

As in the case of the Rouse relaxation time, it is possible to express τ^z in term of the intrinsic viscosity either by using Eq. (24) or the Kirkwood-Riseman theory [24, 47]:

$$[\eta] = \emptyset \cdot R^3/M . \tag{36}$$

The constant $\emptyset$, called the Flory-Fox viscosity parameter, has an experimental value of $\approx 2.5 \times 10^{23}\,\text{mol}^{-1}$ [44]. The corresponding value for τ^z is then:

$$\tau^z \approx 0.95 \cdot \eta_s[\eta]\,M/RT , \tag{37}$$

where R is the molar gas constant.

The numerical factor (0.95) should not be taken too seriously, because it varies with the approximations used to derive Eqs. (35) & (36). In practice, values ranging from 0.42 to 1 can be found in the literature.

The Zimm model predicts correctly the experimental scaling exponent $\tau_1 \approx M^{3/2}$ determined in dilute solutions under θ-conditions. In concentrated solution and melts, the hydrodynamic interaction between the polymer segments of the same chain is screened by the "host" molecules (Eq. 28) and a flexible polymer coil behaves much like a free-draining chain with a Rouse spectrum in the relaxation times.

2.3.2 Bead-Spring Chain in a Good Solvent

The Rouse and Zimm models are valid only under θ-conditions. To extend their range of applicability into good solvent conditions, several improvements have been proposed to include excluded volume effects. Dynamical scaling, however, provides probably the simplest approach to the problem [30].

The dynamic scaling argument supposes that when the geometrical parameters of the chain (i.e. N and l_p) are changed from N into N/λ and l_p into $l_p\lambda^\nu$, any physical quantity (A), either static or dynamic, related to the molecular size will be transformed into $\lambda^x A$. The parameter "ν" is the exponent in Eq. (9) and is equal to $1/2$ in θ-solvent and $\approx 3/5$ in good solvent.

When applied to the relaxation time of a polymer, dimensional analysis of Eq. (22) shows that the following scaling transformation should be written for τ^R:

$$\tau^R \sim (\xi l_p^2/k_B T) \cdot f(N) = (\xi l_p^2 \lambda^\nu/k_B T) \cdot f(N/\lambda) . \tag{38}$$

For the equality to be true for any arbitrary value of λ, $f(N)$ must be of the form $f(N) = \text{numerical constant} \cdot N^{1+2\nu}$ which yields:

$$\tau^R \sim N^{1+2\nu} . \tag{39}$$

In a similar fashion, the scaling transformation of τ^z as given from Eq. (35) is the following:

$$\tau^z \sim (\eta_s l_p^3/k_B T) \cdot f(N) = (\eta_s l_p^3 \lambda^{3\nu}/k_B T) \cdot f(N/\lambda) \tag{40}$$

which permits us to derive the scaling law for τ^z:

$$\tau^z \sim N^{3\nu} . \tag{41}$$

As coils become more expanded, the hydrodynamic interaction decreases: this is reflected in the Zimm relaxation time which approaches the Rouse value in good solvent.

The scaling argument provides only the exponent but not the absolute numerical value for the constant. Therefore, for quantitative results, it should be completed by some more refined technique like the afore-mentioned renormalization group method [49].

2.3.3 Large Deformation Models

The Rouse and Zimm models consider only minute deformations of the molecular coil in the presence of a constant velocity field. In the presence of a velocity gradient, each bead sample has a different fluid velocity resulting in different drag forces which must be incorporated into the system of equations (21).

For the simplest case of a linear dumbbell in a homogeneous velocity gradient of strain rate $\dot\varepsilon$, the force balance equation is the following:

$$\xi \dot{\vec{R}} = \dot\varepsilon \Lambda \vec{R} - 2H_o \vec{R} + \text{Brownian forces} \tag{42}$$

with Λ = dimensionless velocity gradient tensor and $\text{tr } \Lambda = 0$ (incompressible fluid) (cf. Sect. 4, Fig. 21).

Neglecting Brownian motion, Eq. (42) is further simplified into:

$$\overset{\circ}{\vec{R}} = \overset{\circ}{\varepsilon}\Lambda\vec{R} - \vec{R}/(2\tau) \tag{43}$$

in which the ratio H_o/ξ is identified as the relaxation time of the dumbbell (τ).
Equation (43) predicts that unbound exponential growth in R will occur if:

$$(2\overset{\circ}{\varepsilon}\tau)\,\delta^+ > 1, \tag{44}$$

where δ^+ is the largest real eigenvalue of Λ.

Flows in which condition (44) is realized are referred to as "strong" flows, while flows which do not satisfy this condition are known as "weak" ones.

For large distortions, two approximations underlying the preceding calculations should be reconsidered.

1) The entropic restoring force was modeled with a linear elastic spring. This approximation is not so critical in a "weak flow" where molecular coils are modestly distorted, particularly with a high MW polymer which remains Gaussian up to an appreciable deformation ratio. In strong flow, however, the Hookean spring leads to an infinite extension of the chain. To remedy this unrealistic situation, a more appropriate force law like the FENE-spring law, should be employed (Eq. 15).

2) Since the hydrodynamic interaction decreases as the inverse distance between the beads (Eq. 27), it is expected that it should vary with the degree of polymer chain distortion. This is not considered in the Zimm model which assumes a constant hydrodynamic interaction given by the equilibrium averaging of the Oseen tensor (Eq. 34).

In an attempt to describe the behavior at large chain deformations, de Gennes [7] incorporated into the dumbbell model the FENE spring law along with a variable bead friction coefficient which increases linearly with the interbead distance:

$$\xi(r) = \xi_{eq} \cdot (1 + \alpha\beta r). \tag{45}$$

ξ_{eq} is the frictional coefficient of the undeformed dumbbell, α a dimensionless parameter which scales with $N^{1/2}$ and $\beta^2 = 3/(2Nl_p^2)$. Remember that the frictional properties of a chain in the coiled state are modeled with a solid sphere (Eq. 30). For a partly stretched coil, the analogous model would be a cylinder with length L and diameter d. In the presence of a steady elongational flow of strain rate ($\overset{\circ}{\varepsilon}$), the increment in drag force (df^H) on a segment of the cylinder of length (dx) located a distance x from the center (Fig. 9) is given by the results of slender body hydrodynamics:

$$df^H = \eta_s(2\pi/\ln (L/d))\,v\,dx \tag{46}$$

where $v = \overset{\circ}{\varepsilon} \cdot x$ is the velocity of the solvent relative to the cylinder at position x.

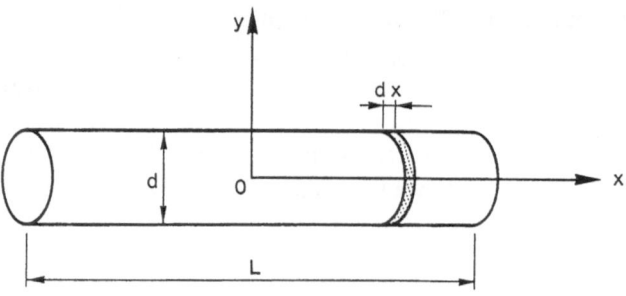

Fig. 9. Stress distribution on a solid cylinder in elongational flow (the flow direction is x)

The form factor $(2\pi/\ln (L/d))$ changes slowly with the aspect ratio (L/d) and can be regarded as constant (k). The total drag on the cylinder is obtained by integration:

$$f^H = k\eta_s \int_{-L/2}^{+L/2} v \, dx = k\eta_s \int_{-L/2}^{+L/2} \mathring{\varepsilon}x \, dx = \tfrac{1}{4} k\eta_s \mathring{\varepsilon}L^2 \tag{47}$$

Since the drag on a dumbbell is equal to:

$$f^H \text{ (dumbbell)} = \tfrac{1}{2} \xi L \mathring{\varepsilon} \tag{48}$$

the equivalence between Eq. (45) and Eq. (47) is obtained by identifying $\xi(r)$ with $1/2 \, k\eta_s L$.

The quadratic dependence of f^H on the end-to-end distance L has a simple physical significance: as the cylinder is stretched, it presents a greater average distance from the center, and hence experiences more drag. In terms of the chain unravelling process, this result means that the hydrodynamic shielding of interior beads from solvent drag is stripped away as the chain is uncoiled. Because of the extra drag produced by this stretching, the gripping action of the flowing fluid increases in efficiency as the chain starts to uncoil. For a large part of the stretching, the entropic force f^S increases more slowly with r $(f^S \approx (r/r_{eq})^{1+\nu})$ than the drag force $(f^H \approx (r/r_{eq})^2)$ and it is expected that a run-away effect will be observed at some stage of coil deformation, leading to a rapid extension of the chain. This intuitive argument for the coil-to-stretch (CS) transition can be put into a more quantitative form with the use of the FENE-dumbbell effective potential (U) [50]:

$$U = -\int (f^H + f^S) \, dr = \int \xi_{eq} \cdot (1 + \alpha\beta r) \, dr + \int H_o r/[1 - (r/l_{max})^2] \, dr$$

$$= -\tfrac{1}{4} \xi_{eq} \mathring{\varepsilon}(r^2 + \tfrac{2}{3}\alpha\beta r^3) - \tfrac{1}{2} H_o l_{max}^2 \ln [1 - (r/l_{max})^2]. \tag{49}$$

Different formulations for Eq. (49) are found in the literature, but they give similar results for the coil-to-stretch behavior of the polymer chain [7, 30, 51].

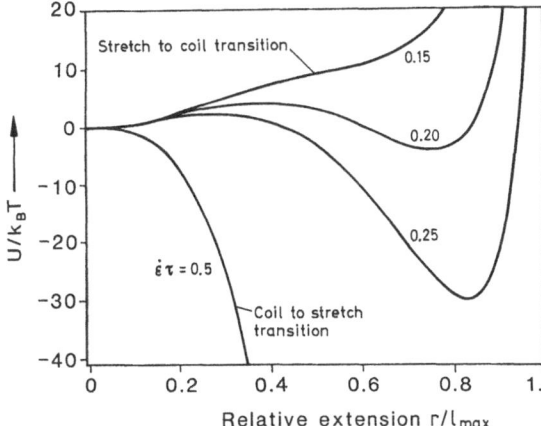

Fig. 10. Free energy versus relative extension (r/l_{max}) for a dumbbell oriented in a longitudinal velocity gradient

Plotting U as a function of L (or equivalently, to the end-to-end distance r of the modeled coil) permits us to predict the coil stretching behavior at different values of the parameter $\dot{\varepsilon}\tau$, where τ is the relaxation time of the dumbbell (Fig. 10). When $\dot{\varepsilon}\tau < 0.15$, the only minimum in the potential curve is at $r = 0$ and all the dumbbell configurations are in the coil state. As $\dot{\varepsilon}\tau$ increases (to 0.20 in the Fig. 10), a second minimum appears which corresponds to a stretched state. Since the potential barrier (ΔU) between the two minima can be large compared to $k_B T$, coiled molecules require a very long time, to the order of $\tau \cdot \exp (\Delta U/k_B T)$, to diffuse by Brownian motion over the barrier to the stretched state: at any stage, there will be a distribution of long-lived metastable states with different chain conformations. With further increases in $\dot{\varepsilon}\tau$, the second minimum deepens. The barrier decreases then disappears at $\dot{\varepsilon}\tau = 0.5$. At this critical strain rate denoted by $\dot{\varepsilon}_{cs}$, the transition from the coiled to the stretched state should occur instantaneously.

If $\dot{\varepsilon}$ is now decreased, with the chain in the extended state, the dumbbell nevertheless stays in the stretched state where the potential is the lowest. The transition back to the coiled state occurs only when there is a single minimum on the potential energy curve, i.e. at $\dot{\varepsilon}\tau = 0.15$. Since the critical strain-rate for the stretch-to-coil transition $(\dot{\varepsilon}_{sc})$ is much below the corresponding value for the coil-to-stretch transition $(\dot{\varepsilon}_{cs})$, the chain stretching phenomenon shows hysteresis (Fig. 11).

The dumbbell relaxation time (τ) in the preceding model is coil deformation dependent. Neglecting Brownian forces, the dumbbell relaxation time is given by $\tau \approx f^H/f^S$. Equation (45) is then tantamount to saying that τ increases approximately in proportion to the root mean square end-to-end separation distance R [52]:

$$\tau = \tau_{lim}/(1 + C/R).\qquad(50)$$

τ_{lim} is the limiting value for a fully stretched chain and should correspond to the Rouse relaxation time; when there is no deformation, τ must equal the Zimm

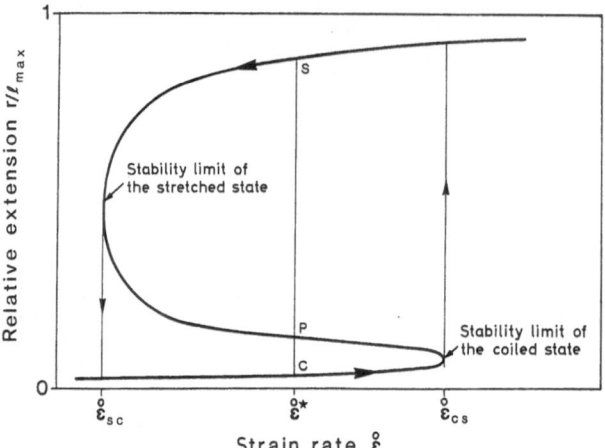

Fig. 11. Hysteresis in the coil-to-stretch (CS) and stretch-to-coil (SC) transitions. For reversible situations, the transition from (S) to (C) should take place at a particular value of the strain rate ($\mathring{\varepsilon}^*$) through an unstable state (P). In practice, hysteresis is expected and should involve the cycle shown by *arrows*

relaxation time. From this requirement, C is found to be equal to:

$$C = (\tau^R/\tau^Z - 1) \cdot R_{eq} . \tag{51}$$

(R_{eq} = rms end-to-end separation distance in the equilibrium configuration)

Using the above argument, it can be seen that $\mathring{\varepsilon}_{cs}$ is governed by the reciprocal value of τ^Z, as expected, it occurs at a much higher value of the strain-rate as compared to $\mathring{\varepsilon}_{sc}$ which depends on the reciprocal value of τ^R.

Using different elongational flow geometries, the CS transition of a flexible polymer chain as well as the phenomenon of hysteresis for the reversed process of SC transition has been confirmed experimentally by the Bristol group [8, 9], the Cal-Tech group [10, 53] and the Paris group [11, 54].

Based on a mechanical model in a time-independent flow, de Gennes' derivation tries to extrapolate it to a time-dependent chain behavior. His implicit assumptions have been criticised by Bird et al. [55]. More recent calculations extending the de Gennes' dumbbell to the bead-spring situation [56] tend, nevertheless, to confirm the existence of a well-characterized CS transition: results with up to 100-bead chains show a critical value of the strain rate at $\mathring{\varepsilon}_{cs} = 0.5035/\tau^Z$ which is just 7% higher than the value predicted by de Gennes.

Most theories predict substantial stretching and alignment of the molecular coils at:

$$\mathring{\varepsilon}_{cs} > A/\tau^Z . \tag{52}$$

Although $A \approx 0.5$ seems to be the most frequently derived number, other values ranging from 1 to 10 can also be found in the literature [57].

2.3.4 Chain Conformation in Non-Steady State Elongational Flow

The bead-spring models reside on the assumption that the rate of stretching is much smaller than the rate at which a portion of the real chain represented by a single spring can sample its distribution of conformations. If this thermodynamic condition is not fulfilled, then the above-considered portion of the chain cannot be considered as an entropic spring regardless of which force law is used: in transient elongational flow with high strain rate, it is thus expected that the bead-spring model is not entirely appropriate. More important, probably, bead-spring models do not account for the noncrossability of the real chain ("phantom" chain) and such effects as internal friction and self-entanglements are completely neglected. A few attempts to model a real chain with the inclusion of non-equilibrium effects have been reported. The elastic dumbbell with internal viscosity was introduced by Kuhn and is simulated by two beads connected by a spring in parallel with a linear dashpot. The inclusion of non-conservative forces adds considerable mathematical difficulties; nonequilibrium kinetic theory is much more involved, even with the simple dumbbell model, and must rely on some form of approximation to obtain useful results [41, 58].

Due to the complexity of the problem, most of the non-steady state models are only empirical in nature; they are generally designed to fit rather than to predict experimental results and are, therefore, limited in scope.

As stated previously, the coil distortion is modest or even absent when $\dot{\varepsilon}\tau^Z < 0.5$. With a purely entropic chain, the system of Eqs. (15), (19) and (20) is sufficient to describe the transient response of the polymer molecule to a supercritical ($\dot{\varepsilon}\tau^Z > 0.5$) elongational flow: it can be predicted, for example, that the f^S will increase approximatively linearly with the chain distortion r until about one-third of the maximum stretch distance l_{max} (Fig. 3). In order to keep the drag force (f^H) from becoming large, the distortion is predicted to be affine with the surrounding medium. This prediction is, however, at odds with the large non-Newtonian effect observed by James and Saringer [59] during the sink flow of dilute PEO solution (Fig. 12). In this experiment, the pressure drop ΔP between the two ports 1 and 2

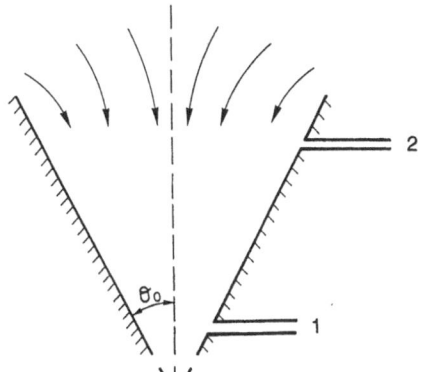

Fig. 12. Sketch of the conical-channel flow experiment of James and Saringer. The pressure drop was measured between ports 1 and 2 (redrawn according to Ref. [59])

was measured as a function of fluid velocity. For a viscoelastic fluid, ΔP is proportional to the elongational stress tensor σ_E which results from the stress contributed by the stretched polymer molecules:

$$\sigma_E = N \cdot \langle f(r) \cdot \vec{r}\vec{r} \rangle , \tag{53}$$

where N is the number of macromolecules per unit volume of solution. For the considered flow geometry (Fig. 12), the residence time was too short for significant distortion of the bead-spring chain. Even so, the measured σ_E surpasses the value predicted from the Gaussian dumbbell by several orders of magnitude. To interpret the results, the authors invoked the formation of aggregates as a possible explanation. Later on, King and James [60] suggested the presence of topological constraints, like self-entanglements or knots [61], which may "freeze" the chain into effectively rigid objects (Fig. 13a). The polymer solution then behaves like a suspension of elongated particles and can generate large stresses as the solvent is forced to flow around these objects [62].

Based on the results of James and Saringer, Ryskin proposed a different version of chain unravelling in transient elongational flow named the "yo-yo" model [63]. In this model, the chain starts to uncoil from its most stressed part, i.e. the central portion (Cf. Eq. (83) in Sect. 5.1) which grows in length at the expense of the outer portions. If the flow becomes weaker, the chain will curl back from its ends like a "yo-yo" toy (Fig. 13b). This representation of chain unravelling was previously formulated in the field of polymer dynamics, by Kuhn [2] then by Frenkel [3], over half a century ago. The Langevin model for a free-draining ideal chain of Rabin [64] and the recent bead-spring calculations of Magda et al. [56] predict that the center of the macromolecule is the most stretched part of the chain, even for a modest extension ratio. These results are, however, at variance with the FENE bead-spring calculations of Wiest, Bird and Wedgewood in non-steady elongational flow [55]. Their results showed that the coil-to-stretch transition takes place in roughly four stages: the equilibrium coil (a), the deformed coil (b), the "locally unravelled" chain (c) and the unfolded chain (d). With the inception of a "strong" flow, the chains which are initially in the equilibrium state (a) are deformed into ellipsoids by rearrangements of the angles without appreciable stretching of the links (b). Later, all the individual links are stretched simultaneously to nearly their maximum extensions; the chain, however has not yet unravelled at this point and remains ellipsoidal in shape (c). Only at the long time limit,

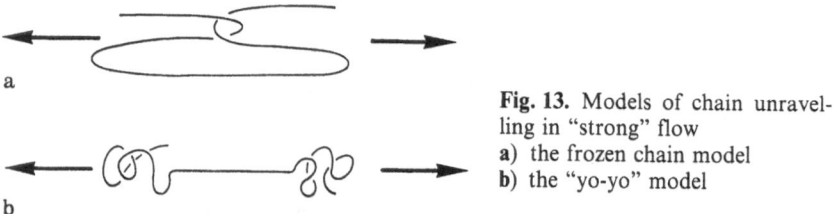

a

b

Fig. 13. Models of chain unravelling in "strong" flow
a) the frozen chain model
b) the "yo-yo" model

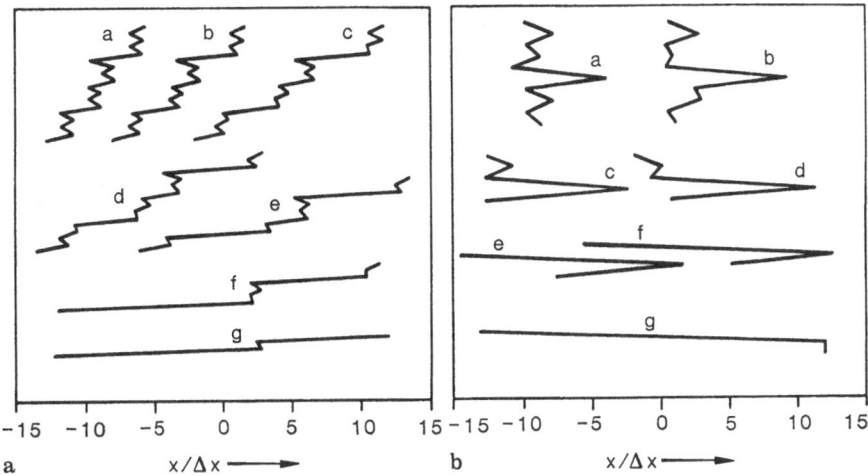

Fig. 14a, b. Development of kinks during chain uncoiling at various stages of deformation; λ = 0 (a), λ = 0.4 (b), λ = 0.8 (c), λ = 1.0 (d), λ = 1.2 (e), λ = 1.4 (f), λ = 1.5 (g) (redrawn according to Ref. [69]). Figure (a) depicts a simulation situation where the chain unraveled quickly, whereas (b) describes a chain which unraveled slowly

when the flow reaches steady state, does the chain unfold completely and assumes a stretched state (d).

In the above models, the development of kinks during chain deformation, implicit to the Kuhn's internal viscosity concept and well-known in parraffin-like chains [65], is not taken into account. In many numerical simulations [66], it was noticed that the presence of a highly folded or kinked state can persist for quite a long time in an elongational flow. This result is in line with some recent direct microscopical observations of DNA which undergoes migration during gel electrophoresis [67, 68]: by using DNA stained with fluorescent dyes, it was observed that "hairpin" conformations can form long-lived metastable states due

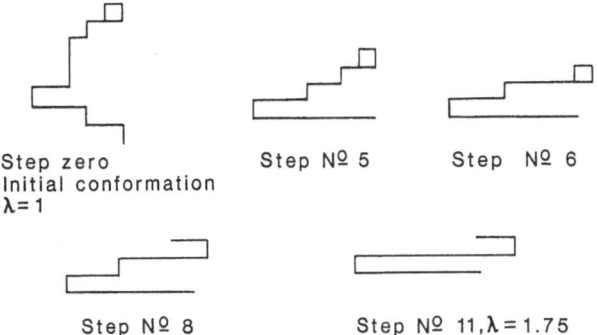

Fig. 15. Simulation of conformational changes in elongational flow by preferred rotation of bonds perpendicular to flow direction (according to Ref. [70])

to partial compensation of the hydrodynamic drag force on each side of the region where the chain reverses its direction.

The existence of kinks was recently explicitly taken into account by Larson (Fig. 14) as a possible model for chain unravelling in the flow [69]. At the same time, Kausch developed a similar model to explain degradation results measured in transient elongational flow (Fig. 15) [70]. With this difference from the Larson model, kinks in the latter model can support compressive stress; chain elastic modulii range from 16 to 110 GPa, depending on the number of defects within the kinked region.

3 Molecular Basis of Mechanochemical Degradation

3.1 Molecular Strain and Reactivity

It is well-known to the organic chemist that the input of external energy, either of thermal or electromagnetic origins, is an efficient way to promote chemical reactions in a system. Mechanical energy, although ubiquitous in every day life, is rarely considered by itself as a bona fide source of chemical activation. The main reason probably stems from the inefficiency of this mode of excitation in chemical systems where most of the applied mechanical energy is easily dissipated into heat and work of bulk deformation, leaving little available as chemical potential. As a matter of fact, many chemical reactions originally thought to be of mechanochemical origins, were later disclosed as thermal reactions activated by the heat produced by viscous friction [71]. Molecular mobility is an efficient means of releasing mechanical stress, so most genuine stress-activated chemical reactions are more likely to be found in the solid state, particularly in crystalline materials where dislocations are easy to create: the explosive decomposition of nitrogen iodide, the formation of F-centers (and its subsequent reaction with humidity to produce H_2) during milling of KBr, the hydrogen embrittlement of some metals under stress or the activation of magnesium foils for Grignard reactions by hammering are just a few examples for this class of reactions. In dilute solutions, mechanochemical reactions are restricted to polymer systems due to the unique propensity of macromolecules to store free-energy upon deformation (Sect. 2.2) and to sustain a high level of stress for a time sufficiently long for chemical reactions to occur. How this deformation energy is transformed into chemical energy, sometime with a surprisingly high efficiency, is still a point to be elucidated (Fig. 16).

Following the application of stress, a molecular system will attempt to minimize the total free-energy by a series of rearrangements of its configuration. According to molecular mechanics, the excess energy of stress (E_s) in a chemical system comes from several sources, viz. non-valence and coulombic interactions, angular deformation, torsional and valence bond strain energies. A low level of stress activates principally the "soft" modes of deformation like breaking of short-range Van der Waals interactions or rotation around torsional barriers. Although the

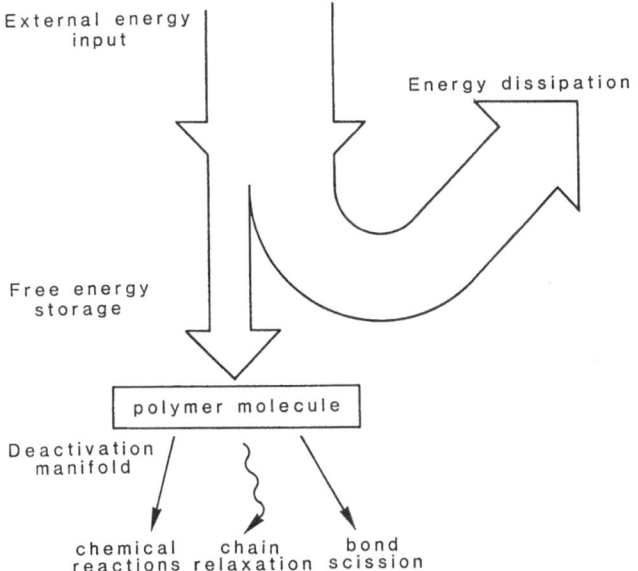

External energy
input

Energy dissipation

Free energy
storage

polymer molecule

Deactivation
manifold

chemical   chain   bond
reactions relaxation scission

Fig. 16. Energy flow diagram in mechanochemical degradation

forces involved during this stage are insufficient to break the valence bonds, they can nevertheless influence the chemical reactivity through the mediation of steric effects as, for example, in the reversible denaturation of dextransucrase by simple shearing [72]. Just after shearing, this enzyme was found to be inactive; after being left overnight, however, the macromolecule recovered most its initial activity after reformation of the secondary and tertiary structures. This supports the idea that the level of shear stress applied in this experiment was just sufficient to break the intramolecular hydrogen bonds and Van der Waals type interactions while preserving the integrity of the molecular backbone.

Rotations around torsional barriers induce changes in chain conformation. For conjugated systems like polydiacetylenes, flow-induced changes in chain conformation can have a profound influence on the photon absorption and electronic conductivity properties of the material [73]. Flow-induced changes in molecular conformation form the basis for several technically important processes, the best known examples are the production of oriented fibers by gel spinning [74], the compatibility enhancement [75] and the shear-induced modification of polymer morphology [76].

Amongst the different modes of molecular deformation, bond angle distortion and valence bond stretching have the most profound influence on the electronic density distribution and hence, on the chemical reactivity. In the range of stresses of ≈ 80 GPa, the valence angles for $C-C$ bonds will become deformed from their equilibrium positions whereas bond stretching and rupture occurs at some still higher values in the vicinity of 740 GPa [15]. The fact that macroscopic force is capable of influencing molecular structure and chemical reactivity is an intriguing fact which, surprisingly enough, has received relatively little theoretical attention [77]. At the present time, a quantum mechanical account for stress-induced chemical reactivity has been given only for the acceleration effect of water during crack propagation of silica glasses. Exposed to the presence of large stresses (up to 15 GPa), the atomic structure of silica becomes distorted from its normal binding configuration. The energy expended to bend the valence bond angle between two oxygen atoms from 105° to 180° (sp^3 to sp^2 hybridization) was calculated to be equal to 350 kJ · mol^{-1} [78]. In the presence of chemisorbed water, the same calculations showed a drastic reduction in the distortion energy which is now lowered to 125 kJ · mol^{-1} following the increase in electron affinity of the silicium atom with bond angle distortion (Fig. 17). As a result, the rate constant of chemisorbed water with a strained silicate structure can be increased by as much as five orders of magnitude with respect to the unstressed state:

$$
H_2O + \quad \begin{array}{c} O \\ \diagdown \diagup \diagdown \diagup \\ Si \qquad Si \\ \diagup \diagdown \diagup \diagdown \\ O \end{array} \longrightarrow \quad \begin{array}{c} OH \quad OH \\ \diagdown | \quad | \diagup \\ Si \qquad Si \\ \diagup \diagdown \diagup \diagdown \\ O \end{array} \quad . \tag{54}
$$

A similar detailed molecular calculation is not available for organic polymers; nevertheless, some qualitative analogy could be made with the preceding results

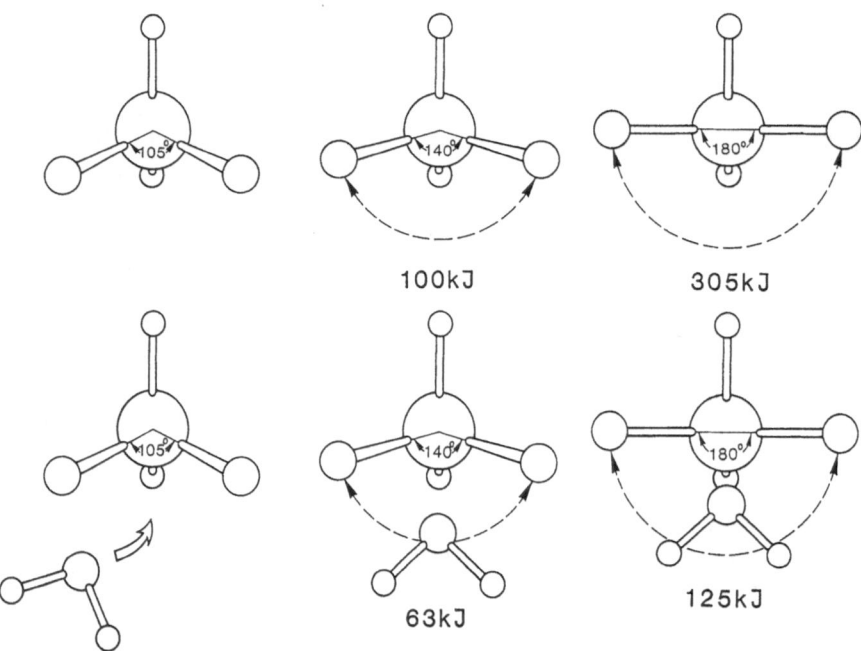

Fig. 17. Bond distortion energy of silicate structure in the absence and presence of water

and with small organic molecules. In carbocyclic compounds where E_s originates principally from valence angle deformation, it is well-known that the ring strain could affect the rate of some chemical reactions in a drastic way. Based on mechanical arguments, it can be inferred that compressive or tensile stresses are effective in reactions which involve distortion of the geometry of the reacting fragments in the direction of the force. For example, if the transition state necessitates distortion of the valence bond angle from θ_0 to θ^*, the valence bond angle contribution to the activation energy is given by

$$E_{ang} = \tfrac{1}{2} k_\theta (\theta_0 - \theta^*)^2 , \tag{55}$$

where k_θ is the elastic constant for bond angle deformation. If however, the same bond angle is already distorted from θ_0 to θ under mechanical stress, the activation energy will be

$$E_{ang} \text{ (stressed)} = \tfrac{1}{2} k_\theta (\theta - \theta^*)^2 \tag{56}$$

resulting in a difference of

$$\Delta E_{ang} = \tfrac{1}{2} k_\theta [(\theta_0 - \theta^*)^2 - (\theta - \theta^*)^2] . \tag{57}$$

It is thus anticipated that compressive stress inhibits while tensile stress promotes chemical processes which necessitate a rehybridization of the carbon atom from the sp^3 to the sp^2 state, regardless of the reaction mechanism. This tendency has been verified for model ring-compounds during the hydrogen abstraction reactions by ozone and methyl radicals: the abstraction rate increases from cyclopropane (c3) to cyclononane (c9), then decreases afterwards in the order anticipated from E_s [79]. The following relationship was derived for this type of reactions:

$$\ln (k_n/k_6) = \alpha \cdot E_s^{1/2}/RT . \tag{58}$$

The indice n indicates the number of carbon atoms in the ring and α is an empirical constant equal to 57 ($J^{1/2} \cdot mol^{-1/2}$).

In semi-cristalline polymers, rate-enhancement under stress has been frequently observed, e.g. in UV-photooxidation of Kapron, natural silk [80], polycaprolactam and polyethylene terephthalate [81]. Quantitative interpretation is, however, difficult in these systems: although the overall rate is determined by the level of applied stress, other stress-dependent factors like the rate of oxygen diffusion or change in polymer morphology could occur concurrently and supersede the elementary molecular steps [82, 83]. Similar experiments in the fluid state showed unequivocally that flow-induced stresses can accelerate several types of reactions, the best studied being the hydrolysis of DNA [84] and of polyacrylamide [85]. In these examples, hydrolysis involves breaking of the ester $O-PO$ and the amide $N-CO$ bonds. The tensile stress stretches the chain, and therefore, facilitates the formation of a transition state in which the bond distance is increased for hydrolysis, thus enhancing the chemical process. Other well-documented examples of stress-induced chemical reactions found in the literature are the acylation of cellulose and the addition of 4-oxipiperidol to rubber chains during mastication [86].

3.2 Breaking Strength of a Polymer Chain

It is common that mechanochemical degradation involves scission of the macromolecule, so one basic question would be to inquire about the level of stress necessary to separate two chemical moieties which have been attached by a covalent bond. Besides the academic interest, the breaking strength of a covalent bond is associated with the ultimate properties of engineering materials and has attracted considerable attention since the beginnings of quantum chemistry.

In the absence of stress, the potential energy for the bond is usually approximated with the Morse function (Fig. 18) [87]:

$$V(l) = D(1 - \exp[-a(l - l_0)])^2, \tag{59}$$

where l is the length of the bond, l_0 the equilibrium separation distance of the atoms, D the bond dissociation energy, $a = \sqrt{k/2D}$ is a parameter which defines the width of the minimum and k is the force constant of the bond in the neighborhood of the equilibrium separation.

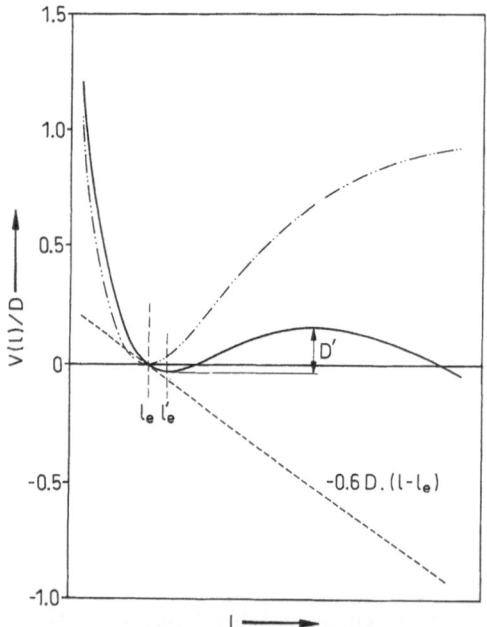

Fig. 18. The Morse potential energy of a bond under equilibrium $(\cdot\cdot - \cdot\cdot)$ and in the presence of an applied force equal to $0.6\,f_b$ (————)

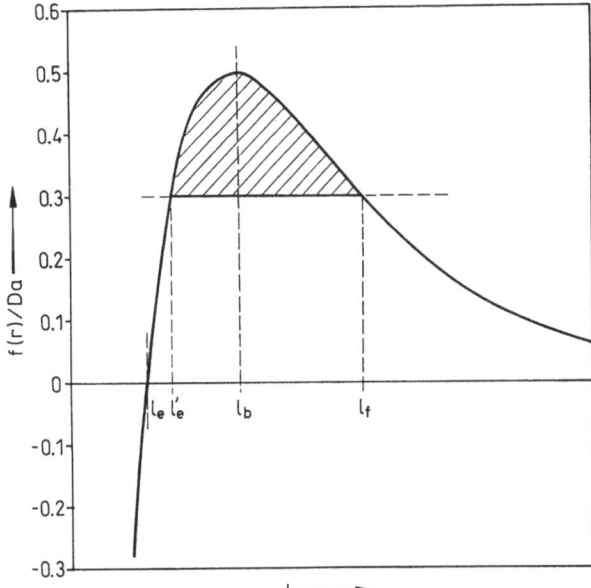

Fig. 19. Derivative of the Morse function. The *shaded area* corresponds to the bond energy under stress

The principle of action and counteraction impose the condition that the acting external force must be balanced by the internal molecular forces. For an isolated diatomic molecule, this internal force (also called the binding force) is given by the derivative of the Morse potential (Fig. 19):

$$f(l) = -dV(l)/dl = -2Da \cdot (\exp[-2a(l - l_0)] - \exp[-a(l - l_0)]).$$
(60)

This function has a maximum at:

$$l_b = l_0 + \ln 2/a$$
(61)

which corresponds to the critical elongation at break and the force to produce this elongation is known as the breaking strength for the bond under consideration:

$$f_b \equiv f(l_b) = \tfrac{1}{2} Da.$$
(62)

The above models consider only one spatial variable which is the bonding distance. It is clear that, for a molecule anything more complex than diatomic, many parameters are needed to define even approximately the potential energy surface. The enormous advances in computational chemistry during the last few years have allowed quantum mechanical calculations on fairly large size molecules. The *first attempt to apply quantum mechanics on deformed polymer chains was made* by Boudreaux [88] who used the CNDO/2 method on a cluster of eight all-*trans* CH$_2$-units combined with an approximate simulation of the periodic environment. The deformation energy and ultimate properties of linear and branched polyethylene in the bulk state have been calculated by this method, giving respectively the values of 3.3 nN (2.7 nN) for the breaking stress and 33% (25%) for the breaking strain of the mentioned species. More recently, Crist et al. [89] pointed out that semi-empirical approaches may not be adequate for molecular configurations far from equilibrium, as for example near the point of rupture, and advocated ab initio calculations as a better alternative. Using the Hartree-Fock self-consistent field molecular orbital scheme, these authors have evaluated the total molecular energy vs elongation for PE, by extrapolation from a series of calculations on odd n-alcanes from C$_1$ to C$_9$ to remove chain-end effects. The computed force-strain curve showed a maximum corresponding to an ultimate strain at $\varepsilon = 0.44$. The tensile strength at this point is 69 GPa, which is equivalent to a breaking force of 12 nN/chain for a perfectly aligned all-*trans* PE molecule at 0 K. At some finite temperature, the actual strength is somewhat lower than the maximum strength due to the influence of thermal fluctuations which stretch interatomic distances in the same way as mechanical stress: in both cases, the increased spacing corresponds to a weaker bonding by the anharmonic potential energy.

It is interesting to note that the simple Morse potential model, when employed with appropriate values for the parameters a and D (a $= 2.3 \times 10^{10}$ m^{-1}, D $= 5.6 \times 10^{-19}$ J as derived from spectroscopic and thermochemical data), gives $f_b = 6.4$ nN and $\varepsilon_b = 20\%$, which are quite comparable to the results obtained with the more sophisticated theoretical techniques [89]. The best experimental data determined on highly oriented UHMWPE fibers give values which are significantly lower than the theoretical estimates ($f_b \approx 2$ nN, $\varepsilon_b = 4\%$), the differences are generally explained by the presence of faults in the bulk sample [72, 90] or by the phonon concept of thermomechanical strength [15].

As the most notable contribution of ab initio studies, it was revealed that the different modes of molecular deformation (i.e. bond stretching, valence angle bending and internal rotation) are excited simultaneously and not sequentially at different levels of stress. Intuitive arguments, implied by molecular mechanics and other semi-empirical procedures, lead to the erroneous assumption that the relative extent of deformation under stress of covalent bonds, valence angles and internal rotation angles ($\Delta r : \Delta\theta : \Delta\Phi$) should be inversely proportional to the relative stiffness of the deformation modes which, for a typical polyolefin, are $100 : 10 : 1$ [15]. A completly different picture emerged from the Hartree-Fock calculations where the determined values of $\Delta r : \Delta\theta : \Delta\Phi$ actually vary in the ratio of $1 : 2.4 : 9$ [91].

3.3 Rate of Stress-Activated Chain Scission

The preceding models permit an evaluation of the ultimate strength of some simple polymers under mechanical stress. Under specific circumstances, it is important to predict the average lifetime of a bond stressed at a level which is below its breaking strength. This problem is deeply associated with the strength of materials and constitutes a classical field of experimentation for the rate theory of fracture.

In 1936, de Boer formulated his theory of a stressed bond which, despite its simplicity, still constitutes the basis for most models of chemical reactivity under stress [92]. In order to fracture an unstressed bond which, in the absence of any vibration, is approximated by the Morse potential of Fig. 18, an energy D must be supplied. If, however, the bond is under tension due to a constant force f_{ext} pulling on either end, the bond rupture activation energy will be decreased by an amount equivalent to the work performed by the mechanical force over the stretching distance from the equilibrium position. The bond potential energy in the presence of stress is given by:

$$V'(l) = V(l) - f_{ext} \cdot (l - l_0) \tag{63}$$

and has a minimum at $l'_0 > l_0$: this is in accord with the intuitive expectation that the bond separation distance should increase in the presence of a tensile stress (Fig. 18). The new activation energy (D') required to break the stressed bond could be calculated from the principle of virtual work performed on the bond in going from l'_0 to l_b:

$$D' = \int_{l'_e}^{l_f} dV'(l)/dl = \int_{l'_e}^{l_f} (f(l) - f_{ext}) \, dl . \tag{64}$$

The ratio (D/D') can be calculated from Eqs. (59) and (64), and is plotted in Fig. 20 as a function of x:

$$D'/D = x \ln \left[(1 - x - \sqrt{1 - 2x})/x \right] + \sqrt{1 - 2x} , \tag{65}$$

where $x = f_{ext}/Da$.

For a large molecule, the internal forces acting on a particular bond are the result of the deformation of some finite number of intramolecular and intermolecular bonds in the vicinity of the bond being ruptured.

Tomashevskii considered this problem by taking a segment of n bonds simultaneously stretched by a rectangular distribution of force in his calculations [93]. The most important results from this model were summarized in the following equation:

$$U(\sigma) \approx (U(0) - C_1\sigma)(1 - C_2T) . \tag{66}$$

In this equation, $U(\sigma)$ is the activation energy for bond rupture in presence of a molecular axial stress (σ), C_1 and C_2 are constant factors. Following the rapid initial decrease of $U(\sigma)$ at low levels of stress, the linearly extrapolated value $U(0)$ to zero stress was found to be significantly lower than D (≈ 0.7 D with $n = 100$).

From a chemical viewpoint, bond scission under stress is a particular case of unimolecular dissociation reaction whose rate is enhanced by mechanical stress.

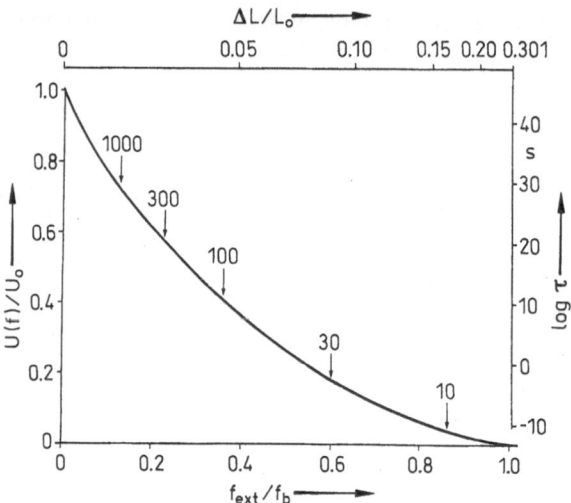

Fig. 20. Bond scission activation energy and lifetime (τ_f) plotted as a function of applied force. The *solid curve* is derived from Eq. (65) based on the Morse potential, the other data are redrawn from Ref. [101]. The upper abscissa gives the overall elastic strain before failure. The numbers indicate the minimum chain lengths which will fail at a particular force

As such, it could be treated with the Eyring's transition state theory. When stated in general terms, the transition state theory is applicable to any physico-chemical process which is activated by thermal energy [94]:

$$k_c = \{(kT/h)\exp(\Delta S^{\#}/R)\}\exp(-\Delta H^{\#}/RT). \tag{67}$$

When comparing Eq. (67) with the empirical Arrhenius equation for chemical kinetics

$$k_c = A \cdot \exp(-E_A/RT) \tag{68}$$

the Arrhenius activation energy, $E_A = (\Delta H^{\#} + RT)$ could be equated with the enthalpy of activation ($\Delta H^{\#}$) while the pre-exponential parameter (A) is linked to the entropy change ($\Delta S^{\#}$) occurring during the activation process:

$$A = (ekT/h) \cdot \exp(\Delta S^{\#}/R). \tag{69}$$

According to the transition state theory, the pre-exponential factor A is related to the frequency at which the reactants arrange into an adequate configuration for reaction to occur. For an homolytic bond scission, A is the vibrational frequency of the reacting bond along the reaction coordinates, which is of the order of 10^{13} to 10^{14} s^{-1}. In reaction theory, this frequency is diffusion dependent, and therefore, should be inversely proportional to the medium viscosity. Also, since the applied stress deforms the valence geometry and changes the force constants, it is expected that A is also a function of the level of stress. The activation enthalpy for the reaction $(\Delta H^{\#})$ is the rate determining factor and, in the unstressed state, should correspond to the thermal activation energy for bond rupture U_0 ($= D - 1/2\, RT$).

The transition state theory permits one to bring the treatment of rate processes within the scope of thermodynamic arguments. By combining de Boer's thermodynamic formulation and the transition state theory, Tobolsky and Eyring developed the rate theory for thermally activated fracture of polymeric threads in 1943 [95]. According to this theory, U_0 is lowered by a quantity $g \cdot \sigma$ in the presence of stress, where g is a stress-dependent coefficient which correlates the applied tension to the work of the force f on the covalent bond during stretching by δl from the equilibrium position. When put into an Arrhenius-like form, the following expression was obtained for k_c

$$k_c = A \cdot \exp\left[-\{U_0 - f(\psi)\}/RT\right], \tag{70}$$

where ψ is the molecular stress.

A detailed discussion about the functional form for $f(\psi)$ can be found in Ref. [15].

The frequencies of molecular vibrations depend on the force constants which are themselves attributed to the bond geometry. It is then not surprising that useful information on bond deformation under stress can come from IR or Raman spectroscopy.

In their pioneering experiment, Zhurkov et al. [96] determined that the IR-frequency shift Δv in a stressed polymer is approximately linear with the applied (macroscopic) stress σ:

$$\Delta v = v_0 \text{ (stressed)} - v_0 \text{ (unstressed)} = \alpha \cdot \sigma, \tag{71}$$

where α is the experimentally observed shift factor.

In an attempt to reproduce the experimental results, Wool and Boyd modeled the polymer as a system of 1-dimensional weakly coupled anharmonic oscillators [91]. Their calculations showed that most of the resulting frequency shifts should be negative and occur primarily due to the anharmonic nature of the atomic potential which results in a decreasing force constant with increasing valence coordinate deformation. With a single exception, a negative displacement was also observed for any of the Raman bands [91]; the highest negative shifts are measured in conjugated systems containing double and triple bonds which are also known to have a highly anharmonic potential. These results seem to conform to the coupled anharmonic oscillators model.

Since the shift of the stretching frequency can be related to a shift in dissociation energy D of the same bond [97], the following relation could be obtained (c = speed of light):

$$D - D' \approx U_0 - U(\sigma) = (2v_0/c) \cdot \Delta v. \tag{72}$$

Finally, it was deduced that [98]:

$$k_c = A \cdot \exp\left[-\{U_0 - \beta \cdot \sigma\}/RT\right]. \tag{73}$$

The factor β, having the dimensions of volume, is identified as the activation volume for the reaction.

Equation (73) is based on the observed shift in vibrational frequency. Since these measurements are usually carried out at a level of stress which is well below the theoretical breaking strength of the chain, they may correspond to the initial portion of the curve in Fig. 20 which also predicts a linear decrease of $U(\sigma)$ on the level of stress.

Several attempts to relate the rate for bond scission (k_c) with the molecular stress (ψ) have been reported over the years, most of them could be formally traced back to de Boer's model of a stressed bond [92] and Eyring's formulation of the transition state theory [94]. Yew and Davidson [99], in their shearing experiment with DNA, considered the hydrodynamic drag contribution to the tensile force exerted on the bond when the reactant molecule enters the activated state. If this force is exerted along the reaction coordinate over a distance δl, the activation energy for bond dissociation would be reduced by the amount:

$$\Delta U = C \eta_s \dot{\gamma} \delta l, \tag{74}$$

where $\dot{\gamma}$ is the fluid shear rate, and C is a parameter which depends on the polymer conformation and friction coefficient per unit length.

From this, an expression for k_c was proposed, which is essentially similar to Eq. 73:

$$k_c = A \cdot \exp\left[-U_0 + C \eta_s \dot{\gamma} \delta l\right]. \tag{75}$$

In a detailed study on shear degradation of DNA, Adam and Zimm found a complex dependence of k_c on solution viscosity. Considering that the macromolecules can rupture only after tumbling had brought them into the right configuration, these authors proposed to include solution viscosity into the pre-exponential factor: $A \sim (1/\eta_s)$ [84].

In a recent version of the Tobolsky and Eyring formulation, the rate of mechanochemical degradation was considered as a Thermally Activated Barrier to Scission (TABS) process. The elastic energy function $f(\psi)$ was explicitly considered in terms of the frictional hydrodynamic drag force acting over the entire macromolecule [100]. A more detailed account of this model will be presented in Sect. 5.1.

An interesting improvement from the classical treatment of the bond under stress was proposed by Crist et al. [101]. Considering the chain as a set of N-coupled Morse oscillators, these authors determined the elongation and time to failure as a function of the axial stress. The results, reported in Fig. 20, show a decreasing correlation between the total elastic strain before failure and the level of applied force with the chain length. To break a chain within some reasonable time interval (for example $< 10^{-3}$ s) requires, however, the same level of stress ($\approx 0.7 \, f_b$) as found from the simpler de Boer's model.

4 Realizations of Transient Elongational Flow

4.1 General Considerations

In rheology, the flow of dilute polymer solutions is distinguished by the molecular response of the embedded macromolecules. One important classification, based on the dynamics of the dumbbell model, divided flow into "weak" or "strong" domains, depending on whether the beads end-to-end distance reaches some equilibrium average value or is extended exponentially without limit. From its nature, a "strong" flow is inherently inhomogeneous: the presence of walls automatically introduces shearing, therefore, pure elongational flow needs to be bound by free surfaces. Moreover, the same fluid element cannot be stretched indefinitely since the total strain imposed on the fluid is bound by experimental geometry. This implies that regions which satisfy the strong-flow conditions are always restricted to a small portion of the flow domain. In elongational flows which possess a center of symmetry, as in the Taylor four-roll mill (Fig. 21) [102] or in the opposed jets device (Fig. 22) [103, 104], the dwell time and the total strain are large in a narrow region at and near the outflow axis where the flow is "stagnant". In the majority of the remaining streamlines, the transit time of the fluid element in the high strain-rate region is short when compared to the time scale necessary for coil extension and the flow is termed as "transient"

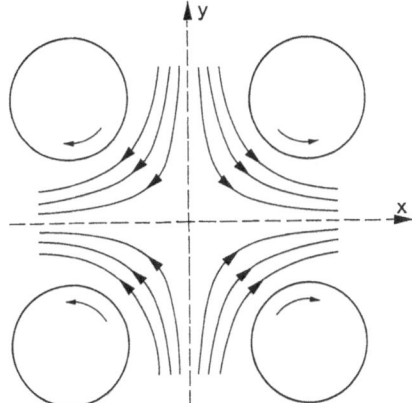

Fig. 21. Schematic drawing of the Taylor four-roll mill with the created flow field. The 2-dimensional velocity gradient can be represented by the tensor Λ:

$$\Lambda = \frac{1}{2} \begin{vmatrix} (1 + \lambda) & (1 - \lambda) & 0 \\ -(1 - \lambda) & -(1 + \lambda) & 0 \\ 0 & 0 & 0 \end{vmatrix} \quad -1 \leq \lambda \leq +1$$

Depending on the relative velocity in opposite pairs of the rollers, flow can be either purely rotational ($\lambda = -1$), simple shear ($\lambda = 0$) or hyperbolic straining ($\lambda = 1$, shown on this figure)

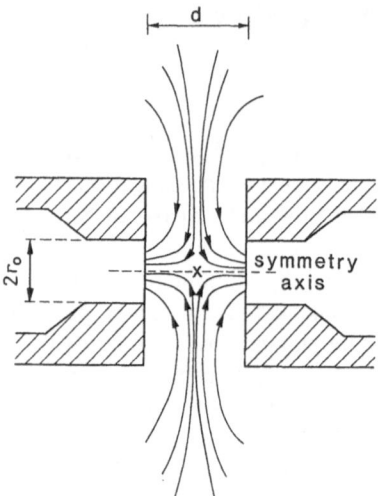

Fig. 22. Schematic diagram of the opposite jets device with some of the associated streamlines (the "x" marks the location of the stagnation point). It has been determined that a ratio of $d/(2\,r_0) \approx 1 - 1.4$ constitutes the optimum geometry for extensional viscosity measurements [104]

according to this criterion. In unsteady flow (as for instance in transient elongational flow), the "strength" of a flow can be defined locally by dividing the flow field into regimes in time and space which can be approximated as regions of homogeneous steady flow. Each regime can then be considered as strong or weak according to the dumbbell mechanics or some other related standards [105].

The above description refers to a Lagrangian frame of reference in which the movement of the particle is followed along its trajectory. Instead of having a steady flow, it is possible to modulate the flow, for example sinusoidally as a function of time. At sufficiently high frequency, the molecular coil deformation will be dephased from the strain rate and the flow becomes "transient" even with a "stagnant" flow geometry. Oscillatory flow birefringence has been measured in simple shear and corresponds to some kind of frequency analysis of the flow

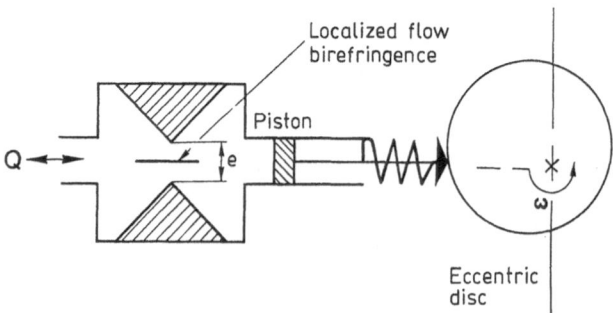

Fig. 23. Oscillatory convergent-divergent flow (depth of the cell = 14 mm, contraction width e = 0.9 to 1.8 mm)

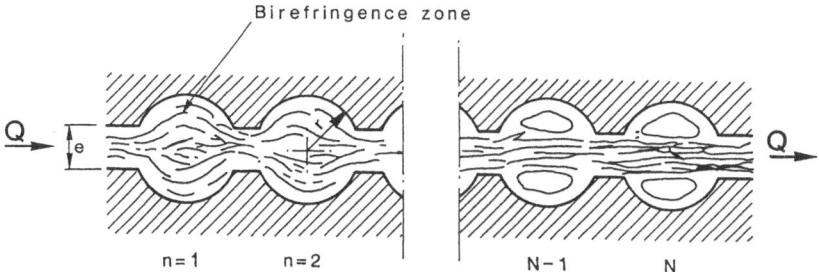

Fig. 24. Flow through a periodical set of convergent-divergent geometry (in the original design, N = 5 to 20, r = 5 mm and e = 0.5 or 1 mm)

birefringence [106]. This method has provided important information on the chain dynamics and on polymer-solvent interactions. With respect to the elongational flow, two interesting designs have been described by Cressely and Hocquart (Figs. 23 and 24) [107] which exploit the hysteresis effect in coil stretching to orient flexible polymer chains even for short residence time. It seems at the present that a theoretical account of polymer response to a periodical sollicitation is still much less developed in elongational flow than in simple shear flow.

4.2 Experimental Arrangements

As opposed to stagnant elongational flow which necessitates highly specific flow geometries, transient elongational flow is readily obtained with some simple arrangements which will be described below.

4.2.1 Abrupt Contraction Flow

Flow through a convergent channel, either two-dimensional or axisymmetric, are probably the most investigated geometries (due to their connection with the technical die entry flow problem). Most of the other convergent flows could be derived from the abrupt contraction flow which is depicted schematically in Fig. 25.

Smith et al. [13] are amongst the first to employ an abrupt contraction flow to investigate chain conformation. A modified version of Smith's design was used by Merrill and Leopairat [108], and later on by the Lausanne group [109] (Fig. 26)

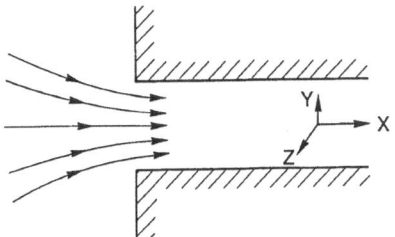

Fig. 25. Sketch of an abrupt contraction flow

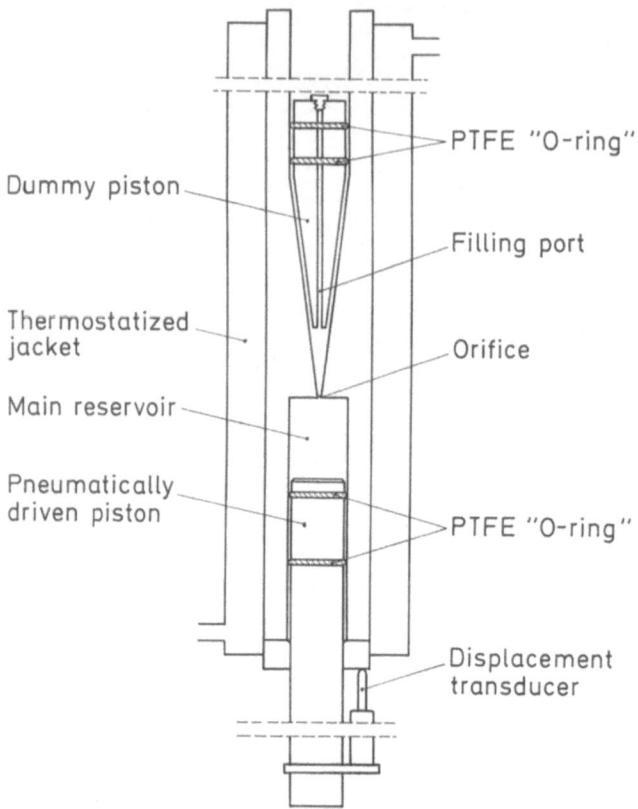

Fig. 26. Schematic representation of the degradation apparatus used in transient elongational flow (orifice diameter = 0.50 mm, reservoir diameter = 21.3 mm, course of the piston = 70 mm)

which permits strain rates in excess of 4×10^5 s^{-1} to be obtained. In experiments with high molecular weight polymers ($>2 \times 10^6$) a much lower range of strain rates is required to break the chains; thus, a cell fitted with glass windows could be used instead of a steel walled cylinder (Fig. 27). In essence, the latter design corresponds to that of the opposed jets in which one arm of the outlet was blocked; without the symmetry center, the stagnation point is destroyed and flow becomes transient along all streamlines. The introduction of transparent windows is highly versatile since it allows flow visualization and birefringence measurement to be performed at the same time as flow-induced degradation.

A large number of the results presented in this review were obtained with the device presented in Fig. 26 and its operating principle will be explained in some detail. The basic set-up consists essentially of a stainless steel syringe closed at one end by a narrow circular orifice and surrounded by a thermostatically controlled jacket. Apart from minor mechanical details used to improve flow field uniformity, the degradation apparatus is similar to the device described by Merrill

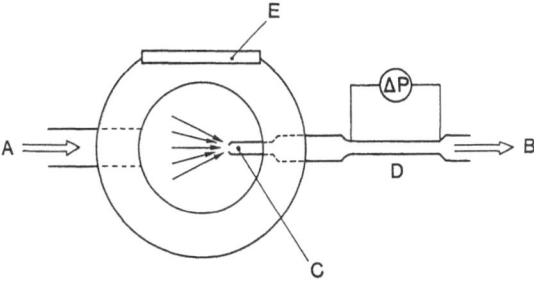

Fig. 27. Abrupt contraction cell for flow visualization, birefringence and degradation measurements A: inlet (from a peristaltic pump of a pressurized reservoir; B: outlet (atmospheric pressure or partial vacuum); C: interchangeable metallic nozzle with a sapphire tip; D: capillary flow meter; E: glass window for flow visualization; ΔP: pressure drop (from pressure transducers)

and Leopairat. Except where noted, the flow geometry was kept constant for all experiments with a piston diameter of 2.13 cm, an orifice diameter of 0.050 cm and a conical outlet with a divergent angle of 14°. The stroke of the piston was 7.0 cm.

The working principle is the following: the polymer solution is introduced into the upper reservoir from where it is sucked very slowly (to avoid degradation) into the main reservoir. Following the application of pressure to the pneumatic actuating cylinder, the confined liquid is pushed at constant speed through the orifice. The sudden variation in diameter through the constriction accelerates the fluid and creates a flow-field of extreme strain-rate ($\dot{\varepsilon}$) but with a residence time as short as a few microseconds. The conical divergent outlet was used to suppress backflow of the high velocity exiting jet, and to prevent degradation from high shear buildup at the walls. To make a smooth transition, the convergent inlet was connected to the divergent outlet by a short channel 0.2 cm long; within reasonable limits, it was found that its exact length had no influence on the degradation process [108].

The velocity of the piston, monitored by a displacement transducer, was found to be constant to better than 2% over the complete stroke. In some experiments, a pressure transducer was added to evaluate the amount of mechanical energy input to the system. The highest strain-rate attainable is determined by the force applied on the piston and by the solution density (in inviscid flow, the pressure drop is proportional to the kinetic energy of the fluid or $\approx 1/2\,\varrho\bar{v}_0^2$, with $\bar{v}_0$ = average volumetric velocity at the orifice); in methyl acetate, a maximum value of $\dot{\varepsilon} \approx 4 \times 10^5\,\mathrm{s}^{-1}$ was obtained when the pressure inside the reservoir compartment reached 200 bar. After a single passage through the orifice, the liquid was recovered for analysis.

4.2.2 Conical Flow

By using a tapered inlet instead of an abrupt constriction, different variations of convergent flow could be envisioned: axisymmetric flow in a conical channel constitutes a standard example of an irrotational sink flow for a Newtonian liquid at a high Reynolds number [105]. To avoid the recirculation vortex, it is common in rheology for investigators to prefer the conical inlet in place of the abrupt constriction: it has been shown experimentally that the recirculation eddy can be suppressed by using a tapered inlet with a conical angle less than 40° [110]. Within the realm of transient elongational flow, using a conical inlet has the added advantage of providing a simple means of modifying the velocity gradient distribution by a proper selection of the entry angle (Cf. experimental results given in Sect. 6.4).

4.2.3 Wedge Flow

Changing the profile of the nozzle permits one to control, to some extent, the uniformity of the flow field. Wedge flow across a hyperbolic channel (Fig. 28) provides a constant average velocity gradient. The axial velocity distribution across the flow tube, is, however, far from being constant. In order to improve the homogeneity of the velocity front at a low Reynolds number, a third-order polynomial profile has been proposed [111]. Wedge flow can produce only moderate velocity gradients which are generally too weak to induce chain degradation. This type of flow is interesting, however, in some applications which require a well-defined flow field, as for instance in the determination of molecular coil expansion in flow by light scattering [112], or in some recirculation experiments where the low strain rate could be compensated by a longer residence time.

4.2.4 Compressional Flow

Reconsidering the opposed jets geometry, instead of sucking the liquid through the orifices, another possibility of creating elongational flow would be to impinge the two high-velocity liquid jets precisely along their common centerline. The flow thus created is biaxially elongational in the plane of symmetry, similar to the one obtained during blow molding. Along the symmetry axis of the jets, flow is "stagnant" elongationally but with a negative strain rate as found previously in flow across a divergent outlet. The same type of flow can be obtained by impinging a free jet onto a solid surface (Fig. 29). Compressional flow has been used to determine rheological [113] and adsorption properties [114] of polymers in solution. Nevertheless, no degradation results have been reported yet in this type of flow. The lack of influence of the type of outlet on the degradation yield, mentioned in Sect. 4.3, suggests that compressional flow is less efficient than elongational flow in inducing mechanochemical degradation. In the solid state, it has been

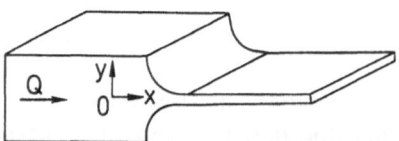

Fig. 28. Hyperbolic entrance flow

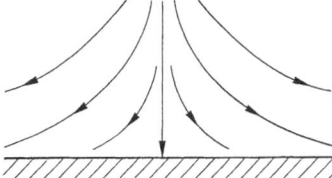

Fig. 29. Free impinging jet on a solid surface

noted in a few instances that the behavior of a chain is completly different whether it is under tensile or compressive stresses [15, 115]. It would be interesting to verify whether the same trend is also observed in solution where chain mobility and environment are distinct from the solid state.

In analogy to the impinging free-jet, transient elongational flow could be created in the vicinity of an appropriate obstacle as, for example, in front of a cylindrical body (Fig. 30) or a spherical particle. Chain extension can be achieved in such a flow and this has been verified by an electrochemical technique [116]. In addition to turbulence, this type of flow may account for part of the degradation observed during high speed stirring of polymer solutions; it may also be relevant to the process of separation and degradation during Gel Permeation Chromatography analysis.

4.2.5 Ultrasonic Irradiation

When a liquid is exposed to ultrasonic irradiation, a variety of phenomena can be observed originating from streaming and cavitation. High intensity sounds in liquids are accompanied by wave radiation pressure which results in steady-state

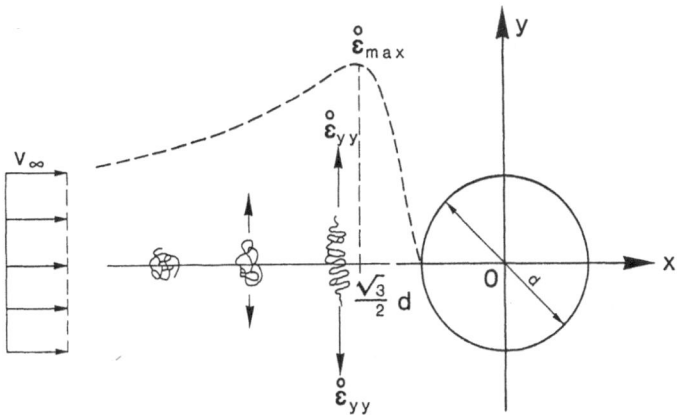

Fig. 30. Flow in the vicinity of a cylindrical obstacle along with the associated elongational gradient $\dot{\varepsilon}_{yy} = -\dot{\varepsilon}_{xx}$ (— — —). The deformation behaviour of a macromolecule convecting toward the cylinder is also sketched

flow known as acoustic streaming. This flow occurs near the walls of the containing vessel or of other obstacles in the liquid, and in the bulk if the sound field is nonuniform. The role of macrostreaming is minor in sonochemistry, excepted from insuring a rapid mixing of the reactants during the irradiation time [117]. On a smaller scale as regards the order of the sound wavelength, the effect of irradiation with high frequency acoustic sound is to alternatively increase and decrease the local pressure. During the compression cycle, the average distance between the molecules is decreased, whilst it increases during decompression. If the applied negative pressure is sufficiently large to overcome the cohesive forces in the liquid, the continuity of the medium will break down, cavitation will occur and bubbles will be formed: dissolved gases tend to escape into the voids and ultrasonic irradiation is a well-proven method for outgasing solutions. Theoretical estimates of the acoustic pressure necessary to cause cavitation in water have given $P_c = 1500$ atm, which is far above the experimental values of ≈ 20 atm. The main reason for the difference comes from the presence of weak spots (dissolved gases, particulate matter) which act as potential nuclei for cavitation and lower the tensile strength of the liquid. Once produced, the cavitation bubble will be forced to contract in the succeeding compression cycle of the wave. Depending on the size of the bubble and the insonation frequency, a violent collapse of the cavity can occur producing shock waves of the order of a thousand atmospheres and local temperatures reaching several thousand Kelvin [118]. Most of the understanding of cavitation came from the work of Lord Rayleigh who, in 1917, derived the basic equations for the adiabatic collapse of a spherical cavity in an incompressible Newtonian fluid. The motion of the cavity wall is given by the following differential equation [118]:

$$\varrho(\ddot{r}r + \tfrac{1}{2}\dot{r}^2) = (P_a - P_h) - 2\sigma/r - 4\eta_s(\dot{r}/r) + (P_h + 2\sigma/r_e - P_v)(r_e/r)^3 , \quad (76)$$

where $\dot{r}$ and $\ddot{r}$ are respectively the velocity and the acceleration of the cavity wall, r_e the equilibrium radius of the bubble, σ the surface tension and ϱ the density of the liquid, P_v the liquid vapour pressure, P_h the atmospheric pressure and P_a the applied acoustic pressure.

Integration of the above equation permits one to estimate the lifetime of a bubble as a function of its initial radius and irradiation frequency. For example, at a frequency of 20 kHz typical in sonochemistry, a cavity of 150 µm formed in water will implode in ≈ 3 µs, adiabatically pressurizing the gases inside the cavity to 500 atm at a temperature of 5800 K (the temperature of the sun's surface), whilst heating the liquid in the immediate surrounding to 2700 K. Both the shock waves and the heat released from the collapse contribute to a unique chemical environment in sonochemistry. In dilute solution, the role of released heat is probably of minor importance for polymer degradation: since the "hot" regions are highly localized and should be quenched in less than 1 µs, the polymer molecules do not have time to diffuse and to reach these spots in such a short time interval. Polymerization reactions may, however, be initiated in these high temperature regions as observed during bulk irradiation of liquid monomers.

The role of cavitation in ultrasound degradation has been confirmed repeatably: in most experiments where cavitation was prevented, either by applying an external hydrostatic pressure, by degassing the solution, by reducing the sound intensity or the temperature, polymer chain scission was also largely suppressed [117].

Elongational flow is created during the formation as well as during the collapse of a cavitation bubble. In an elegant experiment, Odell and coworkers [119] simulated the growth of the bubble by forcing a gas through small pores surrounded by polymer solution. Some chain scission was observed, but it was of minor importance in comparison to the total degradation determined after ultrasonic irradiation. It is plausible to assume that most of the chain scission originated from the elongational flow present during and following the collapse stage. Such a flow field is necessarily transient and must be of extremely high strain rate to be consistent with the low limiting molecular weight found in ultrasonic degradation (30000 [120, 121] as compared to ≈ 300000 with the described transient flow apparatus working at maximum fluid velocity). Recent calculations by Ryskin confirmed the presence of a high strain rate flow field during the cavitational collapse; in addition, it was shown that the presence of added polymer does not affect the growth of the bubble but influences the final stage of the collapse (Fig. 31) [122].

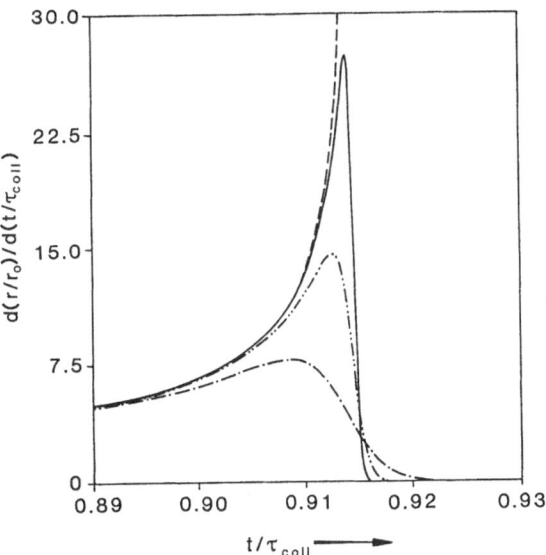

Fig. 31. Bubble wall velocity vs time during cavitational collapse for different values of the parameter λ defined as $\lambda \approx 0.4 c \, [\eta] \, \eta_s/(r_0^{1/2} \, (P_h - P_v)^{1/2})$. λ permits us to account for the viscous and inertia effects of the polymer solution (redrawn according to Ref. [122]); (-----): $\lambda = 0$ (inviscid fluid); (———): $\lambda = 10^{-6}$; (- ·· -): $\lambda = 10^{-5}$; (- · - ·): $\lambda = 10^{-4}$; r_0 = initial bubble radius; τ_{coll} = collapse time scale, for an inviscid fluid, collapse occurs at $t = 0.915 \, \tau_{coll}$

4.3 Flow Field Analysis

In the early investigations of mechanochemical degradation in solution, flow field analysis was almost absent or carried out at a superficial level: no consideration was ever made of the possible role of the rotational component of a flow on chain scission and it took a long time, for example, to realize that most of the observed degradation in capillary tubes did not occur near the walls but in reality at the front of the inlet where flow was elongational [123]. Recent findings have confirmed that the kinetics of polymer degradation is sensitive to subtle modifications of chain conformation, which in turn depends on the details of the pervading flow field. Hence, special attention should be given to the design of the exit nozzle and to flow field calculations to ensure meaningful interpretation of the results. In a previous work, the finite element program POLYFLOW [124] was used extensively for flow field calculations and as a computer-assisted tool for the proper design of the exit nozzle. This flow modeling program is capable of handling steady and time-dependent laminar flows of both Newtonian and viscoelastic fluids at low Reynolds number. The only requirement, that of fluid incompressibility, is easily satisfied in the present context.

4.3.1 Newtonian Flow Approximation

Viscoelasticity is the very essence of polymeric liquids. Flow of such a fluid is highly complex and even with a well-established constitutive equation, flow modeling at the high Reynolds numbers prevalent in mechanochemical degradation may prove impossible to handle due to a lack of proper computational techniques. Fortunately, viscoelasticity is concentration dependent and working at high dilution can relieve many of these complications. In the range of molecular weight (0.3 to 3×10^6) and the concentration commonly used (5 ppm, except for the concentration dependence experiments where a much higher polymer content was employed) the solution should behave approximatively as a Newtonian fluid, whose flow properties are described by the Navier-Stokes equation. We will examine at the end of this chapter some possible errors entailed in this approximation.

A common approximation in many flow field computations at high fluid velocities is to consider that inertial forces dominate the flow and to neglect viscous forces (inviscid approximation). Since solvent viscosity is a variable in some of the experiments discussed here, the above approximation may be not be valid throughout and viscous forces are explicitly considered in the flow equations. Results of computations showed, nevertheless, that even with viscous solvents such as bis-(2-ethyl-hexyl)-phtalate with $\eta_s = 65$ mPa $\cdot$ s, viscous forces do not affect the flow field unless the fluid velocity drops below a few m $\cdot$ s^{-1} at the orifice. This limit is generally more than one order of magnitude lower than the actual range used in the present investigations.

The flow geometry is time-dependent with the constant advance of the piston during the experiment. The flow field, however, was not significantly perturbed by this displacement except when the plunger reached a position a few millimeters before the end of its stroke. The spatially steady flow hypothesis is thus a good approximation; to improve the constancy of the flow field further, a collar was added in a later design to stop the piston movement 0.6 cm before touching the orifice.

Flow field calculations are conveniently performed in dimensionless variables defined in terms of the orifice parameters:

- dimensionless radial coordinate: $r' = r/r_0$
- dimensionless axial coordinate: $x' = x/r_0$
- dimensionless flow velocity: $v'_x = v_x/\bar{v}_0$ ($\bar{v}_0$ = average velocity at the orifice)
- dimensionless strain rate: $\dot{\varepsilon}'_{xx} = \delta v'_x/\delta'_x = \dot{\varepsilon} \cdot (r_0/\bar{v}_0)$.

The streamlines calculated for different values of the stream function are shown in Fig. 32. Some flow recirculation is visible near the corners, but this is negligible as compared to the viscoelastic situation [110].

The computed velocity field has several distinctive features associated with a transient elongational flow:

1) As stated above, elongational flows are inherently restricted to a small portion of the flow domain with streamlines entering and leaving via regions of much reduced flow strength. This is particularly true for flow passing through an abrupt constriction where most of the fluid acceleration occurs within a distance of ≈ 1 orifice radius (r_0) in front of the convergent inlet. This sudden acceleration is reflected in the impulse-like shape of the strain-rate (Fig. 33) which can reach

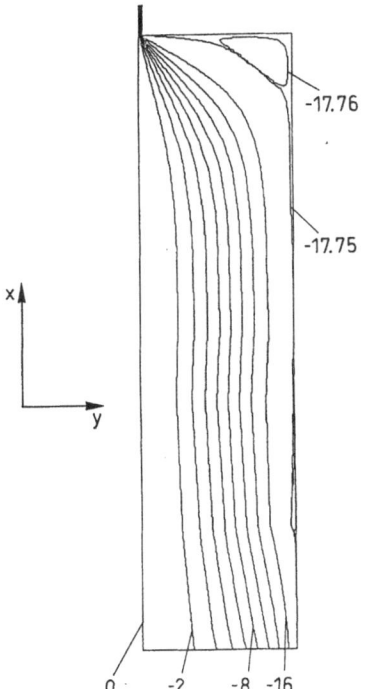

Fig. 32. Streamlines in abrupt contraction flow computed at a midtravel distance of the piston at $x = -40$ mm (the origin of the abscisse is taken at the orifice entrance). Due to axial symmetry, only half of the flow tube is shown. The dimensionless stream function $(\psi/(\bar{v}_0 r_0^2))$ is set arbitrarily to 0 along the centerline. The isolines are indicated in steps of -2, except for the one closest to the walls which has a value of -17.75. Recirculation is visible at a value of the dimensionless stream function equal to -17.76

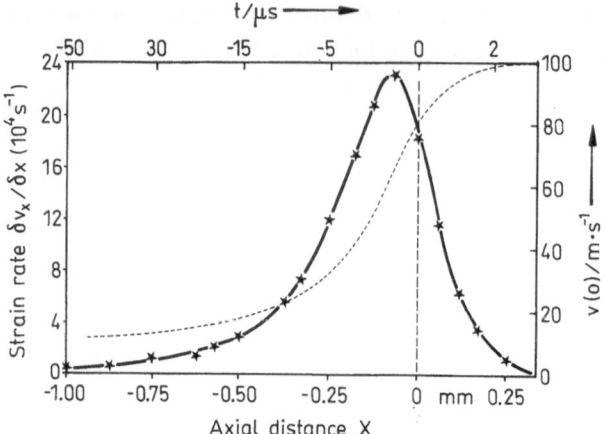

Fig. 33. Axial velocity distribution (right scale ordinate) (--------) and elongational strain rate ($\dot{\varepsilon}_{xx}$) (————) calculated along the centerline of the flow tube in abrupt contraction flow. Values are given for $r_0 = 0.25$ mm and $\bar{v}_0 = 100$ m $\cdot$ s^{-1}. The upper abscissa indicates the transit time of a fluid element passing along the centerline

extremely high amplitudes but only over a limited region of space of the order of r_0. In terms of dimensionless variables, the maximum strain rate along the centerline (denoted by $\dot{\varepsilon}'(0)$) is given by:

$$\dot{\varepsilon}'(0) \equiv \dot{\varepsilon}'_{xx}(max) = 0.56 \quad at \quad x' \approx 0.4 \,. \tag{77}$$

The residence time of a fluid element entering the region of high strain rate, along any streamline, is typically of the order of $r_0/\bar{v}_0$ or ≈ 2–10 μs under present flow conditions. These converging flows are termed "transient" due to this short residence time.

2) In creeping flow with the inertia term neglected, the velocity distribution rapidly reaches a steady value after a distance of $\approx r_0$ inside a capillary tube. At this stage the velocity distribution showed the typical parabolic shape characteristic of a Poiseuille flow. In the case of inviscid flow where inertia is the predominant term, it takes typically (depending on the Reynolds number) a distance of 20 to 50 diameters for the flow to be fully developed (Fig. 34). With the short capillary section ($\approx 4\, r_0$) in the present design, the velocity front remains essentially unperturbed and the velocity along the symmetry axis, i.e. v_x ($y = 0$), is identical to $\bar{v}_0$.

With incompressible liquids, the volumetric average velocity at the orifice ($\bar{v}_0$) is, therefore, given by:

$$\bar{v}_0 = v_p \cdot (r_p/r_0)^2 \,, \tag{78}$$

where r_p is the piston radius.

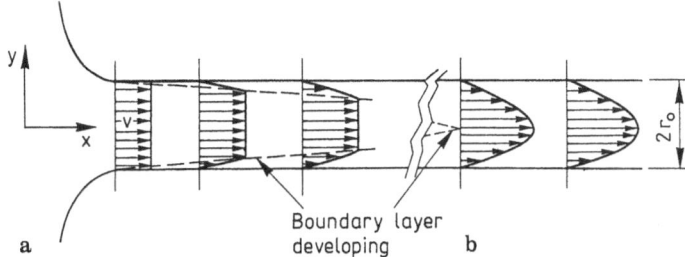

Fig. 34a, b. Laminar flow in a capillary tube. The boundary layer is developing in (**a**), while it is fully developed in (**b**)

3) As with any elongational flow, the velocity field is not homogeneous. Flow is purely elongational only along the central streamline with a maximum in strain rate as given in Fig. 33. Another streamline off the symmetry axis will show a different strain rate function, so that for a given flow rate, there exists a distribution of maximum strain rates across the flow tube (Fig. 35). A kinetics study taking this variation of strain rate into account showed that the result is a broadening in the degradation yield curve as a function of average fluid strain rate [125]. From the dynamics of the dumbbell (Sect. 2.3), it was determined that the degree of coil extension is sensitive not only to the strain rate but also to the relative magnitude of the extensional flow component with respect to the rotational

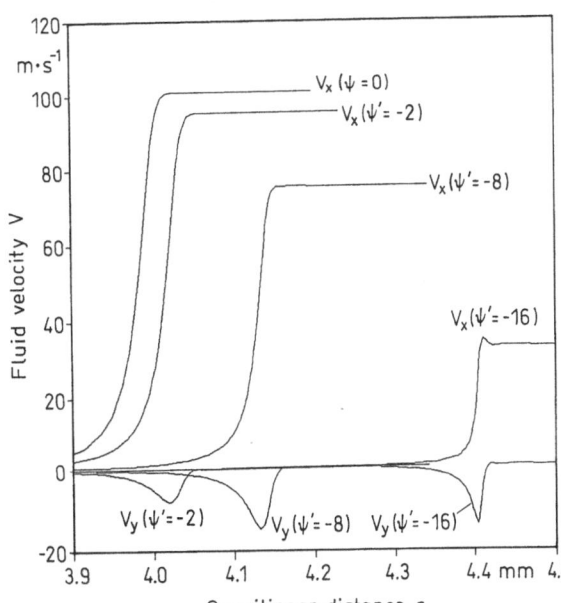

Fig. 35. Velocity distribution along different streamlines in abrupt contraction flow (s is the curvilinear distance along the streamline from the entrance, as shown in Fig. 32).

126 Tuan Q. Nguyen, H.-H. Kausch

component. In terms of the velocity gradient, this means that the rate-of-strain tensor $\dot{\gamma} = \nabla v + (\nabla v)^T$ must be greater than the vorticity tensor $\omega = \nabla v - (\nabla v)^T$ in order to deform the molecular coil:

$$\|\dot{\gamma}\| > \|\omega\| . \tag{79}$$

The Giesekus criterion for local flow character, defined as $\Phi = (\|\dot{\gamma}\| - \|\omega\|)/(\|\dot{\gamma}\| + \|\omega\|)$, has a value of $+1$ in pure extensional flow, 0 in simple shear flow and -1 in solid body rotation [126]. The mapping of Φ across the flow domain provides probably the best description of flow field homogeneity; current calculations in that direction are being performed in the authors' laboratory.

4) A divergent outlet was used to avoid backflow convection and excessive shear buildup after the contraction. As a result of the flow deceleration, a fluid element experiences a uniaxial compression shown as negative strain rates on Fig. 36. With the 14° conical outlet (Fig. 26), the amplitude of the negative strain-rates attains as much as one-third of the maximum value of the extensional strain-rate. In a few experiments, the conical outlet was either removed to allow the outcoming jet to exit freely into a collecting reservoir, or the 14° conical section was replaced with a 5° tapered outlet, thus reducing the range of negative strain-rates by a factor of 8 (Fig. 36). In each case, subsequent analyses did not show any significant change in the scission yield as compared to the standard design, which tends to confirm that most of the degradation took place in the elongational flow region before the orifice and not as a result of biaxial extension or turbulence in the downstream section.

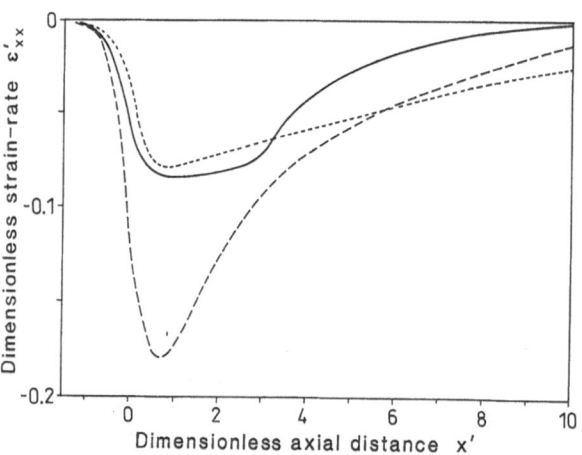

Fig. 36. Dimensionless compressional strain rate ($\dot{\varepsilon}_{xx}'$) calculated along the centerline of the flow tube for the different exit geometries; (———): free jet ($\overline{v}_0 = 35 \text{ m} \cdot \text{s}^{-1}$); (— — —): 10° conical outlet; (-----): 5° conical outlet

4.3.2 Convergent Flow of Viscoelastic Fluids

Practically any liquid, including liquid Argon, is viscoelastic under extreme shear rates. Polymer solutions, on the other hand, are viscoelastic even at modest rates of deformation. It is thus logical to ask what are the limits of experimental conditions, in term of strain rate, polymer molecular weight and concentration, below which it would still be permissible to consider these liquids as Newtonian. It is known that the presence of a high molecular weight polymer in a minute amount (ppm level) could produce detectable rheological effects such as drag reduction and "bathtub vortex" formation. The velocity profile in tube flow is significantly distorted in the presence of dissolved polymer. This effect has been determined by laser-Doppler velocimetry [127] or by dynamic light scattering [128] and the flow field perturbation was explained in terms of the flow energy consumption during stretching of the macromolecule. Recently, a semi-quantitative description of the flow modification by polymers in elongational flow was given by Rabin [129]. It was shown by his calculations that the flow perturbation in the dilute regime is approximately proportional to polymer concentration.

The following qualitative picture emerges from these considerations: in "weak" flow where the molecular coils are essentially undeformed, the polymer solution should behave approximately as a Newtonian fluid. In "strong" flow of a highly dilute polymer solution where the macroscopic velocity field can still be approximated by the Navier-Stokes equation, it should be expected, nevertheless, that in the immediate proximity of a chain, the fluid will be slowed down because of the energy intake to stretch the molecular coil; thus, the local velocity field may deviate from the macroscopic description. In the general case of polymer flow,

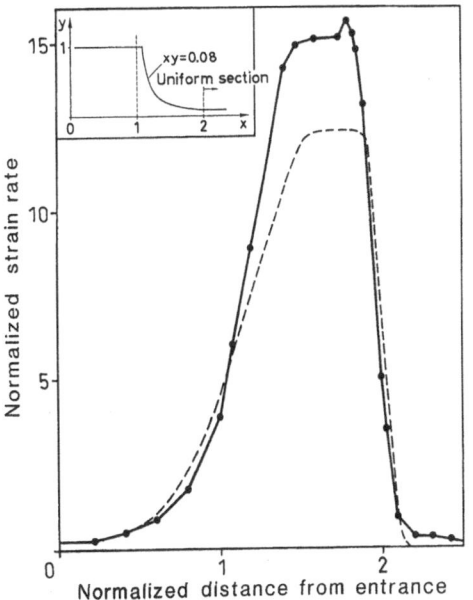

Fig. 37. Strain rate distribution along the centerline in a 2-dimensional hyperbolic flow (the flow geometry is shown as an insert). The solid curve, redrawn according to ref. 131, corresponds to a viscoelastic fluid (the spike at x = 2 is a calculation artefact); the dotted curve is calculated with POLYFLOW for a Newtonian liquid

an appropriate equation of state must be sought for the fluid. The constitutive equation is then solved by the finite element method with appropriate velocity and stress boundary conditions. This proves to be extremely laborious and the results should be checked further by flow visualization techniques. Therefore, it is always advantageous whenever possible, to select experimental conditions where the Newtonian approximation can be employed.

Concluding the discussion of this section, some results obtained for the flow of a dilute polymer solution through a 2-dimensional hyperbolic contraction (Fig. 28) will be used to illustrate a "non-Newtonian flow problem". Using an Oldroyd-B type constitutive equation, which was derived for dilute solutions of Hookean dumbbells [130], Gatski and Lumley calculated by a finite element method the pressure and velocity distributions along the contraction [131]. The computed velocity field revealed an increase of the boundary layer thickness in the contraction region for the viscoelastic fluid. The effect of this is a reduction in the effective contraction width causing higher fluid velocities and a larger strain rate as compared to the Newtonian situation (Fig. 37).

5 Kinetics of Flow-Induced Degradation

The rate of bond scission, as given by Eq. (70) in Sect. 3, depends explicitly on the temperature T, the bond dissociation energy U_0, and the elastic potential function $f(\psi)$. Amongst these parameters, the elastic potential energy is the most difficult to quantify since it depends on a variety of factors, some are inherent to the chemical structure of the polymer, while others are attributed to the surrounding fluid or to the polymer-solvent interactions. The issue is complex and a clear picture on the molecular mechanism of chain scission has not emerged yet, despite of the sizable amount of experimental effort spent in this field over the years [15, 86]. In this section, some important kinetic parameters capable of influencing the rate of degradation will be examined in light of the results obtained in transient elongational flow.

5.1 Mechanisms of Stress Transmission

A polymer responds to the application of stress by chain orientation, disentanglement and bond rupture. Notwithstanding the fact that the relative importance of each mechanism changes with the state of aggregation, structure and molecular weight of the polymer, the extent of degradation is dependent on how the macroscopically applied mechanical force is operative at the molecular level in order to increase the internal energy of the molecule. In a few circumstances, stress could be transmitted by external coupling with an electric field such as during electrophoresis, or with a magnetic field as in the aligment of liquid crystal polymers. By far, the most common method of stress transmission was by mediation of short-range interactions between submolecular entities. Depending on the state of aggregation, the mechanism of stress transmission may be different; in the context of this review, it is practical to distinguish between entangled chains in concentrated solutions or in melts and isolated chains in dilute solution.

5.1.1 Concentrated Solution

In concentrated solution or in the condensed state, interactions between polymer molecules with MW above the critical value for chain entanglement ($M_c \approx 31\,000$ for PS) are additionally effective through the existence of physical crosslinks which act as load-bearing units. The role of entanglements in mechanochemical degradation was first advanced by Bueche in order to explain the positive concentration dependence and the non-random scission in simple shear flow [132]. In the absence of entanglements, little degradation is expected since the coils rotate without extensive deformation whereas the presence of topological constraints hinder the coil rotation. In addition, each entanglement point acts as a stress concentrator and transmits, in a pulley-like fashion, large tensions close to a central portion of the macromolecule, thus faciliting the rupture of the bond (Fig. 38). With this model, Bueche derived a scaling law for chain scission in simple shear as:

$$k_c \sim \overset{\circ}{\gamma} \cdot M^{3.5} . \tag{80}$$

By estimating the frictional force present during the motion of entangled polymer chains in a shear field, Bueche demonstrated that the probability of bond scission along the chain should exhibit a Gaussian distribution with a standard deviation of less than 0.1. These results were subsequently found to be only qualitatively correct: experimentally, the dependence on molecular weight was found to be much weaker and the distribution around "midchain scission" much broader than predicted by the model. More recent models of entangled polymer systems make use of a "modified" form of Graessley's theory in which the viscosity coefficient is split into two parts [133]:

$$\eta = \eta_{\text{ent}} + \eta_{\text{fric}} . \tag{81}$$

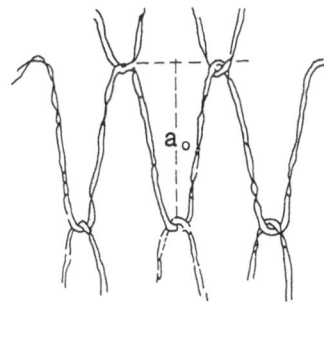

Separation velocity $= 1/2\,\overset{\circ}{\gamma}\, a_o$

Fig. 38. Action of entanglements in simple shear degradation. The tension on the qth link from the center is given by $f_q = f_0(1 - 4q^2/n^2)$, where $f_0 \approx \eta_s \overset{\circ}{\gamma} \cdot f(L)$ is the tension on center link ($f(L)$ = function of chain length) and n = number of links per chain.

The empirical frictional factor (η_{fric}) is independent of shear rate but increases in poor solvent: this permits to account for the dependence of the scission rate constant on solvent quality. The entanglement part (η_{ent}), as given by Graessley's theory which considers the effect of entanglement and disentanglement processes, is a complex function of shear rate:

$$\eta_{ent}/(\eta_0 - \eta_{fric}) = f(\tau_0 \mathring{\gamma}) \tag{82}$$

η_0 is the viscosity at zero shear rate, τ_0 is a shift factor of the order of the Rouse relaxation time and $f(\tau_0 \mathring{\gamma})$ is a universal function independent of MW, concentration and temperature.

5.1.2 Isolated Polymer Chains

In the highly dilute state where each macromolecule is isolated from the other, stress is transmitted by viscous friction between solvent and monomeric segments of the polymer molecule. When a velocity gradient is present, dissymetry in the inelastic exchange of momentum between the flowing solvent and separate portions of the same macromolecule gives rise to molecular tensions and ultimately to chain degradation. The tension in a fully stretched chain is readily obtained from the results of the slender-body hydrodynamics, at the limit of $L = l_{max}$ (Eq. 47). Similarly, the stress distribution can be calculated from the bead-rod model of total length L, fully stretched in the flow direction (x) [3]:

$$f(x) = f_{max} \cdot [(L/2)^2 - (x)^2]. \tag{83}$$

The maximum stress at midchain (f_{max}) is given by Stokes' hydrodynamic drag formula:

$$f_{max} = (3\pi/4)\,\eta_s \mathring{\varepsilon} \cdot a \cdot L_0 (M/M_0)^2, \tag{84}$$

where a is the hydrodynamic radius of the bead, L_0 the distance between two beads, M_0 the molecular weight of a bead and M the molecular weight of the polymer.

From Eq. (84), the level of strain rate ($\mathring{\varepsilon}_f$) necessary to break the polymer chain ($f_{max} \approx f_b$) should scale with molecular weight as M^{-2}.

Applying the TABS model to the stress distribution function f(x), the probability of bond scission was calculated as a function of position along the chain, giving a Gaussian-like distribution function with a standard deviation $\sigma \approx 6\%$ for a perfectly extended chain. From the parabolic distribution of stress (Eq. 83), it was inferred that $f^H < f^B$ near the chain extremities, and therefore, the polymer should remain coiled at its ends. When this fact is included into the calculations of $f(\psi)$ (Eq. 70), it was found that σ is an increasing function of temperature whereas $\mathring{\varepsilon}_f$ increases with chain flexibility [100].

In the very first model of molecular orientation in flow, Kuhn and Kuhn [2] depicted the polymer as having one chain end anchored in space (Fig. 39). In

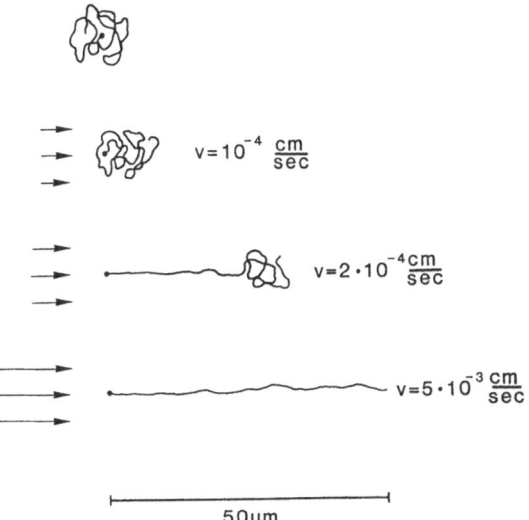

Fig. 39. Uncoiling of a large macromolecular chain with one teathered end (from Ref. [2]).

$v=10^{-4}\frac{cm}{sec}$

$v=2\cdot10^{-4}\frac{cm}{sec}$

$v=5\cdot10^{-3}\frac{cm}{sec}$

50µm

presence of flow, the chain starts to unravel from the position where the viscous friction is the highest, i.e. the free coiled end of the macromolecule. In this model, the velocity gradient necessary to transmit the mechanical stress to the molecule does not stem from the flow field itself but rather from the relative velocity difference between the deforming segment and the fluid velocity. Although few experimental data are available at present, the model of the "chain with tethered end" may be useful to explain flow-induced degradation of adsorbed chains and degradation during GPC analysis which will be examined in Sect. 5.5.

5.2 Chemistry of Degradation

Well before the advent of modern analytical instruments, it was demonstrated by chemical techniques that shear-induced polymer degradation occurred by homolytic bond scission. The presence of free radicals was detected photometrically after chemical reaction with a strong UV-absorbing radical scavenger like DPPH, or by analysis of the stable products formed from subsequent reactions of the generated radicals. The apparition of time-resolved ESR spectroscopy in the 1950s permitted identification of the structure of the macroradicals and elucidation of the kinetics and mechanisms of its formation and decay [15].

In the solid phase, it is possible that the cleavage of the chains is accompanied by the formation of charged species capable of ionizing the surrounding atmosphere, giving rise to the phenomenon of triboluminescence [134–136]. It was determined, for example, that during ball milling of PE and PP in the presence of the ion-scavenger TCNE, as much as 37% of the chain scission occurred through an heterolytic pathway [137].

In the absence of solvation mechanisms, the process of homolytic bond scission in organic compounds requires much less energy than heterolytic bond scission and it is hardly expected that ionic species from organic polymers could be formed in non-polar solvents. Heterolytic cleavage, on the other hand, could be detected in inorganic compounds and it seems now to be well-established that ultrasonic degradation of polydimethylsiloxane proceeds via a ion-radical mechanism.

Bearing an unpaired electron, the fragments formed from homolytic bond scission are highly reactive and are capable of undergoing any of the chemical reactions normally expected from a macroradical:

- diffusion-controlled reactions with paramagnetic species like dissolved oxygen and stable free radicals. This latter reaction forms the basis for the spin-trapping technique frequently used in ESR experiments to prolong the life-time of the macroradicals and permits its identification in solution.
- transfer reactions with radical scavengers by abstraction on labile bonds
- addition reactions on double-bonds
- recombination reactions between macroradicals.

Most of these reactions are well-documented in the literature. The transfer reaction of an hydrogen atom from a β-position transforms a primary radical into a less reactive secondary radical: this type of reaction has undergone detailed investigation during melt extrusion [138]. Due to the high mobility of the chain end bearing the primary radical, this reaction is thought to be the main mechanism responsible for the high rate of radical reactions in glassy polymers below T_g [139].

A large body of literature exists for the addition reaction of mechanoradicals to the vinyl group. This type of reaction was used extensively in the early days of mechanochemistry to synthesize block and graft copolymers with the aim of proving the existence of macroradicals. With an improved understanding of the structure-property relationship in block-copolymers and the availability of inexpensive and reliable generators of high-intensity ultrasound, a renewed interest in mechanochemical synthesis of block copolymers has been witnessed recently [140, 141]. Mechanoradicals were found to follow the same reactivity pattern as conventional macroradicals formed by radical initiation. As a practical consequence, a proper selection of the couple polymer-monomer having the highest rate constant for the propagation reaction is desirable for a successful mechanochemical synthesis: for example, the system poly(vinyl chloride)/methyl methacrylate should be preferred instead of the system poly(methyl methacrylate)/vinyl chloride) for the production of the block copolymer poly(methyl methacrylate-b-vinyl chloride).

Recombination reactions between two different macroradicals are readily observable in the condensed state where molecular mobility is restricted and the concentration of radicals is high. Its role in flow-induced degradation is probably negligible: at the polymer concentration normally used in these experiments (<100 ppm), the rate of radical formation is extremely small and the radicals are immediately separated by the velocity gradient at the very moment of their formation. Thus there is no cage effect, which otherwise could enhance the recombination efficiency.

Unusual reactivities of mechano-radicals have been reported in a few instances. To explain the pseudo first-order kinetics and the high yield of linear block copolymers formed during the mechanochemical degradation of a mixture of two polymers in the solid state, the following mechanism has been proposed [142].

$$P_n^* + R_s - R_t \rightarrow P_n - R_s + R_t^* . \qquad (85)$$

In this reaction, P_n^* is a mechanically activated macroradical which, upon transferring part of its excess energy to a second polymer $R_s - R_t$, induces the degradation of the latter.

A similar mechanochemical chain reaction was proposed to explain the microcrack formation in the glassy state [143]:

$$P_n^* + R_s - R_t \rightarrow P_n H + R_s(-H) + R_t^* . \qquad (86)$$

The newly formed radical (R_t^*) can initiate another sequence of reactions, thus multiplying the number of broken chains while keeping the total radical concentration constant.

The formation of activated species during mechanochemical degradation is, in general, not sufficiently documented both experimentally and with respect to the proposed mechanisms to give a definite proof of their existence. In the dilute state, the rate of energy transfer is high and it is reasonable to assume that any activated species, if present, will be thermalized well before the occurrence of a chemical reaction.

Whether molecular products, such as monomers and oligomers, are formed or not during stress-induced degradation is an unsettled point. The formation of monomers was detected by mass spectrometry during the mechanical fracture of PMMA, PS and PP in the solid state. It was strongly suspected, however, that the recorded signal originated from residual monomer trapped inside the polymer which was subsequently released during the fracture of the sample [144]. In the liquid state, the presence of short fragments was detected by GPC in poly-isobutylene and polyacrylamide solutions degraded at high shear rate [145, 146]. With polyisobutylene solutions, the low MW peak corresponds to the range calculated for entanglement spacings, suggesting that mechanical energy is essentially concentrated at the entanglement coupling points or midway between two such points (Fig. 38) [145].

In transient elongational flow degradation, it was determined in the authors' laboratory, by a detailed mass balance, that main chain scission accounted for >95% of the degradation in dilute solution. Any other type of depolymerization, if present, should then be of minor importance.

Chain scission is the ultimate fate of a stressed bond. At some value below the critical stress for chain rupture, bond angle deformation may result in an increase in reactivity. As stated in Sect. 3.1, mechanically activated hydrolysis of polymers containing ester groups can lead to the scission of the bond: this concurrent reaction should be differentiated from homolytic chain scission, for example by looking at any pH-dependence as was found to be the case during shear degradation of DNA [84].

5.3 Evaluation of Experimental Data

The treatment of experimental data constitutes an essential step in any chemical kinetics study. Although a large part of the present section is based on the investigations in transient flow degradation, the procedure should be general enough to be applicable to other experimental flow arrangements.

5.3.1 Correction for Axial Dispersion in GPC

As previously stated, GPC is the method of choice for studying polymer degradation kinetics. The GPC trace, as given by the detector output, does not provide the true MWD due to various diffusion broadening processes inside the different parts of the equipment. The first step is to correct for instrument broadening if a precise evaluation of MWD is desired. Even with the best columns available, this correction may change the MWD significantly as can be visualized

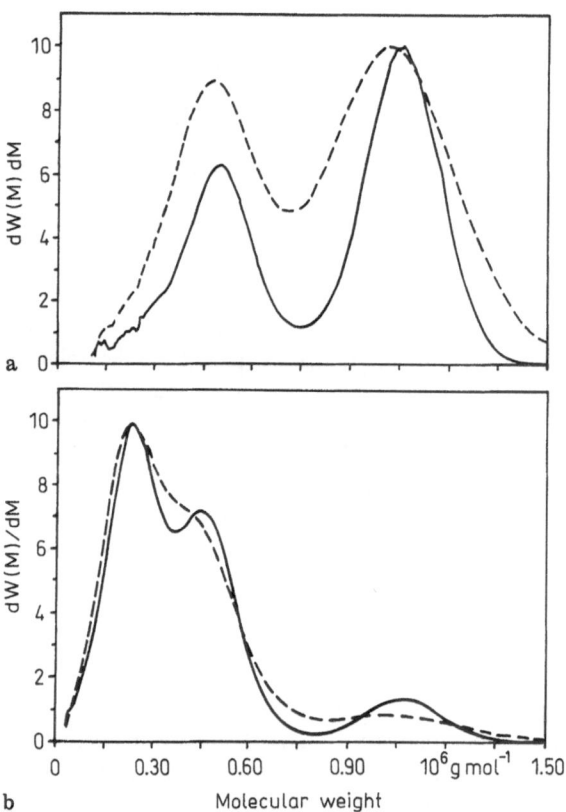

Fig. 40a, b. Effect of instrumental broadening on molecular weight distribution; **a)** PS fraction with $\overline{M_w} = 1.03 \times 10^6$ and $\overline{M_w}/\overline{M_n} = 1.017$ degraded at $\dot{\varepsilon}(0) = 1.7 \times 10^5 \, \text{s}^{-1}$; **b)** Same as in (a), but degraded at $\dot{\varepsilon}(0) = 3.6 \times 10^5 \, \text{s}^{-1}$ $(- - -)$: uncorrected for instrumental broadening; (———): corrected for axial dispersion

from Figs. 40a and 40b [147]. This is particularly true in studies with sharp polymer fractions of high molecular weight where the broadening effect from axial dispersion could be more important than the real width of the MWD itself. For the calibration of axial dispersion, the authors used either the recycling technique with ultra-sharp fractions of PS [148] or the chromophore-bound polymer method [149]. The correction procedure for axial broadening was based on the "trial-and-error" algorithm of Ishige and al. [150]. This method is easily transposed on personal computers but requires, for proper convergence, a good signal-to-noise ratio from the detector response.

5.3.2 Determination of the Degradation Yield

After passage through the orifice, the SEC trace of the degraded polymer solution shows a discernable bimodal distribution. Applying the axial dispersion correction permits one to improve resolution of the degraded peak from the remaining undegraded starting material. Computer substraction of the undegraded SEC trace from the degraded curve provides a sensitive method of quantifying scission yield at low degradation level. This procedure, however, is valid only for low percentages of chain scission since both the MWD of the starting material and of the degraded products varies with the extent of degradation. For scission yields > 15%, the extent of degradation was more accurately calculated from the surface ratio of the degraded peaks to the total area under the SEC trace (Fig. 40).

5.3.3 Scission Probability Along the Chain

The bimodal distribution of the GPC traces (Fig. 40) merely indicates that bond scission is non-random and occurs preferentially near the middle of the chain. In a qualitative way, the width of the degraded peak from the GPC trace is related to the different broadening processes, i.e. instrumental broadening, molecular weight distribution (MWD) of the starting material, and the difference in the scission rate as a function of position along the chain. Based on the observation that the width of the degraded peak is always much broader than the width of the initial undegraded peak, it is frequently quoted that chain scission is not exactly central. Such a statement, however, is misleading as will be shown from the following trivial example. Starting with a monodisperse sample and assuming that the probability of chain scission along the chain follows a Gaussian distribution with a standard deviation σ, the MWD of the degraded fragments with one broken bond will also be Gaussian but with a standard deviation doubled to 2σ. Those with two broken bonds will be even wider with a standard deviation exceeding 4σ. This broadening of the degraded peaks merely reflects the chain halving process; it will be enhanced by the polydispersity effect, even though the chain scission probability distribution function remains exactly unchanged. To determine the exact shape of the probability distribution function for bond scission requires extensive kinetic calculations which will be presented in the following section.

5.4 Kinetics of Bond Scission

Polydispersity is an unavoidable complication when working with synthetic polymers. An (already narrow) anionic PS with a ratio $\bar{M}_w/\bar{M}_n = 1.05$ and following a Schulz-Zimm distribution practically covers molecular weights extending over a decade and even the best refractionated sample with $\bar{M}_w/\bar{M}_n \approx 1.01-1.02$ still contains polymers encompassing a factor of 2 in chain length. For the sake of accuracy, it is always desirable to work with samples having the lowest polydispersity available. Since the rate of mechanochemical degradation strongly depends on the molecular chain length (Sect. 5.6), the MWD of the starting material has a decisive influence on the experimental results and should always be properly taken into account for meaningful evaluation of the kinetic parameters.

The aim of any kinetics study is to determine the individual rate constants from a reaction scheme established in conformity with the available experimental data. More specifically to the transient elongational flow problem, the kinetics calculations should be able to reproduce faithfully:

— the dependence of the scission yield as a function of strain rate (Fig. 41)
— the MWD of the degraded polymer at any given strain rate (Fig. 40a and 40b)
— the existence of a "master curve" when the degradation yields of different MW samples are plotted on a relative strain rate scale (Fig. 42) [147].

Starting with the results of GPC analysis, two approaches have been successfully applied to the problem of polymer degradation, using either the differential or the integral expression of the kinetics equations:

a) in the first technique, a number of GPC traces are recorded at successive degradation times or degradation yields. Each MWD is divided in a number of

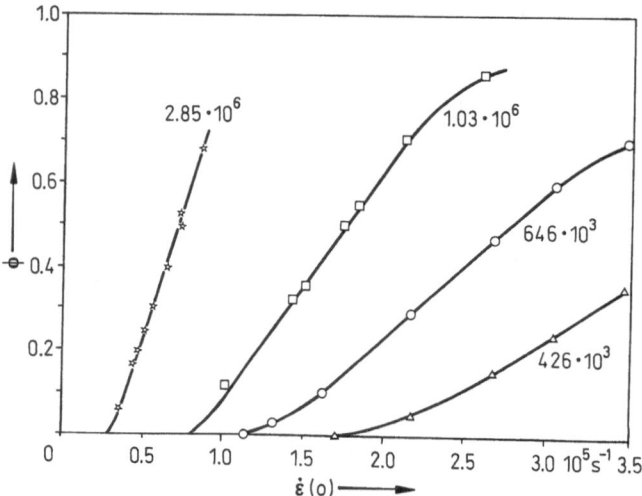

Fig. 41. Dependence of the degradation yield (Φ) on strain rate ($\dot{\varepsilon}(0)$) and initial polymer molecular weight (as indicated at the curves)

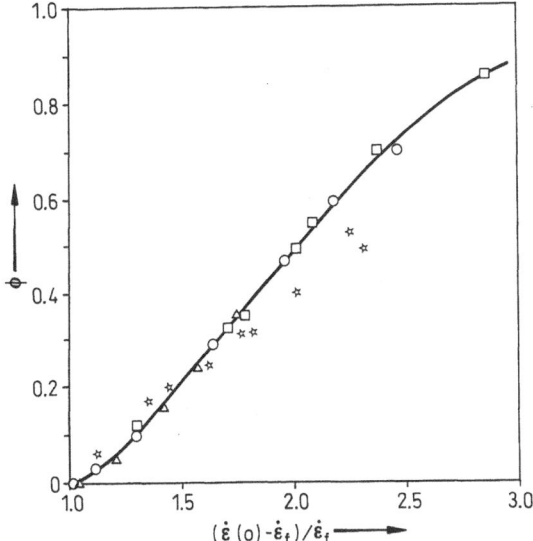

Fig. 42. Dependence of the degradation yield (Φ) on relative strain-rate ($\dot{\varepsilon}(0)/\dot{\varepsilon}_f$)) and polymer molecular weight (same symbols as in Fig. 41)

regular MW intervals and the time derivatives are calculated for each slice from two consecutive chromatograms. The obtained numerical values are then inserted into the corresponding kinetic equations to solve for the unknown rate constants [151]. This method is exact in the sense that no fitting parameter is needed. However, it necessitates highly reproducible MWD determination and a repeated number of experiments performed under identical degradation conditions.

b) the second technique requires numerical integration of the differential kinetic equations, a task which no longer represents a major limitation with the widespread availability of powerful personal computers. The rate constants were used as fitting parameters until a good agreement was obtained between the calculated MWD and the experimental curves [152]. This method is generally preferred in flow-induced degradation. It is less time demanding and nevertheless capable of yielding accurate information about the chain scission kinetics.

Macromolecules have a large number of repeating links and a complete kinetics scheme should include all the individual rate constants for each reacting bond. In most polymer systems, such a detailed analysis is not only cumbersome but unnecessary. For a reliable description of the degradation kinetics, it is generally sufficient to consider rate constants for a representative number of "quasi-monomeric units", each one containing a large multiple of the true monomeric entities. Basedow [153] divided the normalized number molecular weight distribution of the starting material (after correction for axial dispersion) into a discrete set of equidistant molecular weights ("brute force lumping"). The number fraction of molecules having an i-unit chain length is n_i ($1 \leq i \leq r$). With sharp polymer fractions, a value of $r = 100$ was found to be adequate to insure proper convergence of the calculated curves.

The change of n_i with time was calculated according to first-order kinetics. It is given by a system of r linear differential equations and $(\frac{1}{2}) \cdot r(r - 1)$ variables:

$$dn_i/dt = -\sum_{j=1}^{i-1} k_{i,j} \cdot n_i + \sum_{j=1}^{r-i} (k_{i+j,j} + k_{i+j,i}) \, n_{i+j} \,. \tag{87}$$

The individual rate constants $k_{i,j}$ refer to the scission of a chain of i-units in length into two fragments with j and $(i - j)$ subunits respectively.

The system of Eq. (87) can be written in a more general matrix notation:

$$d\bar{n}/dt = \mathbf{A} \cdot \bar{n} \,. \tag{88}$$

$\mathbf{A}$ is called the degradation matrix and $\bar{n}$ is a vector which includes all the species with different degrees of polymerization. Starting with Eq. (88), Basedow et al. devised a general mathematical treatment of chain scission kinetics which does not require numerical integration [153]. The only constraint in their analysis is the absence of radical recombination. This condition is easily fulfilled in practice either by adding some radical scavenger like DPPH or Galvinoxyl to the solution, or simply by not degassing the polymer solution prior to experimentation.

According to the experimental conditions, several simplifications can be considered for reducing the number of unknown parameters:
— for linear molecules, the probability of bond scission is symmetric about the chain center, therefore

$$k_{i,j} = k_{i,i-j} \tag{89}$$

— a midchain scission is found in almost any mechanochemical degradation. In stagnant elongational flow, theoretical arguments supported by experimental results indicate that the probability of chain scission as a function of bond position can be approximated by a Gaussian distribution function of standard deviation σ having the maximum at the center of the chain. Observations in transient elongational flow give results similar to stagnant flow, even if the chains are only partly extended. From these results, it is possible to postulate empirically the following functional dependence for the individual rate constants $k_{i,j}$.

$$k_{i,j} = K_i(\sigma_i \sqrt{2\pi})^{-1} \cdot \exp\{-(j - i/2)^2/(2\sigma_i^2)\} \tag{90}$$

The normalization coefficient K_i is the global scission rate constant given by the contribution of all the individual rate constants for each fracture site:

$$K_i = \sum_{j=1}^{i-1} k_{i,j} \,. \tag{91}$$

Under specific circumstances, alternative forms for $k_{i,j}$ have been proposed like the parabolic or the truncated Gaussian probability distribution function for example [154].

— as a further simplification, it was established that the width of the scission probability distribution function is relatively insensitive to polymer molecular weight [147]. With narrow MWD starting material, a good approximation was to assume that the relative standard deviation (σ_i/i) is independent of chain length:

$$\sigma_i = R \cdot i. \tag{92}$$

By combining Eqs. (87), (89), and (90), most of the unknown variables are eliminated and R is the only parameter left to be adjusted from the experiments. Whenever needed, a change of R with chain length can easily be included in the calculations.

In flow-induced degradation, K_i is strongly dependent on the chain length and on the fluid strain-rate ($\dot{\varepsilon}$). According to the rate theory of molecular fracture (Eqs. 70 and 73), the scission rate constant K_i can be described by the following equation [155]

$$K_i = (i - 1) \cdot A \exp\left[(-U_0 + \beta \cdot \psi_i)/RT\right]. \tag{93}$$

The statistical weight $(i - 1)$ in the pre-exponential factor accounts for the fact that the number of fracture sites under stress is proportional to the number of bonds present in the macromolecule.

The Stokes' viscous drag equation predicts a proportionality between the molecular stress (ψ_i) with the product of solvent viscosity (η_s) and fluid strain-rate ($\dot{\varepsilon}$):

$$\psi_i \sim M_i^x \cdot \eta_s \dot{\varepsilon}. \tag{94}$$

The exponent x is an empirical parameter to be determined from experiments. For a fully extended chain in stagnant elongational flow, x is equal to 2 whereas a value of 1 was found under transient flow conditions (Sect. 5.4).

According to some recent results (Sect. 5.5), the dependence of K_i on $\dot{\varepsilon}$ and η_s is more involved than suggested by Eq. (94). The dependence of K_i on η_s is much weaker than a direct proportionality and the correct flow parameter to be used in Eq. (94) should be the local fluid kinetic energy ($\sim v^2$) rather than the strain rate ($\dot{\varepsilon}$). For a constant flow geometry, however, the two variables v and $\dot{\varepsilon}$ are interchangeable. At the present stage and in order not to complicate unduly the kinetic scheme, it will be assumed that the rate constant K varies with the MW and strain rate as:

$$K_i = (i - 1) \cdot A \exp\left[(-U_o + \alpha \cdot \dot{\varepsilon}(i - 1)/RT)\right]. \tag{95}$$

The coefficient α includes the solvent viscosity dependence and accounts for the negative activation energy found in mechanochemical degradation.

Under steady-state conditions, as in the Couette flow, the strain rate is constant over the reaction volume for a long period of time (several hours) and the system of Eq. (87) could be solved exactly with the matrix technique developed by Basedow et al. [153]. Transient elongational flow, on the other hand, has two distinctive features, i.e. a short residence time (a few μs) and a non-uniform flow field, which must be incorporated into the kinetics equations. In transient elongational flow, each rate constant is a strong function of the strain-rate which varies with time in the Lagrangian frame moving with the center of mass of the macromolecule: the local value of the strain rate for each spatial coordinate must be known before Eq. (87) can be solved.

5.4.1 Monodisperse Fraction

For the purpose of illustration, let us consider the degradation behavior of a hypothetical monodisperse polymer fraction flowing along the central streamline.

First, the function $\mathring{\varepsilon}(t)$ computed from $\mathring{\varepsilon}(x)$ (Fig. 33), is divided into a number of time intervals which are sufficiently short to justify the approximation of a constant average strain-rate within each period. Only the region of space where the strain rate is significantly different from zero, i.e. from $-4r_0 \leqq x \leqq +r_0$ in the case of abrupt contraction flow (Fig. 33), will contribute to the degradation and needs to be considered in the calculations. The system of Eq. (87) is then solved locally using the previously mentioned matrix technique [153].

The fraction of unbroken chains flowing along the centerline of the contraction is given by:

$$d(\ln n) = -K[\mathring{\varepsilon}(x))/v(x)] \, dy = -K(\mathring{\varepsilon}(t)) \, dt \,, \tag{96}$$

where v is the local fluid velocity along the flow direction (x). Denoting n_0 as the number of macromolecules initially present, the scission yield (Φ) can be obtained by substraction:

$$\Phi = 1 - n/n_0 = 1 - \exp - (\int K(\mathring{\varepsilon}(t)) \, dt) \,. \tag{97}$$

From experimental results [147], it is known that a PS with $M = 1.03 \times 10^6$ is degraded in decalin at $\mathring{\varepsilon}_f = 80000 \, s^{-1}$. This value can be put into Eq. (95) to calculate K and the degradation yield Φ in Eq. (97).

The finite transit time (τ_r) in the high strain-rate region has important consequences on the scission yield curve. Since bond scission is a first-order process (Eq. 96), the degradation yield in a single pass experiment is given approximately by:

$$\Phi \approx 1 - \exp(-K\tau_r) \tag{98}$$

Degradation is appreciable only if $K > 1/\tau_r$: this occurs whenever the coefficient $\alpha \cdot M \cdot \mathring{\varepsilon}$ offsets the dissociation energy term $-U_0$. It is physically unrealistic that an elementary chemical process can have negative activation energy, therefore the permissible maximum for K is controlled by the pre-exponential factor $A \cdot (i - 1)$.

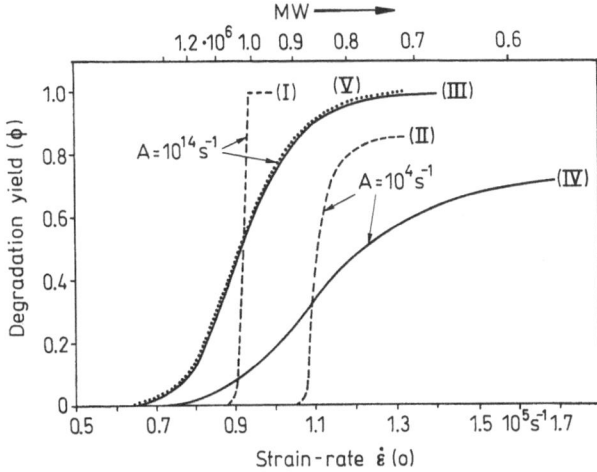

Fig. 43. Dependence of the degradation yield on strain rate for two polymer samples of different ratio of weight-to number-average molecular weights $(\overline{M_w}/\overline{M_n})$, calculated with the pre-exponential factor A as indicated at the curves (I): monodisperse fraction with molecular weight $M = 1.03 \times 10^6$; pre-exponential factor $A = 10^{14}\,s^{-1}$ (II): monodisperse fraction with molecular weight $M = 1.03 \times 10^6$; pre-exponential factor $A = 10^4\,s^{-1}$; (III): poly-disperse fraction with weight-average molecular weight $\overline{M_w} = 1.03 \times 10^6$, ratio of weight- to number-average molecular weights $\overline{M_w}/\overline{M_n} = 1.017$; pre-exponential factor $A = 10^{14}\,s^{-1}$; (IV): polydisperse fraction with weight-average molecular weight $\overline{M_w} = 1.03 \times 10^6$, ratio of weight- to number-average molecular weights $\overline{M_w}/\overline{M_n} = 1.017$; pre-exponential factor $A = 10^4\,s^{-1}$; (V): cumulative molecular weight distribution plotted as a function of molecular weight M on an inverse scale (upper abscissa)

It is expected that the value of the coefficient A lies within the usual range found for unimolecular decomposition, i.e. of the order of 10^{13} to $10^{15}\,s^{-1}$. For $A > 10^8\,s^{-1}$, Eq. (97) combined with the strain-rate function $\dot{\varepsilon}(x)$ predicts an extremely sharp transition from no bond scission to quantitative degradation whenever the coefficient $\alpha \cdot M \cdot \dot{\varepsilon}$ reaches a value of 0.85 to 0.95 U_0 (depending on the exact value of A) at room temperature. In the improbable case of $A < 5 \times 10^4\,s^{-1}$, K is now $< 1/\tau_r$ and the degradation yield after a single passage is limited even at very large strain rates (Fig. 43).

5.4.2 Polydisperse Fraction

The calculations are applied to a polymer sample following a Schulz-Zimm distribution with $\bar{M}_w = 1.03 \times 10^6$ and $\bar{M}_w/\bar{M}_n = 1.017$. These values are representative for the polymer fractions used in most of the experiments in transient elongational flow [147, 155]. To visualize the evolution of the degradation, it is convenient to make a distinction between the polymer fraction N_i from the starting material which remains intact and the fraction N_i^* newly formed following the degradation process

$$n_i = N_i + N_i^* . \tag{99}$$

It is clear that, having the same chain length, these macromolecules are in reality experimentally indistinguishable. One could however think of a labelling technique to make N_i^* different from N_i, for example by using a chromophore-bound radical scavenger which added selectively to the macroradicals issued from the chain scission. Equation (87) can be split into a system of two equations (100) and (101)

$$dN_i/dt = -K_i(\dot{\varepsilon})\, N_i\,, \qquad 2 \leqq i \leqq r\,. \tag{100}$$

The rate of formation of the degraded fragments, with the symmetry relation (82), is given by:

$$dN_j^*/dt = -K_j(\dot{\varepsilon})\, N_j^* + 2 \sum_{i=j+1}^{r} k_{i,j}(\dot{\varepsilon}) \cdot (N_i + N_i^*)\,,$$

$$1 \leqq j \leqq r - 1\,. \tag{101}$$

The change of the MWD with time is obtained by resolution of the system of Eqs. (100) and (101). In analogy with the monodisperse polymer fraction, a large pre-exponential factor selectively depletes from the reaction medium all polymer molecules with MW greater than $U_0/\alpha \cdot \dot{\varepsilon}$. The MWD of the degraded polymer clearly reflects this "all-or-nothing" process by the abrupt drop in the region of high molecular masses (Fig. 44). The dependence of the scission yield on strain-rate is given in this case, with a good approximation, by the cumulative MWD of the original polymer when plotted as a function of $1/M$ (Fig. 43). As a consequence of this behavior, two polymer samples different in MW but following a similar molecular mass distribution function will have their degradation yields directly superposable when plotted as a function of the relative strain-rate $(\dot{\varepsilon}(0)/\dot{\varepsilon}_f)$ (Fig. 42).

It should be stressed that the observed critical strain-rate for bond fracture $(\dot{\varepsilon}_f)$ in the case of a polydisperse fraction refers to the *longest chain* present in the sample. This quantity is significantly different from the critical strain-rate $(\dot{\varepsilon}_f^*)$ defined with respect to an average molecular mass whose value could be determined only after careful consideration of the degradation kinetics.

5.4.3 Degradation Across the Flow Tube

As mentioned in Sect. 4.3, each streamline across the degradation apparatus contributes to a different strain rate function. In order to reproduce the experimental curves, it is necessary to repeat the above calculations and integrate over all the fluid elements passing through the orifice. Such a task is however more complicated than it may appear at first thought: far from the symmetry axis, the flow is no longer purely elongational. With the increase in vorticity, particularly near the wall, chain extension and rupture will be inhibited. Then the knowledge of the strain-rate function alone is insufficient to calculate the scission kinetics.

To avoid these complications, we will rely on an empirical approach to estimate the effect of radial strain-rate distribution on degradation yields.

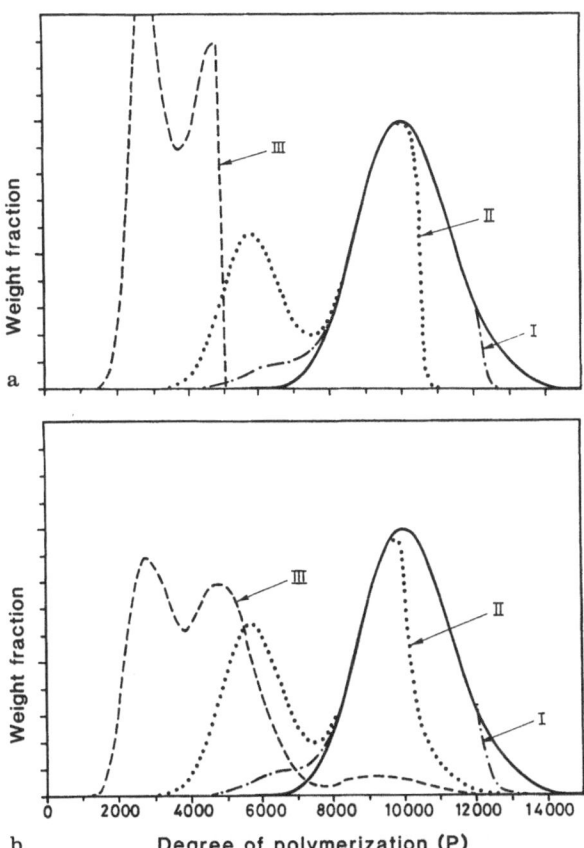

Fig. 44a. Theoretical molecular weight distribution of a polymer sample degraded along the central streamline at different strain rates, calculated with a pre-exponential factor $A = 10^{14} \, s^{-1}$; (I) strain rate $\mathring{\varepsilon} = 75000 \, s^{-1}$; (II) strain rate $\mathring{\varepsilon} = 88000 \, s^{-1}$; (III) strain rate $\mathring{\varepsilon} = 190000 \, s^{-1}$. **b** Theoretical molecular weight distribution of a polymer sample degraded along the central streamline at different strain rates, calculated with $A = 10^4 \, s^{-1}$; (I) strain rate $\mathring{\varepsilon} = 100000 \, s^{-1}$; (II) strain rate $\mathring{\varepsilon} = 120000 \, s^{-1}$; (III) strain rate $\mathring{\varepsilon} = 300000 \, s^{-1}$ (*Solid line*: polymer before degradation, *dotted line*: degraded polymer)

First, we will demonstrate from simple geometric arguments that the existence of a strain rate distribution across the tube gives further support to the previously described existence of a "master curve" (Fig. 42).

Since a critical strain-rate ($\mathring{\varepsilon}_f$) must be surpassed to break the chemical bond, only macromolecules flowing close to the central streamline could be degraded near the scission threshold. Another streamline passing through the orifice at a distance r from the symmetry axis will reach the required strain rate at a higher fluid velocity (Fig. 35) in order to meet the condition:

$$\mathring{\varepsilon}(r) = C(r) \cdot \mathring{\varepsilon}(0) \geq \mathring{\varepsilon}_f. \tag{102}$$

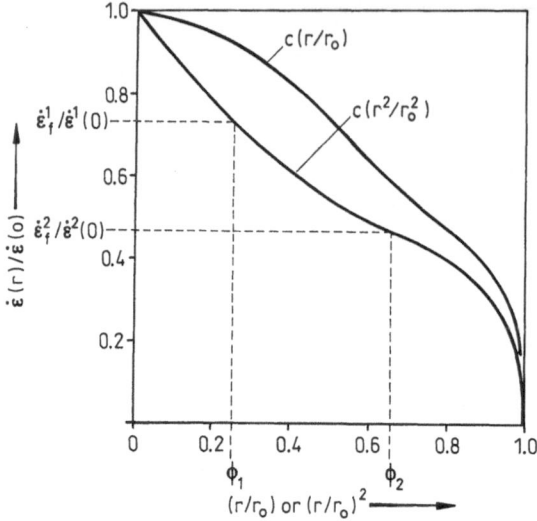

Fig. 45. Radial strain rate distribution function and its influence on the degradation yield Φ (r_0: orifice radius, r: axial distance from the orifice center) $C(r^2/r_0)$: variation of the strain rate as a function of $(r/r_0)^2$ obtained from the reciprocal of the degradation yield curve of a monodisperse polymer fraction (see Fig. 46) $C(r/r_0)$: radial distribution of the strain rate in the plane of the orifice, calculated from the function $C(r^2/r_0^2)$

The fraction of macromolecules which have been degraded (Φ) is equal to the volume of all the fluid elements which satisfy Eq. (102) divided by the total flow across the orifice. Assuming a constant velocity profile at the orifice, Φ in this case is just equal to $(r/r_0)^2$, with r_0 being the radius of the orifice.

In order to reach the same scission yield $\Phi = (r_k/r_0)^2$ for 2 different MW, denoted below by the superscript "1" and "2", the following equalities should be satisfied (Fig. 45):

$$\dot{\varepsilon}^1(0) \cdot C(r_k/r_0) = \dot{\varepsilon}_f^1 \, ,$$

$$\dot{\varepsilon}^2(0) \cdot C(r_k/r_0) = \dot{\varepsilon}_f^2 \, ,$$

or

$$\dot{\varepsilon}^1(0)/\dot{\varepsilon}_f^1 = \dot{\varepsilon}^2(0)/\dot{\varepsilon}_f^2 = 1/C(r_k/r_0) \, . \tag{103}$$

The relation (103) will hold for any value of the scission yield, providing the macromolecules do not cross the boundary between the different streamlines, either by molecular diffusion or by flow turbulence. For samples having similar polydispersities and MWD, all the degradation data could be superimposed onto a single curve when plotted on a relative abscissa scale ($\dot{\varepsilon}(0)/\dot{\varepsilon}_f$).

The function $C(r/r_0)$ can be determined experimentally from the degradation yield curve of a truly monodisperse sample since the function $C(r^2/r_0^2)$ is simply

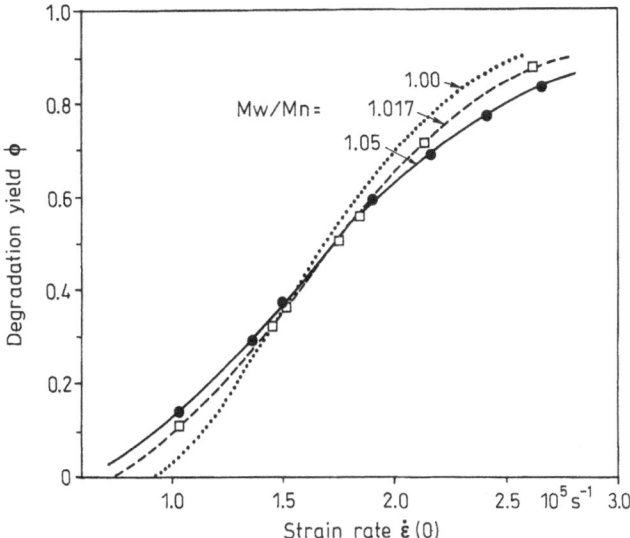

Fig. 46. Dependence of the degradation yield Φ on strain rate $\dot{\varepsilon}(0)$, and on ratio of weight- to number-average molecular weight $(\overline{M_w}/\overline{M_n})$ of polymer as indicated at curves. The *dotted line* is extrapolated for a monodisperse polymer fraction

the reciprocal of the function $(\dot{\varepsilon}/\dot{\varepsilon}_f) = F(\Phi)$. Such a sample is not available for a synthetic polymers and $F(\Phi)$ must be extrapolated by trial-and-error from the degradation yield curves measured at different polydispersities (Fig. 46), until a good fit is obtained between the calculated and experimental values.

No appreciable change in the parameter R with the degradation yield is detected experimentally [147, 155]. Therefore, a constant value for R, independent of the streamline position, can be used for the calculations. The final results, obtained by solving the system of Eq. (87) and taking into account both the axial and radial distribution of the strain-rate, are given in Fig. 47. The presence of the $C(r/r_0)$ term tends to smooth over the calculated MWD which is now in closer conformity with the experimental results (Fig. 40). In particular, the sharp depletion in the high MW tail, observed along a single streamline (Fig. 44) has disappeared. The question remains whether the gradual change in the measured curves solely reflects the averaging process over the radial coordinates or stems from other unaccounted characteristics of degradation in flow. This problem may be resolved by changing the flow geometry leading to a different distribution of the strain rate.

5.4.4 The Empirical Approach

The parameter R in Eq. (92) can be derived empirically from the experimental SEC traces with a minimum of computational effort and without regard to the details of the degradation kinetics by using the following arguments. Once a macromolecule is fractured, the moieties are immediately driven into another region of space. Due to the decrease in MW, a considerably higher strain rate

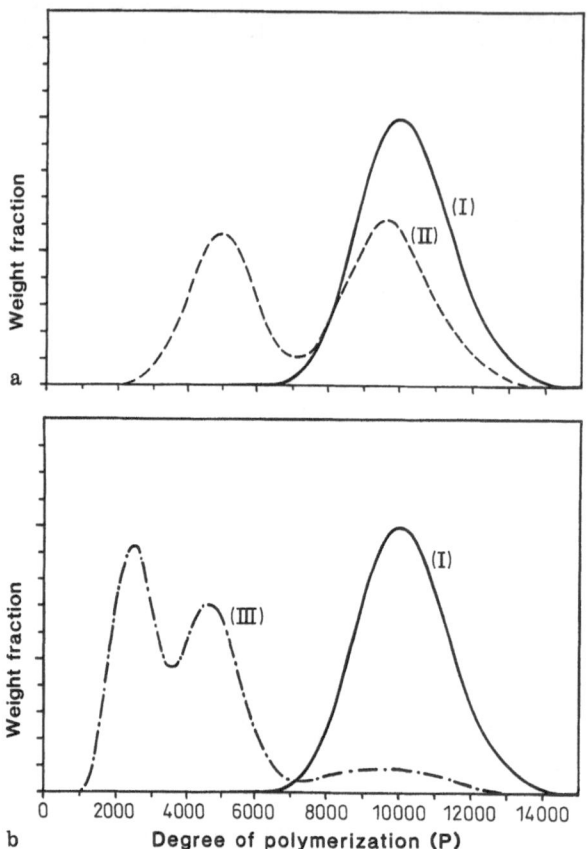

Fig. 47 a, b. Calculated distribution of the degree of polymerization P, taking into account both the axial and radial distribution of the strain rate $\dot{\varepsilon}(0)$, for a polymer sample degraded at strain rate $\dot{\varepsilon}(0) = 1.5 \times 10^5 \text{ s}^{-1}$ (**a**) and at strain rate $\dot{\varepsilon}(0) = 3.5 \times 10^5 \text{ s}^{-1}$ (**b**); (I): before degradation; (II): degraded at a strain rate $\dot{\varepsilon}(0) = 1.5 \cdot 10^5 \text{ s}^{-1}$; (III): degraded at a strain rate $\dot{\varepsilon}(0) = 3.5 \cdot 10^5 \text{ s}^{-1}$

will be necessary to break these fragments a second time. For these reasons, only a limited number of broken bonds per initial chain need to be considered during the flow time across the degradation zone. In addition, the degradation process is spatially inhomogeneous and bond scission can be appropriately treated as a series of discrete events.

Based on the master curve of Fig. 42, it is possible to evaluate graphically for each M_i the fraction of molecules which have been degraded at a given fluid velocity. The MWD of the degraded fragments which result from the scission of molecular mass M_i is a gaussian centered at $M_i/2$ with standard deviation equal to 2 R. Some of the longest fragments can be degraded further, provided the pervading flow field meets the required condition for bond rupture (Eq. 102). In view of the spike-like shape of the strain-rate function, it is unlikely that more

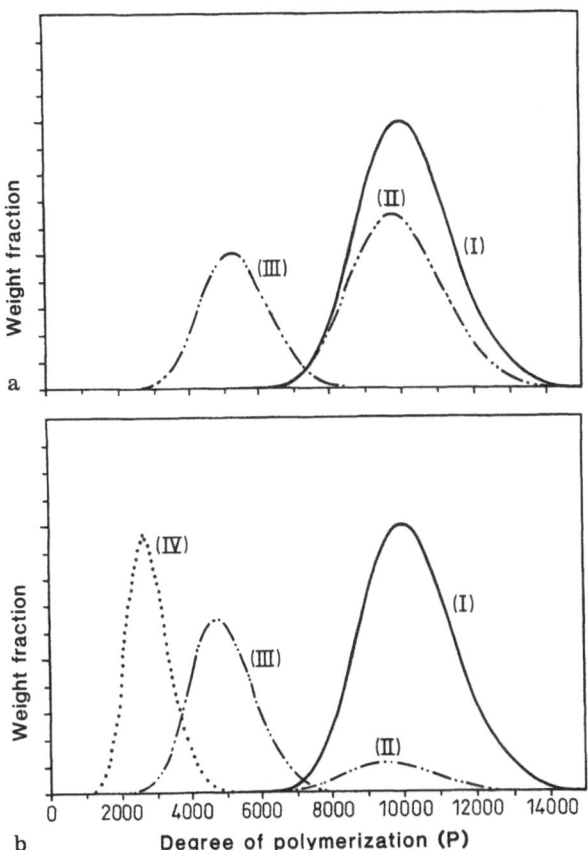

Fig. 48a, b. Distribution of the degree of polymerization P, calculated with the empirical technique, for a polymer sample degraded at strain rate $\dot{\varepsilon}(0) = 1.5 \times 10^5$ s^{-1} (**a**) and at strain rate $\dot{\varepsilon}(0) = 3.5 \times 10^5$ s^{-1} (**b**); (I): before degradation; (II): part of polymer undegraded after passage through the orifice; (III): part of polymer with one broken bond per molecule (IV): part of polymer with two broken bonds per molecule

than 2 scission events per chain can be realized during the flow time (this hypothesis is actually confirmed by the preceding detailed calculations). Since the transit time is even shorter for the degraded fragments than for the initial polymers, the fraction of chains broken twice will follow another degradation yield curve which must be adjusted as to fit the experimental data. The results obtained by using this simple empirical technique are in perfect accord with the experimental MWD (Fig. 48).

5.5 Initial Molecular Weight

The molecular weight of the polymer can influence the rate of flow-induced degradation in at least two respects: first, molecular mobility (characterized by the chain relaxation time) decreases with the chain length, permitting a larger molecular coil to be deformed at a lower fluid strain rate. Secondly, a larger macromolecule by virtue of its physical size, can accumulate more strain and store more elastic energy. These considerations lead to an increase in bond scission susceptibility with chain length. As mentioned in Sect. 5.1, highly expanded chains are predicted to degrade according to the M^{-2} scaling law with a sharp probability for midchain scission. These expectations have been confirmed experimentally in opposed-jets flow. In transient elongational flow, on the other hand, chain expansion is limited due to the short residence time and it is conceivable that the molecular coil remains essentially non-free draining up to the point of rupture. This state of partial chain uncoiling should be detected as a deviation from the $\dot{\varepsilon}_f \sim M^{-2}$ law which is characteristic for perfectly oriented polymers.

Using the device schematized in Fig. 26, polystyrene fractions of different MWs are degraded under θ-conditions in dekalin. To minimize the polydispersity effect, only ultra-narrow fractions are utilized for the experiments (Table 1). The degradation yields as a function of strain rate, measured for each MW, were presented in Fig. 41. From these experimental data, the critical strain rates for chain fracture ($\dot{\varepsilon}_f^*$) are determined according to the kinetics scheme previously developed (the values found correspond to a scission yield of $\approx 7\%$ on the degradation curves). When the different $\dot{\varepsilon}_f^*$ are plotted on a double-logarithm scale as a function of molecular weight, a straight line is obtained with a slope -0.95 (Fig. 49). Such a weaker dependence of $\dot{\varepsilon}_f$ on M was predicted on the basis of slender-body hydrodynamics [12]. Considering a partly deformed molecular coil as an impenetrable ellipsoid, Rabin suggested that the frictional force should scale with the square of the longest axis or $\sim R_g^2$ (Eq. 47). From this relation, the following scaling law was derived:

$$\dot{\varepsilon}_f \sim M^{-2\nu}. \tag{104}$$

Although the derived scaling law seems to conform to the experimental results under θ-conditions where $\nu = 0.50$, the slender body hydrodynamics model is unsatisfactory in many respects.

– superimposed on Fig. 49 are the values of $\dot{\varepsilon}_f$, measured for the same system PS-decalin but under stagnant flow conditions, which distinctly show a slope of -2 as predicted from Eq. (78) [174]. Due to the stronger dependence of $\dot{\varepsilon}_f$ on the chain length ($\sim M^{-2}$), the curve in stagnant flow intersects the transient

Table 1. Principal characteristics of the polymer fractions used

M_w	M_n	M_w/M_n	$\tau_z(\mu_s)$	τ_r/τ_z	σ
2.86×10^6	2.80×10^6	1.020	182	0.29	0.12
1.03×10^6	986×10^3	1.017	37.5	0.31	0.070
646×10^3	639×10^3	1.011	18.6	0.34	0.052
426×10^3	421×10^3	1.012	9.8	0.36	0.055

σ is the standard deviation from midchain scission

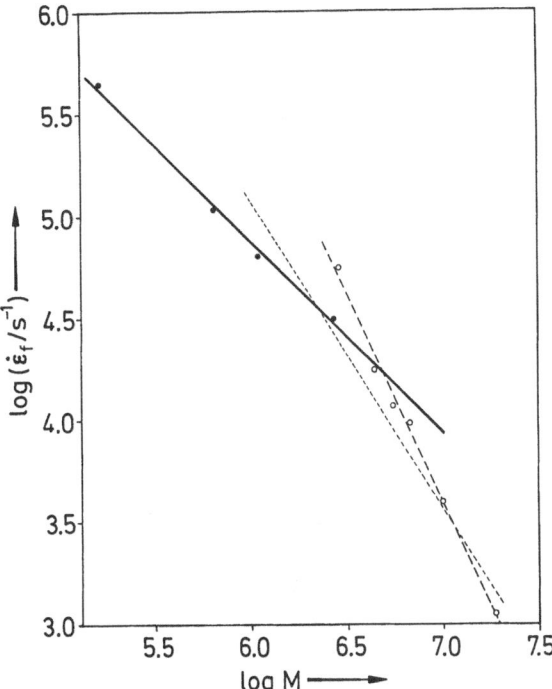

Fig. 49. Dependence of critical strain rate for chain scission ($\dot{\varepsilon}_f^*$) on initial polymer molecular weight for the system PS in decalin; (———): transient elongational flow; (– – –): stagnant elongational flow (data from Ref. [174]) The *dotted line* (– – –) with a slope of −1.54 corresponds to the experimental values of $\dot{\varepsilon}_{cs}$ determined by the Bristol group (Ref. [9])

flow line at $M \approx 3–4 \times 10^6$. This means that below this cross-over molecular weight, scission in transient flow will occur at a lower strain rate than under stagnant flow. This conclusion is in complete contradiction to the hydrodynamic screening concept of non-free draining coils where the gripping action of the flow increases rapidly with the degree of coil expansion regardless of the molecular weight.

— the slender body model implies that $\dot{\varepsilon}_f \sim M^{-6/5}$ for expanded coils in good solvent and $\sim M^{-2/3}$ for globules below the θ regime. The limited data reported for 1-methyl-naphtalene do not seem to support this prediction (Sect. 5.8).

5.6 Bond Strength and Chemical Structure

The bond dissociation energy is a prime variable in mechanochemistry since it delineates the level of stress a chain can sustain before rupture (Eq. 62). The role of weak bonds in mechanochemistry has been investigated mainly in the solid state. Electron spin resonance has shown that rupture of polysulfides and polyamides generally occurs at the carbon-carbon bond in β-position to the heteroatom where the bond energy is the lowest. The possibility of resonance structures in the original and degraded polymer is important for the location of bond scission: it is known for example that polyisoprene degrades from the middle link in $=C-C-C-C=$ where bond energy is reduced by as much as $80 \text{ kJ} \cdot \text{mol}^{-1}$ by resonance stabilization of the generated allyl macroradicals [156]. In vulcanized natural rubber containing different types of cross-links ($C-C$, $C-S-C$, $C-S_x-C$), mechanochemical degradation was shown to take place mainly (although not exclusively) through the sulfur atoms due to the weaker $S-S$ bonds [156].

Much fewer experiments are available in solution where the few reported data are generally more concerned about the effect of molecular structure than about bond dissociation energy. In simple shear, it is generally agreed that chain flexibility dominantly influences the rate of bond scission, with the most rigid polymers being the easiest to fracture [157]. The results are interpreted in terms of the presence of "good" and "poor" sequences in the chain conformation.

The role of long side groups in mechanochemical degradation is unclear as yet. During ultrasonic irradiation of polyalkyl acrylates, polymer stability was found to be independent of side chain length [158]. This is different with the results reported in simple shear flow [159] where the presence of long side groups protected the main chain from mechanochemical degradation. This effect was tentatively assigned to the shielding effect of the polymer skeleton from solvent action. In turbulent flow, it was similarly found that branched structures can confer some additional resistance to bond scission during drag reduction [160, 161]. The presence of lateral groups can influence both the chain flexibility and the monomer friction coefficient (ξ); the possibility that ξ may be strain rate dependent, to allow for flow orientation of the side branches, has not been explored yet. In stagnant elongational flow, it was found that polyethylene oxide (PEO) could resist a higher strain rate before rupture than polystyrene. The difference was explained in terms of a lower monomeric friction coefficient combined with a higher chain flexibility in PEO [162].

The possible role of weak links on the service lifetime of polymer materials has attracted considerable interest over the years. It seems that a systematic investigation of the relation between bond energy or chemical structure and the rate of chain scission should be most useful in elongational flow which provides an almost ideal environment to study uniaxial stretching of isolated chains. Unfortunately, only a single study in this direction is available at present [163]. In the cited reference, a degradation study in transient elongational flow was undertaken comparing bisphenol-A polycarbonate and a similar material copolymerized with a small percentage of sterically hindered bisphenol monomer (Fig. 50). The labile bonds in the comonomer are destabilized by resonance and steric hindrance resulting in a bond dissociation energy of only $140 \text{ kJ} \cdot \text{mol}^{-1}$ or 45% of the value for a normal $C-C$ bond. Random copolymerization ensures that the weak links are statistically distributed along the chain, thus avoiding any bias of the location of the broken bonds; its small number (3%) will minimally perturb the chain

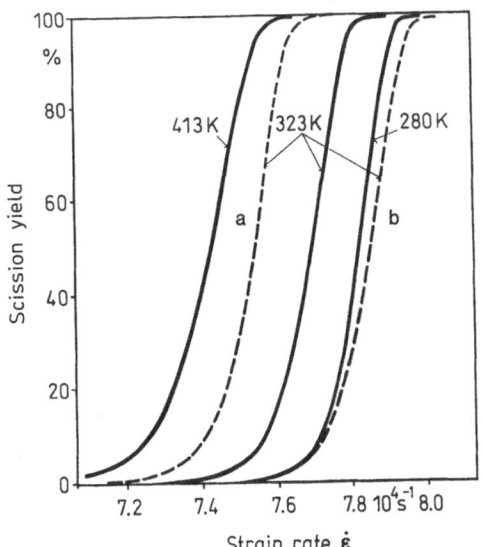

Structure of bisphenol-A polycarbonate copolymerized with 3% by weight of labile groups

structure with respect to the homopolymer. Degradation results, summarized by the graphs in Fig. 50 (shown in the Appendix), reveal a weak effect of the bond dissociation energy on the critical strain rate for chain fracture ($\dot{\varepsilon}_f$). With a decrease in bond dissociation energy by over a factor of 2, the reduction in $\dot{\varepsilon}_f$ was only of about 10 to 30%. Although the relevant quantity for this comparison should be the bond strength (f_b) rather than the bond dissociation energy (U_0), the two quantities are related (Eq. 62); thus, it is hard to believe that the smallness of the observed difference in $\dot{\varepsilon}_f$ can be entirely accounted for by this approximation. In the described experiment, the type of the broken bond was not identified, for example by ESR technique. This leaves opened the question whether the process of degradation is selective concerning the weakest bonds, or rather affects undiscriminately any bond of the main chain, in which case the weak dependence of $\dot{\varepsilon}_f$ on bond dissociation energy can be better understood.

The preceding results on polycarbonate are at variance with the ultrasonic degradation of poly(vinyl pyrollidone) prepared with peroxide linkages where the rate of chain cleavage was determined to be 5000 times faster at the $-O-O-$ than the $-C-C-$ bonds [164].

Fig. 51. Dependence of the scission yield on strain rate and temperature. Curves with a *continuous line* are calculated with the pre-exponential factor $A = 10^{12}$ s^{-1}; curves with *broken lines*, (a) and (b), are obtained using respectively $A = 10^{11}$ s^{-1} and 10^{13} s^{-1}

5.7 Temperature

From the Arrhenius form of Eq. (70) it is intuitively expected that the rate constant for chain scission k_c should increase exponentially with the temperature as with any thermal activation process. It is practically impossible to change the experimental temperature without affecting at the same time the medium viscosity. The measured scission rate is necessarily the result of these two combined effects; to single out the role of temperature, k_c must be corrected for the variation in solvent viscosity according to some known relationship, established either empirically or theoretically.

Under stagnant elongational flow, it was determined that polymer chains degrade according to the TABS model of bond scission under stress (Eqs. 64 and 70, Cf. also Sect. 5.1.2) [165]. The calculated data for PS further predicted that no chain stretching should be possible above $T \approx 160\ °C$. At this temperature, the thermal energy is just sufficient to overcome the potential energy barrier for bond scission and any additional mechanical energy input will immediately serve to break the chain instead of deforming it.

Experiments in transient elongational flow demonstrated, on the contrary, a quasi-absence of temperature effect on the chain scission kinetics [109]. This seemingly remarkable result stems from the transient nature of the flow as can be verified from the following qualitative argument. With a typical average residence time in the high strain rate region of the order of $\tau_r \approx 10^{-6}$ s and a preexponential factor of $10^{12}\ s^{-1}$, the degradation yield as given from Eqs. (70) and (98), is:

$$\ln (n(t)/n_0) \approx 10^{12}\tau_r \cdot \exp\left[-U_0 + f(\psi)/RT\right]. \tag{105}$$

In order to reach a degradation yield of $> 50\%$, the following inequality must be satisfield:

$$f(\psi) > U_0 - 14RT. \tag{106}$$

Since U_0, which is $347\ kJ \cdot mol^{-1}$ for PS, is orders of magnitude larger than the thermal energy under normal experimental conditions, the effect of changing T has only a minimal influence on the value of $f(\psi)$ in Eq. (106).

By using the kinetic equations developed in Sect. 5.2, the degradation yield as a function of strain rate and temperature can be calculated. The results, with different values of the temperature and preexponential factor, are reported in Fig. 51 where it can be seen that increasing the reaction temperature from 280 K to 413 K merely shifts the critical strain rate for chain scission by $< 6\%$.

In order to observe any temperature dependence in transient flow degradation, it would be necessary to prolong considerably the effective residence time of the polymer coil. This can be accomplished either by recirculating the solution or by using an oscillatory flow equipment as described in Sect. 4.1 (Figs. 23 and 24).

5.8 Solvent Viscosity

Regardless of the state of aggregation, it is invariably observed that the rate of mechanochemical degradation at a constant strain rate decreases with increasing temperature. This statement is universal to the point that a negative temperature coefficient was considered as a prime criterion for mechanochemical degradation. The reason for this apparent negative activation energy should be sought in the indirect effect of temperature, which, in addition to increasing the probability of fracture of a stressed bond, promotes at the same time the mobility of the chain and of the surrounding medium, thus lowering the rate of energy input by this combined effect. In most of the current theories of chain dynamics in solution, the frictional contact between polymer segments and solvent molecules, known as the "monomer-solvent friction coefficient" (ξ) is assumed to be directly proportional to the bulk viscosity of the solvent (Eqs. 22, 47). A chain in a more viscous solvent is more easily oriented ($\dot{\varepsilon}_{cs} \sim 1/\tau_1 \sim 1/\eta_s$) and can be fractured at a lower strain rate. Solvent viscosity, thus, plays an active role in channeling part of the kinetic energy of the flow into the chain elastic energy. In this respect, it is therefore, as strain rate and molecular weight are, one of the central parameters that control the kinetics of degradation in flow.

In a few studies, solvent viscosity was varied as a result of change in temperature [109, 165]. In transient flow, the direct effect of temperature on the scission rate was shown to be minimal (Sect. 5.7). Even so, it is desirable to look for a system where the solvent viscosity can be studied independently of the other kinetics parameters [166]. Ideally, the solvents used should satisfy the following criteria:

— they should cover the broadest range of viscosity at a given temperature (preferably room temperature)
— they should have similar θ-temperatures to keep the solvation power constant (it is advantageous to work under θ-conditions where the scaling laws for polymer molecular properties are well-established to keep the number of parameters to a minimum)
— to recover the polymer for GPC analysis, they should have a sufficiently low boiling point to allow its complete removal by vacuum distillation without thermal degradation of the solute (T < 120 °C).

The physical properties of some solvents which meet these criteria are listed in Table 2. The degradation results using these solvents are reported in Table 3 and Figs. 52 and 53.

The most outstanding feature revealed by the experimental data is the relative insensitivity of $\dot{\varepsilon}_f$ to a large change in solvent viscosity: increasing η_s by two orders of magnitude in going from methyl acetate to dimethyl phtalate merely reduces $\dot{\varepsilon}_f$ by a factor of 3.4, approximately equivalent to $\dot{\varepsilon}_f \sim \eta_s^{-0.25}$ (Fig. 53). This weak dependence of $\dot{\varepsilon}_f$ on η_s, already noticed but left unexplained in a previous work on the effects of temperature [109], is entirely unexpected from the current model of chain scission through frictional loading which assumes a proportionality between the molecular stress and solvent viscosity (Eq. 47). It is also at variance with the results obtained in stagnant flow where the relation $\dot{\varepsilon}_f \sim \eta_s^{-1}$ was verified [165].

Table 2. Physical constants of some selected solvents for PS

Solvent	η_s [mPa s]	ϱ [g cm^{-3}]	T [°C]	b.p. [°C]
Methyl acetate	0.329	0.902	43.5 (θ)	57
Cyclohexane	0.772	0.765	34.5 (θ)	81
Decalin	2.769	0.890	14.8 (θ)	189–191
1-Methyl-naphtalene	2.793	1.014	27.0	241–245
Bis-(methyl)-phtalate	32.10	1.200	10.0	283–288
DOF	65.00	0.985	22.0 (θ)	386
Bis-(decyl)-phtalate	10.46	0.918	72.5 (θ)	–

[a] η_s and ϱ are measured at the temperature given in column 4
[b] the addition (θ) indicates the θ-temperature
[c] boiling point at atmospheric pressure, as given in the literature

Table 3. Molecular properties of polystyrene in the different solvents

Solvent	τ_z [μs]	$\dot{\varepsilon}_f^*$ [10^5 s^{-1}]	[η] [dl×g^{-1}]	K [dl×g^{-1}]	a	R_g [nm]	(fluid)	σ
Methyl acetate	8	3.48	0.737	72×10^{-5}	0.50	26.1	2.1	0.08
Cyclohexane	23	2.84	0.863	85×10^{-5}	0.50	27.6	5.3	0.07
Decalin	82	2.06	0.812	80×10^{-5}	0.50	27.0	7.1	0.07
1-Methyl-naphtalene	240	2.06	2.44	87×10^{-6}	0.74	39.0	15.3	0.04
Bis-(methyl)-phtalate	1400	0.92	1.14	67×10^{-7}	0.87	30.2	31.4	0.06
Bis-(methyl)-phtalate(a)	9200	0.25	2.77	67×10^{-7}	0.87	40.6	37.7	0.11
Bis-(methyl)-phtalate(b)	–	–	1.17	67×10^{-7}	0.87	–	–	–

[a] polystyrene with M $= 2.86 \times 10^6$, $\bar{M}_w/\bar{M}_n = 1.02$
[b] measured at T $= 25$ °C
a = exponent in the Mark-Honwink equation

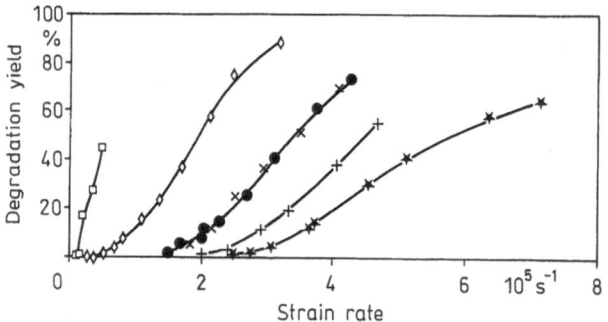

Fig. 52. Degradation yield as a function of strain-rate $\dot{\varepsilon}(0)$ and solvent viscosity ($\dot{\varepsilon}(0)$: maximum elongational strain-rate along the centerline); Except for the high MW series, all the data refer to the same PS fraction with $\overline{M_w} = 1.03 \times 10^6$, $\overline{M_w/M_n} = 1.017$; $-\square-$: bis-methyl phtalate ($\overline{M_w} = 2.86 \times 10^6$, $\overline{M_w/M_n} = 1.02$); $-\diamond-$: bis-methyl phtalate ($\eta_s = 32.1$ mPa · s); $-\bullet-$: decalin ($\eta_s = 2.77$ mPa · s); $-\times-$: 1-methyl-naphtalene ($\eta_s = 2.79$ mPa · s); $-+-$: cyclohexane ($\eta_s = 0.772$ mPa · s); $-*-$: methyl-acetate ($\eta_s = 0.329$ mPa · s)

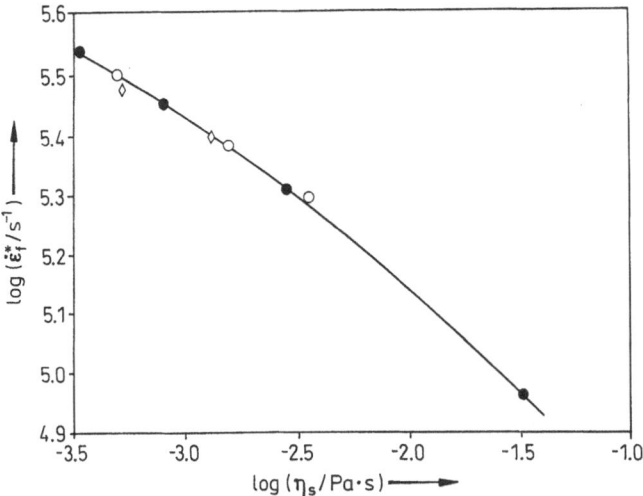

Fig. 53. Dependence of the critical strain-rate for chain scission ($\dot{\varepsilon}_f^*$) on solvent viscosity (η_s) data from this work: $-\bullet-$ data from Ref. [109], where η_s was changed concomitantly with the solvent temperature: $-\circ-$: decalin at 7, 22 and 140 °C; $-\diamond-$: dioxane at 22 and 90 °C

To determine the influence of solvent quality on the degradation kinetics, some comparative experiments were performed in a good solvent (1-methylnaphtalene) and in a θ-solvent (decalin), at temperatures where the two solution viscosities perfectly match each other (Table 2). Since polymer dimensions and molecular relaxation times are much larger in 1-methyl-naphtalene than in dekalin (Table 2), it is expected that the coil should also start to deform, then to break at a lower critical strain-rate. The experimental data tend to confirm otherwise: the degradation curves in both solvents are exactly superposable (Fig. 52), suggesting that solvent quality has little influence on the degradation mechanism. A fortuitous coincidence cannot be rejected and it is desirable in a future research to be able to extend the range of investigations to polymers having different molecular weights, and eventually to the collapsed state in a highly dilute solution.

From the weak dependence of $\dot{\varepsilon}_f$ on the surrounding medium viscosity, it was proposed that the activation energy for bond scission proceeds from the intramolecular friction between polymer segments rather than from the polymer-solvent interactions. Instead of the bulk viscosity, the rate of chain scission is now related to the "internal viscosity" of the molecular coil which is strain rate dependent and could reach a much higher value than η_s during a fast transient deformation (Eqs. 17 and 18). This representation is similar to the large loops "internal viscosity" model proposed by de Gennes [38]. It fails, however, to predict the independence of the scission yield on solvent quality (if this proves to be correct).

5.9 Polymer Concentration

Flow-induced degradation in solution is a complex function of polymer concentration which can alter the rate of chain scission in several respects. It can modify:

- the bulk flow properties (viscoelasticity)
- the stress transmission efficiency (entanglements)
- the static and dynamic coil dimensions (hydrodynamic screening and topological hindrance)
- and the chain relaxation time.

Under quiescent conditions, polymer solutions are divided into four categories depending on the average distance separating the centers of mass of the molecular coils: the dilute, the semi-dilute (or semi-concentrated), the concentrated and the entangled state.

Polymer solutions are considered as dilute when the degree of coil interaction is negligible. Under quiescent conditions, contacts are practically absent when the polymer concentration is below some limiting value (c*) defined by [167]:

$$c^* \approx 3M/(4\pi R_g^3 N_A) = 1.46/[\eta], \qquad (107)$$

where N_A is Avogadro's number and $[\eta]$ is the intrinsic viscosity of the polymer solution.

As the polymer concentration increases, polymer-solvent interactions are replaced by more frequent interchain dynamic contacts. A topological network structure can be formed only if the polymer molecular weight exceeds some critical value $M_c \approx 2M_e$ (M_e = molecular weight between entanglements ≈ 18000 g mol^{-1} for PS) and if the time-average number of entanglements per chain exceeds 1.5. From the last consideration, it could be seen that intermolecular entanglements occur at much higher concentration than molecular interpenetration, typically in the range of $[\eta] \cdot c \approx 4$–10 [16].

In elongational flow, the entanglement regime was observed at much lower concentrations, even below $[\eta] \cdot c = 0.1$ [169]. The effect was initially thought to be the result of coil expansion in flow, but was later discarded in favor of the lifetime for entanglement formation, under dynamic conditions of flow.

As mentioned in Sect. 5.1 the polymer concentration can alter the efficiency of stress transmission by the formation of entanglements. In the absence of turbulence, it is generally believed that degradation in simple shear flow is absent below the critical concentration for chain entanglement [132, 170]. In elongational flow, the influence of concentration seems to be more involved than suggested solely by the stress transmission mechanism. In a preliminary investigation, it was found that the scission yield decreased with increasing polymer concentration, which is contrary to the findings in simple shear flow [171]. This result was later confirmed indirectly in some experiments in which the polymer concentration was increased as a result of phase separation [109, 172]. In a series of experiments, the effect of polymer concentration was investigated more quantitatively by determining not only the degradation yield but also the scission probability along the chain (Figs. 54 and 55). From Eq. 107, the overlap concentration for the 1.08×10^6 PS sample in dekalin is 12000 ppm. Even if the experimental concentration range of 5 to 500 ppm is well within the dilute regime, a fast decrease in the scission yield was actually observed down to the ppm level (Fig. 54). This decrease in the scission yield was accompanied by a broadening of the degradation peak. Although central

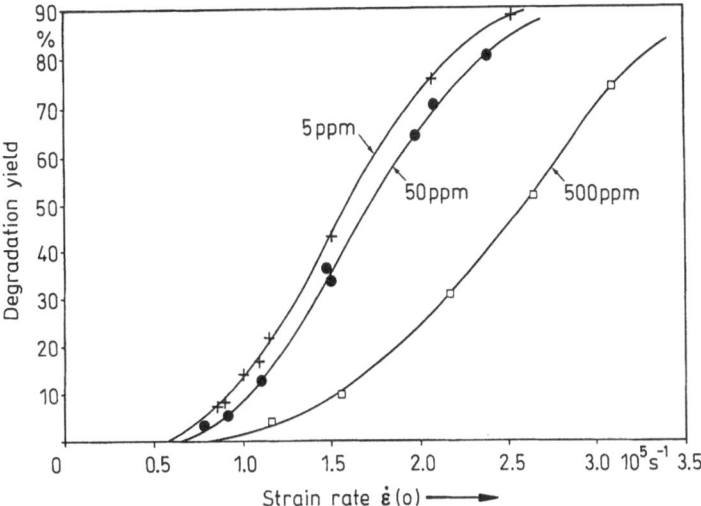

Fig. 54. Degradation yield as a function of strain-rate $\dot{\varepsilon}(0)$ and polymer concentration for a 1.08×10^6 PS with $\overline{M_w}/\overline{M_n} = 1.05$

scission is clearly visible at all concentrations, the standard deviation for the probability of midchain scission increases gradually from $\sigma = 0.07$ at 5 ppm to 0.16 at 500 ppm (Fig. 55). A similar trend was reported from degradation results under stagnant elongational conditions [173]. The more randomized scission can be expected due to the increased number of entanglements at higher concentration.

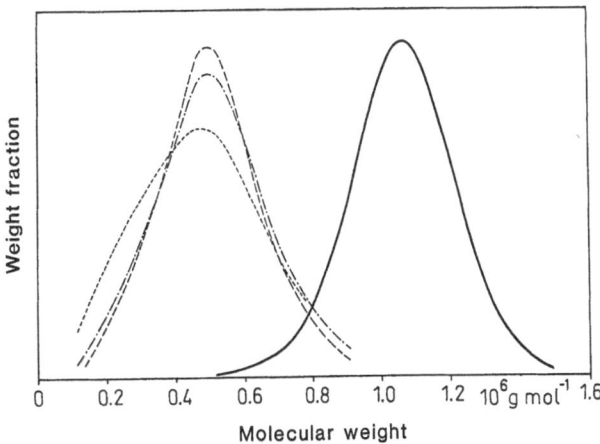

Fig. 55. Molecular weight distribution of the virgin (————) and degraded PS samples as a function of polymer concentration: (— — —): 5 ppm; ($\cdot$ — $\cdot$ — $\cdot$): 50 ppm; (-- ---): 500 ppm; (the degraded peaks are computer-substrated from the undegraded fraction and normalized to a degradation yield of 50%)

The reduction in scission yield, however, is more difficult to visualize with the same model since the amount of transmitted stress should increase with the number of entanglements. In one of the earlier papers, the negative concentration dependence was envisaged as the result of steric hindrance which prevented coil expansion and lowered the molecular stress [108]. With the information presented in this review, it seems more appropriate to rationalize the effect in terms of the stored free energy which can now be redistributed amongst different gripping points such as those formed by entanglements (Cf. Sect. 6). In order to reach the energy for dissociation, a larger total deformation energy will therefore be required.

5.10 Fluid Strain-Rate

As mentioned in Sect. 3.3, most of the current theories on mechanochemical degradation consider the forces generated along the chain as the dominant parameter which controls the fate of a bond under stress. Whenever this force is comparable to the breaking strength of the bond, scission will occur. By using Eq. (47), it is then possible to relate the breaking strength of a chain with the critical strain rate for degradation ($\dot{\varepsilon}_f$) and solvent viscosity (η_s) through a constant factor (C) pertaining to the coil dimensions.

With highly extended chains in stagnant elongational flow, a breaking force in the range of 3–13 nN/chain has been calculated from these relations for the degradation of PS and polyethylene oxide (PEO), which did indeed correspond to the expected value for the breaking strength of the $C-C$ and $C-O$ bonds [174]. The situation is more confusing when chains retain their state of partial uncoiling: from degradation of DNA in capillary flow, Levinthal and Davison deduced an experimentally determined critical stress of 11 nN/chain which they claimed to be in satisfactory agreement with the theoretical prediction of 8.9 nN for the $C-O$ bond [175]. Using similar devices, Harrington and Zimm found on the contrary that the critical stresses for breaking PS and DNA in simple shear flow are two orders of magnitude lower than the theoretical estimates based on bond strength [176]. The same conclusion was reached by Merrill and Leopairat in transient elongational flow where the determined breaking stresses were also two orders of magnitude below the theoretical estimates [108].

Shear rate can be controlled by machine design, whereas the breaking strength is specific to a chemical bond. In order to verify the proportionality between f_b and $\dot{\varepsilon}_f$, the most rational step would be to change the flow geometry and to recalculate f_b for each experimental situation. This procedure is the only means to resolve the discrepancies mentioned previously and it has only recently been applied to transient elongational flow [177]. In this study, the basic design of the degradation equipment was kept constant while the orifice diameter and the nozzle inlet angle were changed within suitable limits (Fig. 56).

A flow field analysis at fixed conical angle and varying orifice diameters confirmed that all the strain rate distribution functions are exactly superposable onto a single curve when plotted against the dimensionless parameters $\dot{\varepsilon}'_{xx} = \dot{\varepsilon}_{xx}/(\bar{v}_0/r_0)$ and $x' = x/r_0$. Three such master curves for different angles are

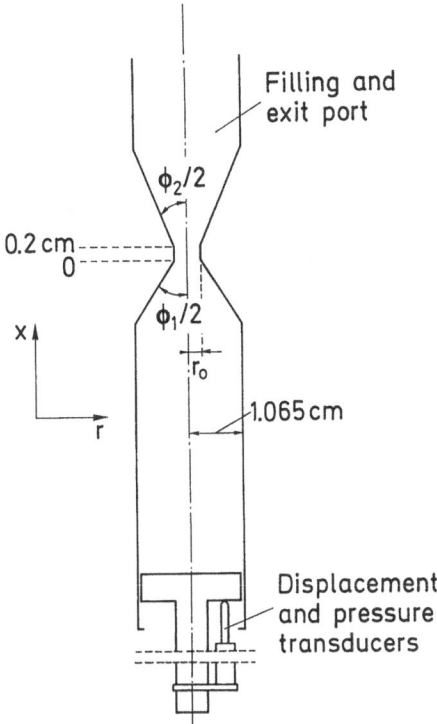

Fig. 56. Schematic representation of the convergent flow apparatus $\Phi_1 = 180°$ (abrupt contraction), 14° or 5° (conical inlet); $\Phi_2 = 14°$ or 5° (conical outlet) (the figure is not drawn to scale; see text for the exact values of the orifice diameter)

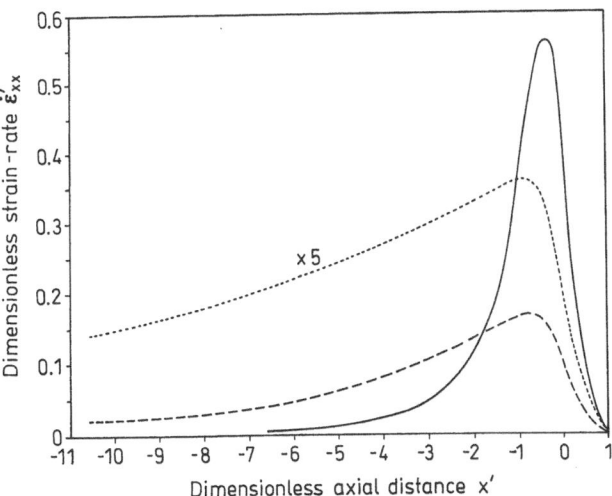

Fig. 57. Entrance elongational strain rate ($\dot{\varepsilon}_{xx}$) calculated along the centerline of the flow tube for the different nozzle geometries (the origin of the abscissa is taken at the orifice entrance); (————): abrupt contraction; (— — —): 14° conical inlet; (–––––): 5° conical inlet

shown in Fig. 57. As compared to the abrupt contraction at identical flow rate and orifice diameter, the conical entrance provides an increase in spatial distribution of the strain rate by a factor of 3 with the 14° inlet and of up to 8 with the angle of 5°. This increase in width is accompanied by a decrease of approximately the same factor in the amplitude of the fluid strain rate:

abrupt contraction: $\Phi_1 = 90°$, $\mathring{\varepsilon}_{xx}(\text{max}) = 0.56\,\bar{v}_0/r_0$,   (108)

conical inlet: $\Phi_1 = 14°$, $\mathring{\varepsilon}_{xx}(\text{max}) = 0.17\,\bar{v}_0/r_0$,   (109)

$\Phi_1 = 5°$, $\mathring{\varepsilon}_{xx}(\text{max}) = 0.07\,\bar{v}_0/r_0$.   (110)

The degradation yields from a 1.08×10^6 PS sample are plotted in Fig. 58 as a function of strain rate and nozzle geometry. The most apparent feature of the reported data is the lack of any coincidence between the different values of critical strain rate for chain rupture ($\mathring{\varepsilon}_f$). In other words, a given fluid strain rate ($\mathring{\varepsilon}$) degrades the same molecules either completely (5° angle) or hardly at all (abrupt contraction). This dispersion in the degradation yields suggests that the scission event is not controlled by the frictional stress (proportional to the fluid strain rate) as implied by the model of chain scission through frictional loading. In order to reconciliate the above data with the model of chain scission through frictional loading, the hypothesis of a nozzle-dependent coil expansion factor has been

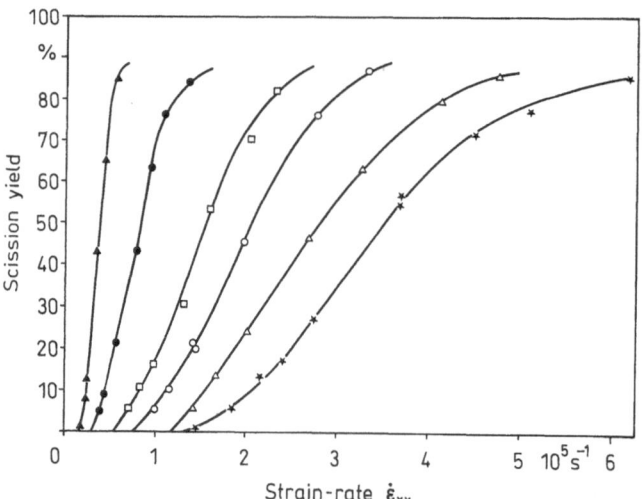

Fig. 58. Degradation yield as a function of maximum entrance strain rate $\mathring{\varepsilon}(0)$ for different nozzle geometries; ($\times$): abrupt contraction with $r_0 = 0.175$ mm; ($\triangle$): abrupt contraction with $r_0 = 0.25$ mm; ($\circ$): abrupt contraction with $r_0 = 0.34$ mm; ($\square$): abrupt contraction with $r_0 = 0.50$ mm; ($\bullet$): 14° conical inlet with $r_0 = 0.25$ mm; ($\blacktriangle$): 5° conical inlet with $r_0 = 0.25$ mm

explored: since f_{max} changes as $\dot{\varepsilon} \cdot L^2$ (Eq. 47), a more expanded coil could also be fractured at a lower strain rate. Such a model is a priori plausible from the results of flow field calculations which show that the residence time, one of the key parameters to determine the degree of coil expansion, changes with the nozzle geometry. Closer scrutiny showed, however, several drawbacks of the hypothetical model. The relaxation time (τ_z) for the 1.08×10^6 PS sample in decalin, calculated from Eq. (37), is 88 µs. The corresponding critical strain rate for the coil-to-stretch transition $(\dot{\varepsilon}_{cs})$, based on the most conservative estimate with $A = 0.50$ (Eq. 52), would be 5700 s^{-1}.

A plausible assumption would be to suppose that the molecular coil starts to deform only if the fluid strain rate $(\dot{\varepsilon})$ is higher than the critical strain rate for the coil-to-stretch transition $(\dot{\varepsilon}_{cs})$. From the strain rate distribution function (Fig. 59), it is possible to calculate the maximum strain (λ_{max}) accumulated by the polymer coil in case of an affine deformation with the fluid element $(\varepsilon_{fl} = v_{sc}/v_{cs} \approx \bar{v}_0/v_{cs})$. The values obtained at the onset of degradation at $\bar{v}_0 \approx 35$ m $\cdot$ s^{-1}, actually go in a direction opposite to expectation. With the abrupt contraction configuration, λ_{max} decreases from 19 with $r_0 = 0.0175$ cm to 8.7 with $r_0 = 0.050$ cm. Values of λ_{max} are even lower with the conical nozzles $(r_0 = 0.025$ cm), varying from 3.3 with the 14° inlet to a mere 1.6 with the 5° inlet. In any case, the values obtained are lower than the maximum stretch ratio for the 10^6 PS which is ≈ 40. It is then physically impossible for the chains to become fully stretched in this type of flow.

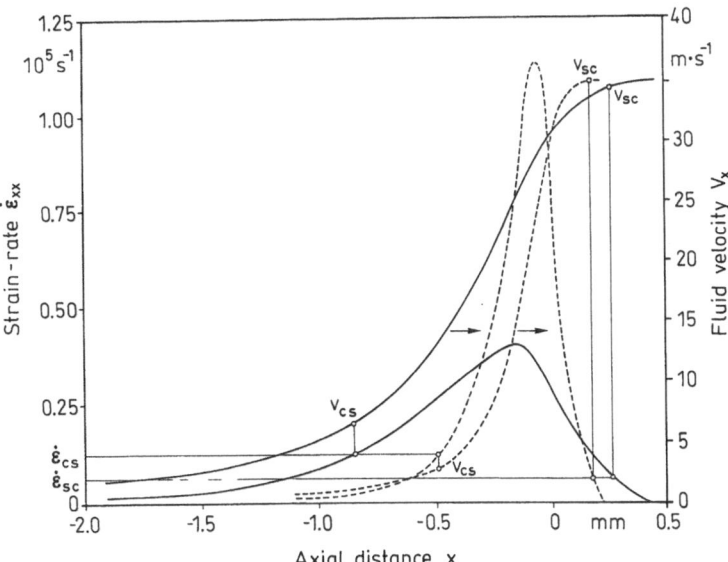

Fig. 59. Relation between the degree of chain extension and the axial velocity profile in abrupt contraction flow (axial velocity curves: right scale ordinate); $(---)$: $r_0 = 0.175$ mm; (———): $r_0 = 0.50$ mm

The results obtained with the 5° inlet have several interesting implications which may stimulate further investigations:

- the maximum strain rate at the onset of degradation ($\mathring{\varepsilon} = 9800 \text{ s}^{-1}$) is either just sufficient to distort the molecular coil or may be even below the threshold depending on the exact value of $\mathring{\varepsilon}_{cs}$ (Eq. 52). It is then legitimate to question whether a polymer chain can still be degraded even if the fluid strain rate is lower than the critical value to induce coil deformation. The correctness of this ascertion could be verified by using a less viscous solvent or a lower molecular weight (to increase $\mathring{\varepsilon}_{cs}$)
- by gradually reducing the angle Φ_1 of the conical inlet, it is possible to make a smooth transition from an elongational flow to a simple shear flow. If it is correct that isolated molecular coils cannot be degraded in "weak flow", then, it can be easily predicted that at some stage the degradation yields must drop sharply since in laminar capillary flow little chain scission is expected.

Figure 58 reveals an inverse dependence of $\mathring{\varepsilon}_f$ on orifice diameter. A similar influence of orifice diameter exists with the fluid strain rate (Fig. 57); this suggests that a better correlation can be obtained by reporting the degradation yields as a function of average orifice velocity ($\bar{v}_0$) for the different nozzle geometries. Figure 60 shows that effectively all the degradation data for the abrupt contraction as well as for the conical inlets can be superimposed onto a single curve. Although some dispersion is still present, it is considerably reduced as compared to the totally separated curves of Fig. 58; in part, the remaining dispersion can be accounted for by the lack of precision in the determination of the orifice diameters.

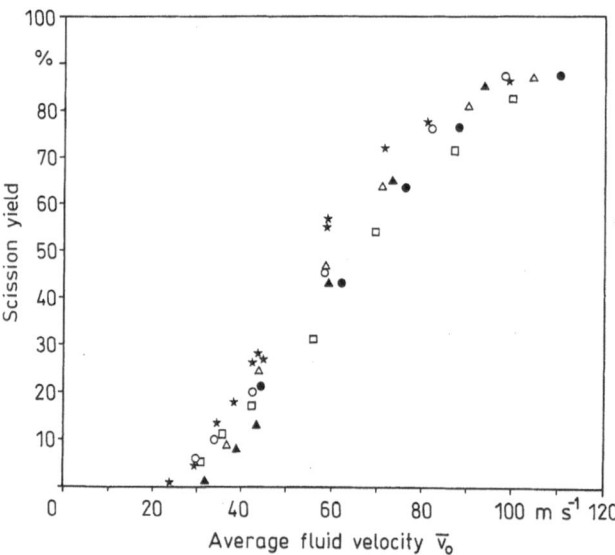

Fig. 60. Degradation yield as a function of volumetric average fluid velocity at the orifice ($\bar{v}_0$) (symbols are the same as used in Fig. 58).

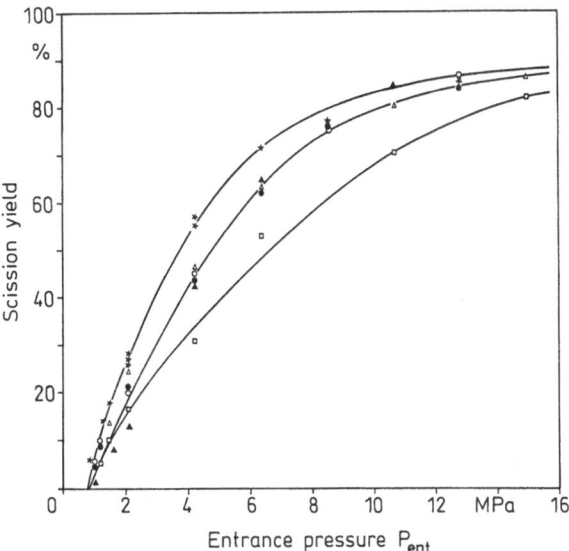

Fig. 61. Degradation yield as a function of experimental entrance pressure drop (symbols are the same as used in Fig. 58) (The curves for $r_0 = 0.25$ mm and $r_0 = 0.34$ mm are experimentally indistinguishable)

In convergent flows, flow-rate and entrance pressure drop are two dependent variables whose relationship is well-documented [178]. Instead of reporting the degradation yields as a function of $\bar{v}_0$, the other alternative would be to report the same results as a function of the entrance pressure drop P_{ent} (Fig. 61): a weak increase in degradation efficiency with decreasing orifice diameter is observed with this type of plot and may have some importance in polymer flow passing through porous media. The most remarkable feature is certainly the perfect coincidence between the data points obtained with the conical inlets and with the abrupt contraction of identical orifice diameters. Although the velocity fields are quite different in each case, the chains break with identical efficiency at a given pressure drop (or energy input): such a coincidence is certainly not fortuitous and conceals a deeper underlying mechanism which still needs to be investigated. An empirical model of chain degradation in transient elongational flow, based on the concept of critical energy for bond scission, will be developed in Sect. 6.

5.11 Other Types of Transient Elongational Flow

As mentioned in Sect. 4.2, transient elongational flow can be found in a variety of experimental situations. Generally, the flow geometries are not well-defined and the flow field can only be estimated qualitatively.

5.11.1 Ultrasonic Degradation

Ultrasonic irradiation is a common method of promoting degradation in solution (Cf. Sect. 4.2.5). The first documented experiment in this field dates back to 1933, when it was discovered that the viscosity of natural polymer solutions decreased on treatment with high frequency sound waves [117]. In the beginning, experiments were hindered by the broad initial molecular weight distribution and a lack of a proper molecular weight characterization techniques. Although midchain scission was frequently quoted as a plausible degradation mechanism, a definite proof in this direction was only a recent accomplishment. In a meticulous series of experiments with ultra-sharp fractions of dextran ($\bar{M}_w/\bar{M}_n = 1.01$), Basedow and Ebert showed conclusively that ultrasonic sonication cut the polymer chains precisely in half. Furthermore, it was determined that the scission rate constant increased linearly with the polymer molecular weight [179]. These same authors also found that the rate of degradation of dextran was not changed when dissolved in media with quite different viscosities like water, glycerol or ethylene glycol [180]. The experimental data reported in Figs. 62 and 63 show a striking resemblance to the results obtained in convergent flow; this lends further support to the conviction that ultrasonic degradation does indeed occur in a transient elongational flow field. As a final point of similitude, good correlation was obtained between the rate of ultrasonic or hydrodynamic energy input with the rate of bond scission [181, 182] (Cf. Sect. 6).

5.11.2 Spraying

The presence of small amounts of dissolved polymer can alter sizably the aerosol particle dimensions when the solutions are sprayed. This antimisting property has received special attention in an effort to develop additives for jet fuel to prevent accidental ignition following crash landing. As in drag reduction, the polymer

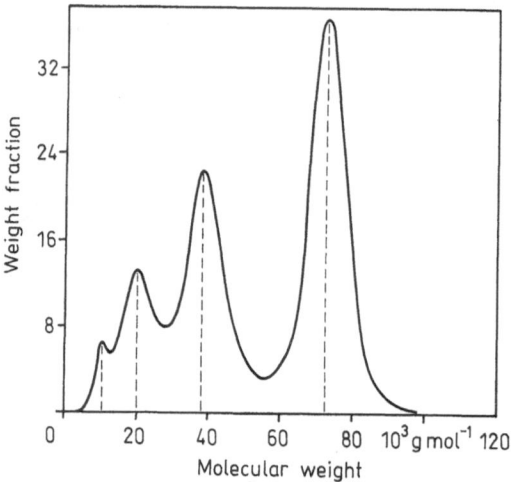

Fig. 62. Molecular weight distribution of a sharp dextran fraction ($\bar{M}_w = 72\,400$, $\bar{M}_w/\bar{M}_n = 1.01$) which has been irradiated ultrasonically (according to Ref. [179])

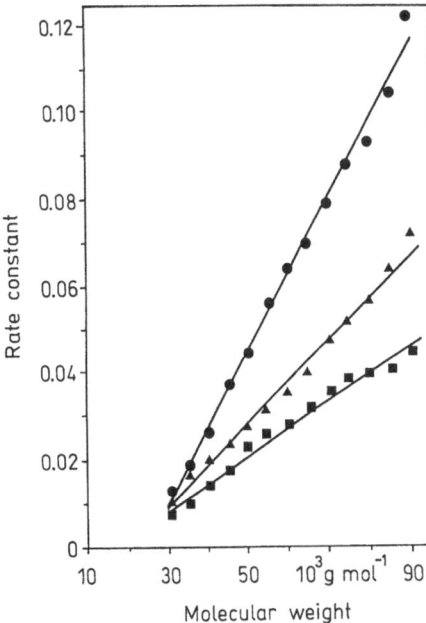

Fig. 63. Scission rate constant for the ultrasonic degradation of dextran as a function of molecular weight (M), in different solvents (according to Ref. [179]); ($\bullet$): formamide; ($\blacktriangle$): 10% MgSO$_4$; ($\blacksquare$): water

additive was found to be degraded after a single passage across the nozzle. The separation of the jet into small particles involves transient elongational flow and presumably, degradation happens during this stage. For polyisobutylene with 4.1×10^6 and 11.9×10^6 MW, it was found that the polymer started to degrade above some critical value of the wind-spraying speed of 30–40 m · s^{-1} [183]. No systematic investigation on the effects of molecular weight and nozzle geometry on the degradation yield has been reported, so a more quantitative comparison with transient elongational flow results is presently not possible.

5.11.3 Turbulent Flow

In the preceding categories of flow, the velocity field is deterministic since it can be calculated (at least in principle) from the constitutive equation of the fluid and the experimental boundary conditions. Turbulent flow, on the other hand, is distinctively unpredictable, as was pointed out a century ago by Osborne Reynolds.

Turbulence is known to occur in pipe flow at a Reynolds number (Re) above 2300. Beyond this stability limit, any disturbance will grow exponentially in time and the flow becomes fully chaotic at Re $\approx$ 4000, where Re is defined by:

$$Re = \bar{v} \cdot D \cdot \varrho / \eta_s \tag{111}$$

with $\bar{v}$: average fluid velocity, D: tube diameter, ϱ: solvent density and η_s: solvent viscosity.

The effect of polymer additives on turbulent flow is at the origin of the important phenomenon of drag reduction and has found other industrial applications such as oil recovery and antimisting action. Drag reduction in dilute polymer solutions

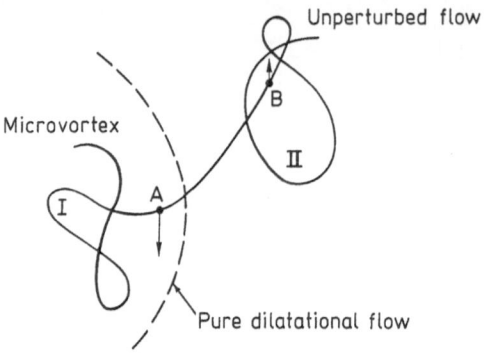

Fig. 64. Possible molecular stretching mechanism between region A and B in turbulent flow (sketched according to Ref. [185])

is observed only in the turbulent regime and is invariably accompanied by chain scission.

Although a molecular mechanism for drag reduction remains elusive, most theories agree that the phenomenon takes its origins from the resistance of polymer molecules to elongation. Turbulence in flow results from the boundary-layer oscillations which create the so-called "hairpin" vortices [184], each one acting as a source of local elongational flow. According to a classical scheme [185], if one of these disturbances was formed next to a polymer molecule, one chain end can get trapped in the core of the "microvortex" causing partial molecular extension under the velocity gradient (Fig. 64). Due to this strain, some energy is absorbed locally from the flow field, altering the energy balance in a direction which favors viscous dissipation to the detriment of turbulence formation [186]. Under conditions of drag reduction, random fluctuations in the velocity field have correlation times much shorter than the characteristic response time (τ^z) of the polymer molecule. The polymer coils are thus in a state of incessant transient stretching and contraction, and degradation should happen in a partly uncoiled conformation. Although chain stretching in turbulent eddies has been proposed as a plausible mechanism for drag reduction since 1950, the study of polymer conformation when exposed to a turbulent fluid in which the velocity component is random or stochastic is still in at a very much primitive stage. One interesting development was given by Jhon et al. [34]. They showed that the effect of a fluctuating velocity field was to screen the interaction between any two points of the chain, leading to a reduction in the connector force between the beads in proportion to the turbulent strength. The consequence was that the effective radii of gyration of polymer molecules in a turbulent vortex become greater than they would be at rest.

Chain degradation in turbulent flow has been frequently reported in conjunction with drag reduction and in simple shear flow at high Reynolds numbers [187]. Using poly(decyl methacrylate) under conditions of turbulent flow in a capillary tube, Muller and Klein observed that the hydrodynamic volume, $[\eta] \cdot M$, is the determining factor for the degradation rate in various solvents and at various polymer concentrations [188]. The initial MWD of the polymers used in their experiments are, however, too broad ($\bar{M}_w/\bar{M}_n = 5!$) to allow for a precise determination of the scaling exponent, so some further experimentation with narrow polymer fractions may be desirable. Another complication arises from the inter-relation between turbulence strength and molecular weight. Drag reduction is known to increase in proportion to the chain length; this effect, if not properly considered, may alter the flow conditions and the kinetics of chain scission. Furthermore, flow entrance effects and high shear build up at the walls are two serious problems which can additionally affect the degradation results.

Apart from the mentioned difference in molecular weight dependence, the degradation kinetics in turbulent flow offer striking resemblances with those in convergent flow. Using a tube 224 cm long and having an inner diameter of 0.70 cm. Horn and Merrill determined that PS degraded in turbulent flow with the same high propensity for midchain scission as in transient elongational flow [189]. In drag reduction, it was similarly established that the amount of degraded polymer: increases in proportion to the input of mechanical energy [190]; is constant or decreases with increasing polymer concentration [191]; and is weakly dependent on solvent viscosity [192]. The main difference, however, between degradation in turbulent flow from degradation in transient flow resides in the strong dependence of k_c on solvent quality which favors chain scission in poor solvents [192]. The issue of this dissimilarity is still unclear and we must wait for more extensive results concerning the effect of solvent quality in transient elongational flow. In an attempt to quantitatively support this idea, the chain has been modeled as a sequence of compact and extended bundles. According to this model, compact bundles are regions of high local rigidity and are more prone to degradation [192].

5.12 Miscellaneous Flows

There are a few instances where a macromolecule in solution can be degraded even in the absence of an elongational flow field. In these circumstances, the chain scission was believed to happen in a partly extended state: this situation is sufficiently related to the degradation in transient elongational flow to be worth mentioning.

5.12.1 Simple Shear Flow

In simple shear flow where vorticity and extensional rate are equal in magnitude (cf. Eq. (79), Sect. 4), the molecular coil rotates in the transverse velocity gradient and interacts successively for a limited time with the elongational and the compressional flow component during each turn. Because of the finite relaxation time (τ^z) of the chain, it is believed that the macromolecule can no more follow these alternative deformations and remains in a steady deformed state above some critical shear rate $(\dot{\gamma}^*)$ given by [193] (Fig. 65):

$$\dot{\gamma}^* = \pi/\tau^z . \tag{112}$$

This state of partial deformation has been determined recently by small angle neutron scattering [194] and has been a favorite subject of investigation in the

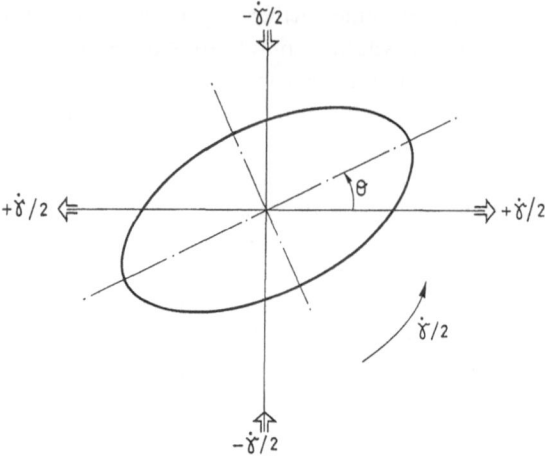

Fig. 65. Molecular shape according to the dummbbell model in a steady two-dimensional shear flow (according to Ref. [186])

field of flow birefringence [195]. A quite extensive body of literature exists describing polymer degradation in simple shear flow [86]; nevertheless, no unanimous concordance about the correct degradation kinetics to be used in the laminar regime seems to have been reached at the present. Early experiments were plagued with entrance effects and flow turbulence so that no reliable kinetic descriptions were available. With improved design, it was recently found that the rate constant for shear degradation of poly(decyl methacrylate) increased linearly with the polymer MW [196]. In a related context, it was similarly determined that the rate constant for the shear-induced hydrolysis of polyacrylamide and dextran showed a direct proportionality to the polymer chain length [197]. In most other circumstances, the MWD of the starting material was too large for any reliable determination on the MW dependence, although there is a general consensus that k_c should increase with the initial MW [198].

In the semi-dilute regime, the rate of shear degradation was found to decrease with the polymer concentration [132, 170]. By extrapolation to the dilute regime, it is frequently argued that chain scission should be nonexistent in the absence of entanglements under laminar conditions. No definite proof for this statement has been reported yet and the problem of isolated polymer chain degradation in simple shear flow remains open to further investigation.

5.12.2 Freeze-and-Thaw Cycles

Mechanical forces developed between adjacent solvent crystals during the freezing of polymer solutions can induce severe degradation. The most detailed studies on freezing degradation have been reported by Abbas and Porter [199] who proved that midchain scission is preponderant at low concentrations whereas random scission is predominant above the critical concentration for chain entanglement. Recent investigations in our laboratory corroborated the preference for midchain scission although with a much broader distribution than found in elongational

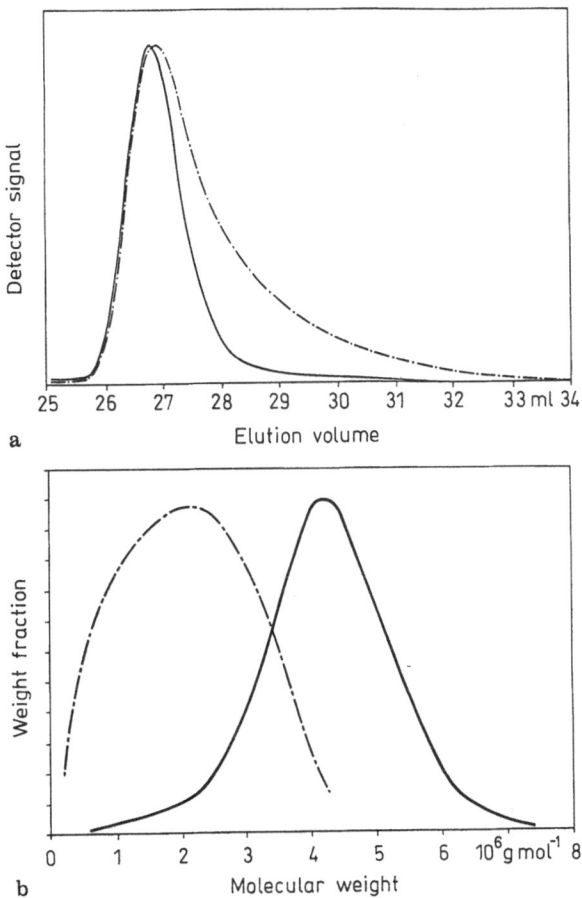

Fig. 66a. GPC traces of a PS sample (initial $\overline{M_w} = 4.34 \times 10^6$, $\overline{M_w}/\overline{M_n} = 1.06$), initial concentration = 200 ppm before and after 50 cycles of freeze-and-thaw degradation (the curves are normalized to the same height). **b** Molecular weight distribution of the virgin (———) and the degraded PS sample (— — —). To better visualize the scission probability along the chain, only the degraded polymer fraction is shown on the curve; the undegraded part has been removed from the original distribution by computer-substraction

flow. A degradation rate constant proportional to the initial polymer molecular weight was also established [200] (Figs. 66 and 67). It is believed, however, that the origins of this scaling law are different from those found in transient elongational flow.

5.12.3 Gel Permeation Chromatography

High molecular weight polymers are known to be degraded during GPC analysis [201–203]. As for any flow-induced degradation, it was determined that bond scission is flow rate and molecular weight dependent. In practice, with an adequate set of columns, degradation can be considered as negligible below $M \approx 4 \times 10^6$ but becomes rapidly intolerable above this threshold value. In one typical

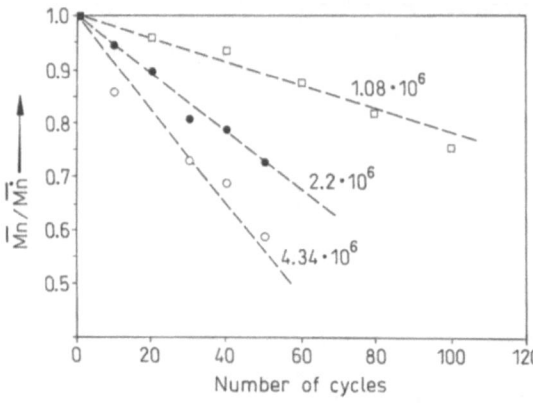

Fig. 67. Kinetics of the degradation by freezing-and-thawing: dependence on the number of cycles and polymer molecular weight

experiment, a 44×10^6 PS specimen was found to be degraded by over one-half (19×10^6) after a single passage through a GPC column [203]. Degradation was more severe with PIB where degradation was detected at a MW as low as 2×10^6 [201].

Because of the complex hydrodynamics associated with GPC systems, it is difficult to arrive at a simple correlation between GPC operational parameters and chain scission kinetics. At least five degradation mechanisms (given below with the associated flow field) may be operative in the different parts of the column, during a standard GPC analysis:
— at the entrance of the column, following an abrupt change in tubing diameters (abrupt contraction flow)
— across the inlet fritted filter (flow through porous media)
— in front of the particles of the stationary phase (flow in the turbulence caused by a spherical obstacle, Fig. 30)
— in the interstices between the particles (convergent-divergent flow, Fig. 24)
— and finally, inside the pores of the stationary phase itself.

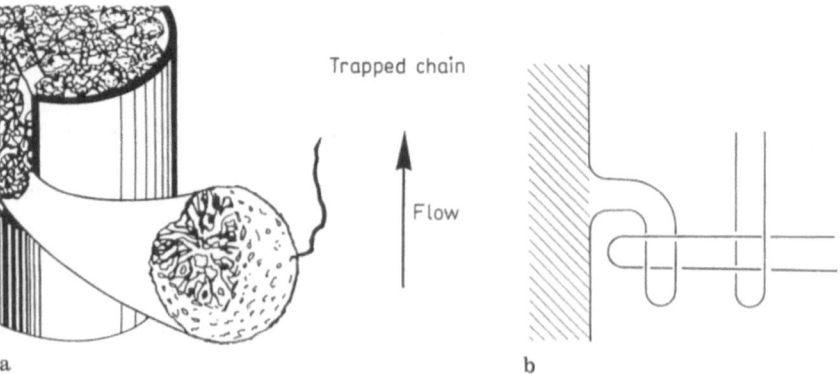

Fig. 68a. Sectional view of a packed GPC column and a single, greatly enlarged gel particle with a trapped chain. **b.** Mechanism of chain trapping by polymer loop and substrate loops interpenetration

The first four mechanisms involve transient elongational flow, whereas a "loop entanglement" model between the stationary phase and the eluted polymer was proposed for the last mechanism [203]. During the process of diffusion in and outside the pores, long flexible macromolecules can form loops which get temporary locked to the walls of the stationary phase. Due to the fast moving solvent outside the pores, a velocity gradient exists with the pore inside, leading to stretching and degradation (Fig. 68). The model is similar in many respects to Kuhn's model of the "chain with a teathered end" (Fig. 39). Using different types of polymers of varying MWs, it was determined empirically that the rate of chain scission increases according to a complex scaling law changing between $M^{0.6}$ and $M^{1.2}$ with the type of polymer and the stationary phase [203].

6 Critical Energy Model of Bond Scission

Most recent theories of mechanochemical degradation stem from the works of Frenkel [3], Kauzman and Eyring [204] and Morris and Schnurmann [205]. In any of these models, the process of chain scission is associated with the frictional forces generated on the chain and transmitted to the bond for chain rupture, in a similar fashion to the de Boer's formulation described in Sect. 2.2. To sustain the high level of stress necessary to overcome the binding strength of the covalent bond (4–10 nN), molecular spring laws predict that the chain (or a large portion of the chain containing the bond to be fractured) must first become highly stretched. These arguments lead to the pictorial representation of a mechano-chemically degraded chain as a macroscopic string frictionally loaded until fracture. Recent degradation results obtained under *stagnant* elongational flow and summarized below, further reinforce this perception of chain scission in flow [165, 174]:

— first, it was shown from birefringence measurements that flexible polymer molecules could be highly extended above a critical strain rate $\dot{\varepsilon}_{cs} \approx 1/\tau^z$
— for given experimental conditions, there exist a critical strain rate, denoted by $\dot{\varepsilon}_f$, below which there is no degradation
— except for very high MW ($> 30 \times 10^6$ PS), $\dot{\varepsilon}_f > \dot{\varepsilon}_{cs}$ which means that the chains are degraded in a highly stretched state (Fig. 49)
— chains are broken precisely into halves
— the critical strain rate for chain fracture ($\dot{\varepsilon}_f$) is proportional to the reciprocal solvent viscosity and scales with the inverse square of polymer molecular weight ($\dot{\varepsilon}_f \sim \eta_s M^{-2}$)
— the frictional force, calculated with the bead-rod model, corresponds to the breaking strength of the covalent bond ($f^H \approx f_b$)
— the degradation yield ($< 0.4\%$ per pass) corresponds to the small fraction of polymer molecules which flow close to the stagnation point and hence, can get highly stretched.

All these features can be rationalized with the simple model of chain scission through frictional loading previously mentioned. The series of experiments performed in transient elongational flow and reported in this review show that polymer chains in a *partly uncoiled state* behave in a quite different and intricate way as compared to chains in a highly expanded state:

— first, a flexible macromolecule can be degraded well before it reaches complete chain uncoiling, as can be assessed from the transient flow field (Fig. 33)
— even in a partially uncoiled state, polymers are fractured above a critical strain rate ($\dot{\epsilon}_f$) with a sharp propensity for midchain scission (Fig. 40)
— the critical strain rate for chain fracture ($\dot{\epsilon}_f$) is weakly dependent on the solvent viscosity $\sim \eta_s^{-0.25}$
— $\dot{\epsilon}_f$ scales with the reciprocal of polymer molecular weight (Fig. 49)
— degradation yield can be almost quantitative (90%) after a single passage (Fig. 41)
— degradation yield is a fairly unique function of (average) fluid velocity almost independent of differences in elongational strain rate introduced by nozzle geometry (Fig. 60)
— degradation yield is a fairly unique function of entrance pressure drop and thus, of energy introduced into the deformed liquid volume element (Fig. 61).

These experimental results point to the fact that a partly deformed coil is easier to degrade than a fully extended chain: even at a lower level of strain rate (Fig. 49), a much larger amount of polymer can be degraded in transient elongational flow (90%) than in opposed jets flow ($<0.4\%$). The frictional force at break, calculated from the slender body hydrodynamics, is orders of magnitude less than the breaking strength of the covalent bond. Using Eq. (47) and the experimental data for degradation in cyclohexane (Table 3), it is possible to evaluate the frictional forces involved during the coil stretching. With $L \approx 690$ nm, $d \approx 130$ nm ($\lambda \approx 5.3$), $\eta_s = 0.772$ mPa $\cdot$ s and $\dot{\epsilon}_f = 1.42 \times 10^5$ s^{-1}, $f_{max} = 2.5 \times 10^{-11}$ N. This value agrees with the result from Merrill and Leopairat [108] but is over two orders of magnitude below the actual breaking strength of a $C-C$ bond.

The preceding facts show that the model of chain scission through frictional loading, when applied to the transient flow conditions, is not only in quantitative disagreement but also in qualitative contradiction with the degradation behavior of partly extended molecular coils. The absence of any correlation between the critical strain for chain fracture ($\dot{\epsilon}_f$) and the fluid strain rate when varying the nozzle geometry (Sect. 5.10) is probably the most intriguing observation in transient elongational flow. The shape of the curves in Fig. 61 suggests the existence of a critical level of energy, corresponding to $P_{ent} \approx 1$ atm or $\bar{v}_0 \approx 35$ m $\cdot$ s^{-1}, below which no degradation occurs regardless of the nozzle geometry. At a given rate of mechanical energy input, the fraction of degraded polymer is practically identical regardless of the nozzle geometry. From these findings, it is tempting to speculate that the principal factor which governs the fate of a stressed macromolecule is not the frictional *force* at break but the amount of elastic *energy* (E_{el}) stored into the deformed molecule.

From the standpoint of thermodynamics, the essential quantity which governs the course of a chemical reaction is the chemical potential of the system or, in the case of interest, the free energy storage within the molecular coil. This quantity, unfortunately, is difficult to evaluate in non-steady flow. At modest extension ratios ($\lambda < 4$), the free energy storage of a freely-jointed bead-spring chain is usually given by the well-known theory of rubber elasticity [32]:

$$E_{el} = 3/2 \, kT \cdot \lambda^2 . \tag{113}$$

It is expected, however, that the Gaussian representation is inadequate in transient elongational flow, even if the chain is only weakly deformed. During a fast deformation, the presence of non-equilibrium effects, like "internal viscosity", noncrossability and self-entanglements will stiffen the molecular coil which is now capable of storing a much larger amount of elastic energy than that predicted from Eq. (113).

Practically any model of mechanochemical degradation resorts to some empirical relationship between the rate of chain scission and an experimental variable which can be directly determined: while fluid shear rate is the most commonly selected parameter, a few investigations adopted the rate of mechanical energy input as the prime degradation criterion. The notion of a critical energy for bond rupture is not new in itself and has periodically surfaced over the years in the field of mechanochemistry. Bestul was the first to give a sound basis to that idea, in an effort to interpret degradation results during capillary extrusion of concentrated polyisobutylene (PIB) solutions [206, 207]. In this model, it was assumed that a polymeric system under shear reaches a pseudo-equilibrium steady-state condition with respect to the temporary stored energy contributed by the shear field. By analogy with the kinetics for thermal activation, the following expression was derived for the probability that a bond attains the activation energy necessary to be ruptured:

$$P_{(E > E^*)} = k_B T/(k_B T - aJ) \cdot \exp\left[(-E^*/k_B T) - aJ(-E^*/aJ)\right] \tag{114}$$

In this equation, E is the sum of thermal and mechanical energies, E^* the mechanical activation energy for bond rupture (approximated as $\approx U_0$), J the rate of application of shear energy which equals the shear rate times the shear stress and a is a coefficient having the dimension of time. This is constant for a given shearing condition but increases with the polymer molecular weight. The product aJ gives the average amount of applied shear energy which is temporarily stored in bonds.

Neglecting the thermal energy contribution, the rate for shear degradation was obtained as:

$$K = k_c/w = B \exp\left(-E^*/aJ\right), \tag{115}$$

where w is the weight percent of polymer and B a constant factor. By plotting $\log k_c$ as a function of $1/J$, Bestul and coworkers found that all of their experimental points fell on straight lines in agreement with the preceding equation.

Similar approaches were proposed during ultrasonic degradation:
- El'tsefon and Berlin [181] determined that the degradation of PS in benzene gave the best correlation with the parameter q defined by

$$q = Ut/Vc , \tag{116}$$

where U is the ultrasonic intensity, t the irradiation time, V the volume of the solution and c the polymer concentration.
— good agreement was found by Doullah [182] between the scission rate constant and limiting molecular weight with the cavitation energy

More recently, Wolf and coworkers [170] found that moderately concentrated polymer solutions in laminar shear flow do not degrade as long as the quality of the solvent remains high; with decreasing solvent quality, the rate of chain rupture increases steeply in the vicinity of the temperature for phase separation then decreases again. In order to explain this complex behavior, Wolf [208] proposed the formation of "grip points" along the chain: degradation will be observed only if the amount of stored energy per grip point is superior to the activation energy E* for bond scission. The maximum in the degradation curve as a function of temperature could now be envisioned as a balance between the reduction in coil volume and the increase in free energy storage per macromolecule.

The present results obtained under transient elongational flow tend also to confirm that the principal factor which governs the fate of a stressed macromolecule is the amount of energy stored in the deformed chain, and not the level of stress. Although no intermolecular entanglements could be formed at the low polymer concentration used in transient flow investigations, the possibility of long-range intramolecular contact points and of local "kinetic rigidity" implicit in the concept of "internal viscosity" must be admitted. Under these conditions, stress accumulation becomes possible leading to bond rupture in a similar way as through the "grip points" considered by Wolf.

7 Prospects and Conclusion

For almost a decade, it has been predicted and verified that flexible macromolecules in solution could be stretched and fractured in a highly extended state. The sharp midchain scission, as well as the dependence of $\dot{\varepsilon}_f$ on $\eta_s^{-1} \cdot M^{-2}$ have been correctly explained with the use of a model of chain scission through frictional loading. The ultimate mechanical properties of highly oriented polymer chains in bulk are also accurately predicted with the above-mentioned model, leaving little doubt to its ability to explain physical properties of polymer molecules in the fully stretched state. Degradation results obtained in convergent flow disturb this seemingly coherent picture by showing that:

$$-\dot{\varepsilon}_f \sim M^{-1},$$

$$-\dot{\varepsilon}_f \sim \eta_s^{-0.25},$$

$$-f^H \ll f_b.$$

The dissimilarity in the kinetic laws for chain fracture observed under different flow conditions reflect the deficiencies of the present theories which should be able to incorporate both dependences (M^{-1} and M^{-2}) into its structure, with either one of the two terms becoming predominant depending on the degree of coil extension at fracture. Recent light scattering results from Menasveta and Hoagland [209] indicate that high MW polystyrene coils (20×10^6) are only modestly deformed by a factor of 2 at the stagnation point in opposed jets flow. From this result, it is predicted that the exponent in the scaling law $\dot{\varepsilon}_f \sim M^{-2}$ should decrease with very high MW polymers, to eventually reach the same value (-1) as found in transient elongational flow.

Experimentally, it was observed that a good correlation could be obtained by reporting the degradation yields as a function of the entrance pressure drop (Fig. 61). This correlation is entirely empirical and, as such, does not support or refute any molecular mechanism of chain scission. It does suggest, however, that the prime variable which controls the rate of chain scission is not the fluid strain rate, or any quantity which is proportional to it like the frictional hydrodynamic forces, but rather the free energy storage over the deformed molecular coil. This energy storage, evidently, must originate from the viscous friction between the polymer segments and the flowing solvent molecules during the stretching process. In light of the experimental results which have been presented, we believe that the most probable hypothesis which can explain the distinct degradation behavior of partly uncoiled relative to highly expanded chains stems from the non-equilibrium conformation of the polymer coil in a rapidly deforming liquid (the fully stretched conformation represents a state of equilibrium in "strong" flow). The rapidly deforming coil stiffens and is capable of storing appreciable amounts of elastic energy. Part of this free energy storage is then diverted into a few specific bonds which participate in the dissociation process. In an affinely deforming liquid, chains are almost statically loaded from the surface of the circumscribed volume element, which may explain the $\dot{\varepsilon}_f \sim M^{-1}$ dependence (since R_g^2 is proportional to the molecular coil surface). The internal relaxation of the chain relieves some of the axial chain stresses and contributes to the small to moderate elongation. By intersegmental interaction, stresses are transferred from the surface to the center of the coil. The remaining problem in the explanation of sharp midchain scission is how the partly deformed chain knows its center during the process of unravelling. On purely geometrical grounds, it can be appraised that bond rupture should be non-random. Since the chain ends are more mobile than the inner parts of the polymer, they should be able to relieve stresses and resist degradation more readily than the middle by reason of symmetry. This argument, however, is insufficient in itself to explain why the "center of the coil" coincides to such precision with the "center of the chain", unless the process of chain unravelling can be described by the so-called "yo-yo" model. Degradation of DNA solutions across a narrow orifice reveal a somewhat broader MWD than with PS [210]. The presence of nicks or weak links in DNA strands, and intramolecular entanglements in these very high MW polymers may, to some extent, account for this difference.

Although significant insight into the process of bond rupture has been gathered from the studies on the scission kinetics, it remains desirable at this stage to carry out further experimentation at a molecular level to get information on the chain conformation at the moment of fracture. As a first step in this direction, birefringence measurements have been attempted recently in the single jet cell

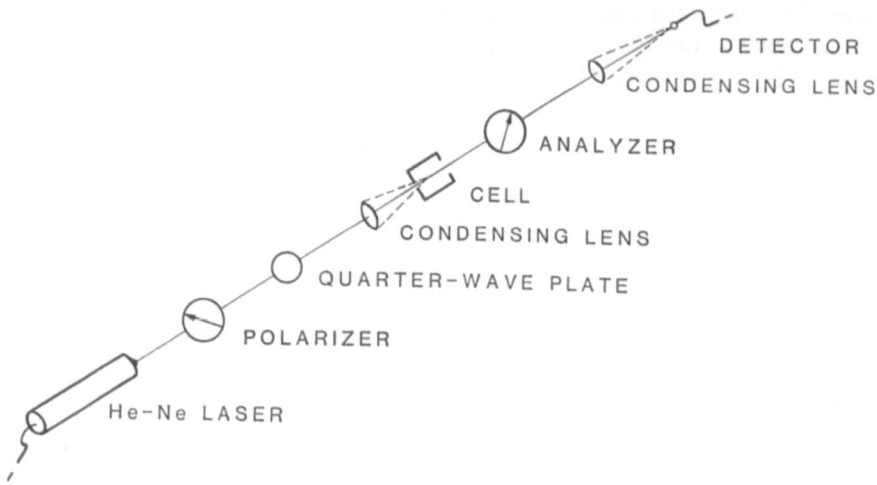

Fig. 69. Schematic of the optical train for birefringence measure ments. For microphotographs, an instant camera replaced the photodetector

(Fig. 27) employing the optical set-up schematized in Fig. 69 [211]. A perceptible birefringence signal was recorded by pushing dilute polyethylene oxide solutions of varying molecular weights (2 to 7×10^6) across a narrow contraction: this confirmed that significant segmental orientation did occur in a transient elongational flow field at approximately $\dot{\varepsilon} = \dot{\varepsilon}_{cs}$, before the occurrence of complete chain stretching. As opposed to the stagnant elongational flow where birefringence was highly localized in a small zone between the jets [9, 212], birefringence in transient elongational flow was found to pervade over the entire flow volume in front of

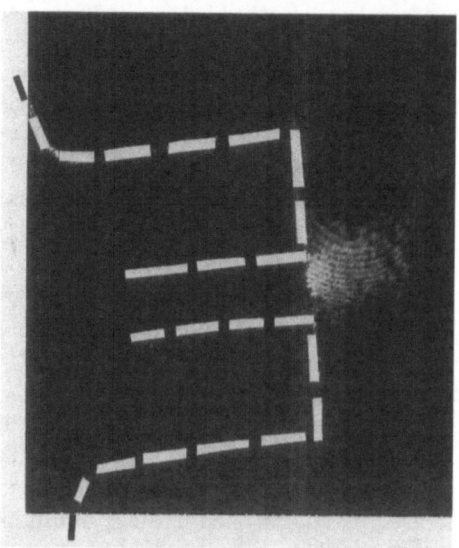

Fig. 70. Optical micrograph of the birefringence zone in a PEO solution (100 ppm, 4.10^6 MW). The dotted line delineated the contour of the nozzle

the jet (Fig. 70); this may explain the high degradation efficiency recorded in this type of flow.

To settle disputes amongst the different models of chain unravelling, it is proposed to investigate both the local and overall dynamics of polymer extension. Birefringence, which probes the dynamics of segmental orientation, should be complemented with other experimental techniques capable of yielding information about the chain geometry, such as light scattering for the degree of molecular coil expansion, and extensional viscosity for the total deformed length of the macromolecule. All these measurements should be supported by GPC analysis which provides a unique means of quantifying degradation yields and to localizing the point of scission along the chain.

Acknowledgements: The authors would like to thank D. Hunkeler (invited professor at EPFL) for reviewing this paper, and JA Odell and A Keller for frequent discussions and comments. The financial support from the Swiss National Science Foundation is gratefully acknowledged.

8 References

1. Staudinger H (1930) Ber Dtsch Chem Ges 63: 3152
2. Kuhn W, Kuhn H (1944) Helv Chim Acta 27, 493
3. Frenkel YaI (1944) Acta Physicochim URSS 19: 51
4. Peterlin A (1966) Pure and Appl Chem 12: 563
5. Frank FC (1970) Proc Roy Soc London A 319: 127
6. Marucci G (1975) Polym Engin Sci 15: 229
7. de Gennes PG (1974) J Chem Phys 60: 5030
8. Pope DP, Keller A (1978) Coll and Polym Sci 256: 751
9. Farrell CJ, Keller A, Miles MJ, Pope DP (1980) Polymer 21: 1295
10. Fuller GG, Leal LG (1980) Rheol Acta 19: 580
11. Chaveteau G, Moan M (1981) J Phys (Paris) 42: L-201
12. Rabin Y (1987) J Chem Phys 86: 5215
13. Smith KA, Merrill EW, Peebles LH, Banijamali SH (1975) In: Wolf C (ed) Polymères et lubrication, Colloques Internationaux du CNRS, no. 233, Paris, p 341
14. Dennisson GL (1976) Master's thesis, Pennsylvania State University
15. Kausch HH (1987) Polymer fracture, 2nd (edn) Springer, Berlin Heidelberg New York, chap 5
16. Forsman WC (1986) Polymers in solution, Plenum Press, New York, p. 91
17. Kramers HA (1944) Physica 11: 1
18. Kuhn W (1936) Kolloid Z 76: 258
19. Flory PJ (1969) Statistical mechanics of chain molecules, Wiley Interscience, New York, chap VIII
20. Aharoni SM (1983) Macromolecules 16: 1722
21. Peiffer DG, Kim MW, Lundberg RD (1986) Polymer 27: 493
22. Tinland B, Rinaudo M (1989) Macromolecules 22: 1863
23. Ricka J, Meewes M, Nyffenegger R, Binkert Th (1990) Phys Rev Lett 65: 657
24. Flory PJ (1973) Principles of polymer chemistry, Cornell University Press, Ithaca, New York, chap XIV
25. Kuhn W (1934) Kolloid-Z 68: 2
26. Solc K, Stockmayer WH (1971) J Chem Phys 54: 2756
27. des Cloizeaux J, Jannink G (1987) Les polymères en solution: leur modélisation et leur structure, Les Editions de Physique, Les Ulis, France, p 269

28. de Gennes PG (1972) Phys Lett 38 A: 339
29. Ohta T, Oono Y, Freed KF (1982) Phys Rev A 25: 2801
30. de Gennes PG (1979) Scaling concepts in polymer physics, Cornell University Press, Ithaca, p 165
31. Park IH, Wang QW, Chu B (1987) Macromolecules 20: 1965
32. Treloar LRG (1975) The physics of rubber elasticity, 3rd (edn) Oxford University Press, London, chap VI
33. Warner HR Jr (1972) Ind Eng Chem Fundamentals 11: 379–387
34. Jhon MS, Sekhon G, Armstrong R (1987) The response of polymer molecules in a flow, Adv. Chem Phys 66: 153
35. Rabin Y, Creamer DB (1985) Macromolecules 18: 302
36. Kuhn W, Kuhn H (1945) Helv Chim Acta 28: 1533
37. Cerf R (1957) J Polym Sci 23: 125
38. de Gennes PG (1977) J Chem Phys 66: 5825
39. Rabin Y, Öttinger HC (1990) Europhys Lett 13: 423
40. Kuhn W (1934) Kolloid-Z 68: 2
41. Bird RB, Curtiss CF, Armstrong RC, Hassager O (1987) Dynamics of polymeric liquids, 2nd (edn) John Wiley, New York, vol 2, p 58
42. Booij HC (1988) J Rheol 32: 47
43. Tanner RI (1988) Engineering rheology, Clarendon Press, Oxford, rev (ed), p 185
44. Doi M, Edwards SF (1986) The theory of polymer dynamics, Clarendon Press, Oxford, chapters 4 & 5
45. Rouse PE (1953) J Chem Phys 21: 1272
46. Debye P, Bueche AM (1948) J Chem Phys 16: 573
47. Kirkwood JG, Riseman J (1948) J Chem Phys 16: 565
48. Zimm BH (1956) J Chem Phys 24: 269
49. Oono Y, Kohmoto M (1983) J Chem Phys 78: 520
50. Larson RG (1988) Constitutive equations for polymer melts and solutions, Butterworths, Boston, p 256
51. Henyey FS, Rabin Y (1985) J Chem Phys 82: 4362
52. Larson RG, Magda JJ (1989) Macromolecules 22: 3004
53. Fuller GG, Leal LG (1980) Rheol Acta 19: 580
54. Ambari A, Deslouis C, Tribollet B (1984) Chem Engin Comm 29: 63
55. Wiest JM, Bird RB, Wedgewood LE (1988) On coil-stretch transitions in dilute polymer solutions, RRC Report no. 116, University of Wisconsin-Madison, May
56. Larson RG, Magda JJ (1989) Macromolecules 22: 3004
57. Müller AJ (1989) Extensional flow of macromolecules in solution, Ph D Thesis, University of Bristol, Physics Department
58. Phan-Thien N, Atkinson JD, Tanner RI (1978) J Non-Newt Fluid Mech 3: 309
59. James DF, Saringer JH(1980) J Fluid Mech 97: 655
60. King DH, James DF (1983) J Chem Phys 78: 4749
61. Broachard F, de Gennes PG (1977) Macromolecules 10: 1157
62. Batchelor GK (1971) J Fluid Mech 46: 813
63. Ryskin G (1987) J Fluid Mech 178: 423
64. Rabin Y (1988) J Chem Phys 88: 4014
65. Bleha T, Gajdos J (1988) Coll Polym Sci 266: 405
66. Rallison JM, Hinch EJ (1988) J Non-Newtonian Fluid Mech 29: 37
67. Boots S (1989) Anal Chem 61: 551 A
68. Schwartz DC, Koval M (1989) Nature 338: 520
69. Larson RG (1990) Rheol Acta 29: 371
70. Kausch HH, Nguyen TQ (Feb. 1991) Molecular coils and their deformation, IUPAC Symposium Polymer 91, Melbourne 10–15
71. Sacher E, Engel PA, Bayer RG (1979) J Appl Polym Sci 24: 1503
72. Tecante A, Choplin L, Lencki RW (1987) Polym Prep 28: 624
73. Viallat A, Pincus PA (1989) Polymer 30: 1997
74. Dikstra DJ, Pennings AJ (1988) Polym Bull 19: 73

75. Takahashi H, Matsuoka T, Ohta T, Fukumori K, Kurauchi T, Kamigaito O (1988) J Appl Polym Sci 36: 1821
76. Buerger DE, Engberg K, Jansson J-F, Gedde UW (1989) Polym Bull 22: 593
77. Ladik JJ (1988) Quantum theory of polymers as solids, Plenum Press, New York London
78. Michalske TA, Bunker BC (1984) J Appl Phys 56: 2686
79. Popov A, Rapoport N, Zaikov GE (1991) Oxidation of stressed polymers, Gordon and Breach Sci Publ New York London Tokyo, p 335
80. Smoilov GG, Tomashevskii EE (1969) Sov Phys — Solid State 10: 2395
81. Vershinina MP, Kvachadze NG, Tomashevskii EE (1977) Sov Phys — Solid State 19: 1382
82. Lloyd BA, Dvries KL, Williams ML (1972) J Polym Sci A 2 10: 1415
83. Gooden R, Davis DD, Hellman MY, Lovinger AJ, Winslow FH (1988) Macromolecules 21: 1212 (1988)
84. Adam RE, Zimm BH (1977) Nucleic Acis Res 4: 1513
85. Basedow AM, Ebert KH, Hunger H (1979) Makromol Chem 180: 411
86. Casale A, Porter RS (1978) Polymer stress reactions, Academic Press, New York, vol 1, p 9
87. Kelly A (1966) Strong solids, Clarendon Press, Oxford, p 6
88. Boudreaux DS (1973) J Polym Sci Polym Phys (ed) 11: 1285
89. Crist B, Ratner MA, Brower AL, Sabin JR (1979) J Appl Phys 50: 6047
90. Penning JP, van der Werff H, Roukema M, Pennings AJ (1990) Polym Bull 23: 347
91. Wool RP, Boyd RH (1980) J Appl Phys 51: 5116
92. de Boer JH (1936) Trans Far Soc 32: 10
93. Tomashevskii EE (1971) Sov Phys — Solid State 12: 2588
94. Krausz AR, Eyring H (1975) Deformation Kinetics, John Wiley New York, p 338
95. Tobolsky A, Eyring H (1943) J Chem Phys 11: 125
96. Zhurkov SN, Vettegren VI, Korsukov VE, Novak II (1969) Sov Phys — Solid State 11: 233
97. Bernstein HI (1962) Spectrochim Acta 18: 161
98. Zhurkov SN, Korsukov VE (1974) J Polym Sci Polym Phys (ed) 12: 385
99. Yew FFH, Davidson N (1968) Biopolymers 6: 659
100. Odell JA, Keller A, Rabin Y (1988) J Chem Phys 88: 4022
101. Crist B Jr, Oddershede J, Sabin JR, Perram JW, Ratner MA (1984) J Polym Sci Polym Phys (ed) 22: 881
102. Pope DP, Keller A (1977) Coll & Polym Sci 255: 633
103. Farrell CJ, Keller A, Miles MJ, Pope DP (1980) Polymer 21: 1292
104. Mikkelsen KJ, Macosko CW, Fuller GG (1988) Xth Internat Congr Rheol Sidney, vol 2, p 125
105. Tanner RI (1988) Engineering rheology, Clarendon Press, Oxford, rev (ed), p 185
106. Lodge TP, Miller JL, Schrag JL (1982) J Polym Sci 20: 1409
107. Cressely R, Hocquart R (1981) Polym Prepr 22: 120
108. Merrill EW, Leopairat P (1980) Polym Eng in Sci 20: 505
109. Nguyen TQ, Kausch HH (1986) Coll & Polym Sci 264: 764
110. Nguyen H, Boger DV (1979) J Non-Newtonian Fluid Mech 5: 353
111. Moreau V (1984) internal report T-84-2, Fluid mechanics laboratory, EPFL, Lausanne
112. Dennisson GL (1976) Master's thesis, Pennsylvania State University
113. Hill JW, Cuculo JA (1976) In: Elongational flow behavior of polymeric fluids, Rev in Macromol Chem 14 B, Marcel Dekker, New York, p 143
114. Besio GJ, Prud'homme RK, Benziger JB (1988) Macromolecules 21: 1070
115. Robinson IM, Yeung PHJ, Galiotis C, Young RJ (1986) J Mater Sci 21: 3440
116. Ambari A, Deslouis C, Tribollet B (1984) Chem Eng Comm 29: 63
117. Sheth PJ, Johnston JF (1978) in Polymer stress reactions, Academic Press, New York, vol 2, p 501
118. Mason TJ, Lorimer JP (1989) Sonochemistry: Theory, applications and uses of ultrasound in chemistry, Ellis Horwood Limited, Chichester, chap 2
119. Odell JA (December 1990) private communication

120. Ederer HJ, Basedow AM, Ebert KH (1981) In: Ebert KH, Deuf Chard P, Jager W (eds) Modelling of chemical reaction systems. Springer, Berlin Heidelberg New York, p 197
121. Van der Hoff BME, Glynn PAR (1974) J Macromol Sci-Chem A 8: 429
122. Ryskin G (1990) J Fluid Mech 218: 239
123. Culter JD, Zakin JL, Patterson GK (1975) J Appl Polym Sci 19: 3235
124. Polyflow, a finite element program for calculating viscous and viscoelastic flows, Polyflow SA, Place de l'Université 16, B-1348 Louvain-la-Neuve, Belgium (1988)
125. Nguyen TQ, Kausch HH (1989) Makromol Chem 190: 1389
126. Giesekus H (1962) Rheol Acta 2: 122
127. Gardner K, Pike ER, Miles MJ, Keller A, Tanaka K (1982) Polymer 23: 1442
128. Leal LG (1986) In: Rabin Y (ed) Studies of flow-induced conformation changes in dilute polymer solutions, Proceedings of the 1985 La Jolla Institute Workshop, Academic International Press, New York, p 5
129. Rabin Y, Henyey FS, Creamer DB (1986) J Chem Phys 85: 4696
130. Bird RB, Curtiss CF, Armstrong RC, Hassager O (1987) Dynamics of polymeric liquids, 2nd (ed), John Wiley & Sons, New York, vol 2, p 71
131. Gatski TB, Lumley JL (1978) J Fluid Mech 86: 623
132. Bueche F (1960) J Appl Polym Sci 4: 101
133. Ballauff M, Krämer H, Wolf BA (1983) J Polym Sci Polym Phys (ed) 21: 1205
134. Zakrevskii VA, Pakhotin VA (1981) Polym Sci USSR 23: 741
135. Fuhrmann J, Scherer GH, Nick L (1987) Makromol Chem 188: 2241
136. Dickinson JT, Crasto AS (1988) ACS Symposium Series no. 367, American chemical society, Washington DC, p 145–168
137. Kinpara H, Hori Y, Shimada S, Kashiwabara H (1985) Polym Comm 26: 142
138. Porter RS, Casale A (1985) Polym Engin Sci 25: 129
139. Ranby B, Rabek JF (1977) ESR Spectroscopy in polymer research, Springer, Berlin
140. Fijiwara H, Goto K (1990) Polym Bull 23: 27
141. Price GJ (1990) Adv Sonochemistry 1: 231
142. Berlin AA Chem Abstr 52: 11456b
143. Zhurkov SN, Zakrevskii VA, Korsukov VE, Kuksenko VS (1972) J Polymer Sci A 2 10: 1509
144. Grayson MA, Wolf CJ (1979) In: Stress mass spectrometry of polymeric materials: a review, ACS Adv Chem Series no. 174, Koenig JL (ed), p 53–80
145. Porter RS, Cantow MJR, Johnson JF (1967) J Polym Sci C 16: 1
146. Abdel-Alim AH, Hamielec AE (1973) J Appl Polym Sci 17: 3769
147. Nguyen TQ, Kausch HH (1988) J Non-Newtonian Fluid Mech 30: 125
148. Grubisic-Gallot Z, Marais L, Benoit H (1976) J Polym Sci A-2 Polym Phys 14: 959
149. Nguyen TQ, Kausch HH (1988) J Chromatogr 449: 63
150. Ishige T, Lee SI, Hamielec AE (1971) J Appl Polym Sci 15: 1607
151. Abbas KB (1980) Polym Engin Sci 20: 703
152. Ballauff M, Wolf BA (1981) Macromolecules 14: 654
153. Basedow AM, Ebert KH, Ederer H (1978) Macromolecules 11: 774
154. Ziff RM, McGrady ED (1986) Macromolecules 19: 2513
155. Nguyen TQ, Kausch HH (1989) Makromol Chem 190: 1389
156. ref 86, vol 1, p 125 and vol 2, p 261
157. Brostow W, Ertepinar H, Singh RP (1990) Macromolecules 23: 5109
158. Malhotra SL (1986) J Macromol Sci Chem A 23: 729
159. Herold FK, Schulz GV, Wolf BA (1986) Polym Comm 27: 59
160. McCormick CL, Hester RD, Morgan SE, Safieddine AM (1990) Macromolecules 23: 2124
161. Deshmukh SR, Singh RP (1986) J Appl Polym Sci 32: 6163
162. Odell J, Keller A (1986) J Polym Sci Polym Phys (ed) 24: 1889
163. Nguyen TQ, Kausch HH (1987) Polym Prep 28: 409
164. Encina MV, Lissi E, Sarasua M, Gargallo L, Radic D (1980) J Polym Sci Polym Lett 18: 757

165. Odell JA, Muller AJ, Narh KA, Keller A (1990) Macromolecules 23: 3093
166. Nguyen TQ, Kausch HH (1990) Macromolecules 23: 5137
167. Fujita H (1990) Polymer solutions, Elsevier, Amsterdam Oxford, chap 6
168. Aharoni SM (1983) Macromolecules 16: 1722
169. Keller A, Odell JA, Miles MJ (1985) Polymer 26: 1219
170. Ballauff M, Wolf BA (1988) Adv Polym Sci 85: 1
171. Merrill EW, Horn AF (1984) Polym Comm 25: 144
172. Horn AF, Merrill EW (1987) Polym Comm 28: 172
173. Muller AJ, Odell JA, Carrington S (April 1991) In: Degradation of polymer solutions
    in extensional flow, Proceedings of the polymer physics: a Conference to mark the
    retirement of A Keller, Bristol UK 3–5
174. Keller A, Odell JA (1985) Coll Polym Sci 263: 181
175. Levinthal D, Davison PF (1961) J Mol Biol 3: 674
176. Harrington RE, Zimm BH (1965) J Phys Chem 69: 161
177. Nguyen TQ, Kausch HH (1991) Influence of nozzle geometry on polystyrene degrada-
    tion in transient elongation flow, to be published in Coll Polym Sci
178. Ref 105, p 329
179. Basedow AM, Ebert KH (1977) Adv Polym Sci 22: 83
180. Basedow AM, Ebert KH (1978) Makromol Chem 179: 2565
181. El'tsefon BS, Berlin AA (1964) Polym Sci USSR 5: 668
182. Doullah MS (1978) J Appl Polym Sci 22: 1735
183. Ilano AL, Williams MC, Grens EA (1986) II, J Appl Polym Sci 32: 3649
184. Hoyt JW (june 1972) J Basic Engin Trans ASME 258
185. Peterlin A (1970) Nature 227: 598
186. Lumley JL (1969) Ann Rev Fluid Mech 1: 367; (1973) J Polym Sci Macromol Rev 7: 263
187. Nguyen TQ, Kausch HH (1986) Chimia 40: 129
188. Muller HG, Klein J, Rottlof A (1981) Makromol Chem 182: 529
189. Horn AF, Merrill EW (1984) Nature 5990: 140
190. Ting RY, Little RC (1973) J Appl Polym Sci 17: 3345
191. Gold PI, Amar PK, Swaidan BE (1973) J Appl Polym Sci 17
192. Brostow W (1983) Polymer 24: 631
193. Dupuis D, Wolff C (1985) Chem Eng Comm 32: 203
194. Lindner P, Oberthür RC (1985) Coll Polym Sci 263: 443; (1988) Coll Polym Sci 266: 886
195. Janeschitz-Kriegl H, Burchard W (1986) J Polym Sci A 2 6: 1953
196. Müller HG, Klein J (1981) Makromol Chem 182: 513
197. Basedow AM, Ebert KH, Hunger H (1979) Makromol Chem 180: 411
198. Ref 86, vol 1, p 104
199. Abbas KB, Porter RS (1976) J Polym Sci Polym Chem (ed) 14: 553
200. Zysman V, Nguyen TQ, Kausch HH unpublished results
201. Huber C, Lederer KH (1980) J Polym Sci Pol Lett (ed) 18: 535
202. Barth HG, Carlin FJ Jr (1984) J Liq Chromatogr 7: 1717
203. McIntyre D, Shih AL, Savoca J, Seeger R, MacArthur A (1984) ACS Symposium series,
    no. 245: 227
204. Kauzmann WJ, Eyring H (1940) J Amer Chem Soc 62: 3113
205. Morris WJ, Schnurmann R (1947) Nature 160: 674
206. Bestul AB (1956) J Chem Phys 24: 1196
207. Goodman P, Bestul AB (1955) J Polym Sci 18: 235
208. Wolf BA (1987) Makromol Chem Rapid Comm 8: 461
209. Menasveta MJ, Hoagland DA (1991) Macromolecules 24: 3427
210. Reese HR, Zimm BH (1990) J Chem Phys 92: 2650
211. Hunkeler D, Nguyen TQ, Kausch HH (Aug 1991) ACS meeting in New York City
212. Cathey CA, Fuller GG (1990) J Non-Newt Fluid Mech 34: 63

Editor: H.-H. Kausch
Received July 2, 1991

Appendix

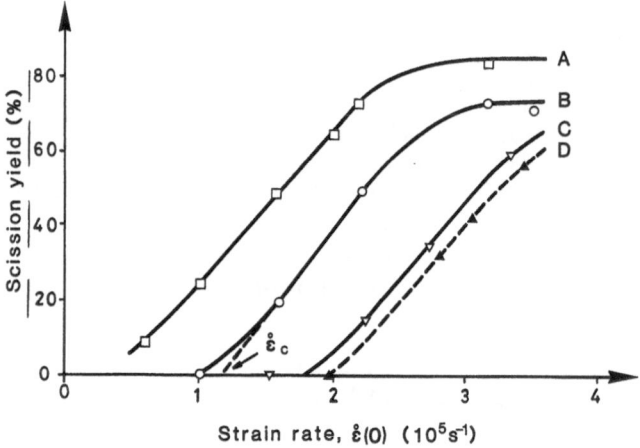

Fig. 50. Yield for chain scission as a function of strain rate for different fractions of polycarbonate (PC) in benzyl alcohol/dioxan (90:10 v.v) at 20 °C. A: normal PC with $M_p = 417000$; B: normal PC with $M_p = 321000$; C: normal PC with $M_p = 256000$; D: PC with weak bonds, $M_p = 217000$; M_p: molecular weight at peak maximum; $\overset{\circ}{\varepsilon}_c$: critical strain rate for chain scission (extrapolated from the linear portion of the degradation curve)

Polymorphism in Polymers

Paolo Corradini and Gaetano Guerra
Dipartimento di Chimica, Università di Napoli via Mezzocannone 4,
I-80134, Napoli, Italy

Some aspects of the polymorphic behavior of polymers, with particular reference to the structural organization at the molecular level are reviewed.

After a temptative structure-based classification of different kinds of polymorphism, a description of possible crystallization and interconversion conditions is presented. The influence on the polymorphic behavior of comonomeric units and of a second polymeric component in miscible blends is described for some polymer systems. It is also shown that other characterization techniques, besides diffraction techniques, can be useful in the study of polymorphism in polymers. Finally, some effects of polymorphism on the properties of polymeric materials are discussed.

Advances in Polymer Science, Vol. 100
© Springer-Verlag Berlin Heidelberg 1992

1 Introduction

Polymorphism is a widespread phenomenon in polymers; almost all crystalline polymers are polymorphic, if the right conditions for the crytallization of the different forms can be found.

In this article some literature studies together with studies conducted recently in our laboratories on the crystalline and molecular structure of polymorphic polymers are reviewed, also with the aim of showing possible influences of the polymorphism on the properties and, as a consequence, on the applications of polymeric materials.

In the second section a classification of the different kinds of polymorphism in polymers is made on the basis of idealized structural models and upon consideration of limiting models of the order-disorder phenomena which may occur at the molecular level. The determination of structural models and degree of order can be made appropriately through diffraction experiments. Polymorphism in polymers is, here, discussed only with reference to cases and models, for which long-range positional order is preserved at least in one dimension.

The following section deals with the crystallization and interconversion of polymorphic forms of polymers, presenting some thermodynamic and kinetic considerations together with a description of some experimental conditions for the occurrence of solid-solid phase transitions.

In the fourth section the influence of comonomeric units on the polymorphic behavior of some polymers is described. In particular it is shown that comonomeric units can produce changes in the solid-solid transition temperatures as well as variations of the crystalline forms, which can be obtained for a given crystallization procedure. In the same section, some recent studies, showing that the polymorphic behavior of some polymers can be altered in particular miscible blends, are also reviewed.

In the penultimate section, it is shown, with some examples, that other characterization technique (in particular FTIR and solid-state NMR), besides diffraction techniques, can be useful in the study of the polymorphism of polymers at the molecular level.

Some relevant effects of the polymorphism on the properties of polymeric materials are shown in the final section. In particular, it is shown that, while the occurrence of transitions between polymorphic forms can be detrimental for some systems, a precise knowledge of the polymorphic behavior and of the physical properties of the single forms can be used advantageously to improve the in use properties as well as the processing conditions of some polymeric materials.

2 Structure-Based Classification of Polymorphism

2.1 Structural Considerations and Diffraction Techniques for Characterization

A classification of the various possible "idealized" ordered states of matter can be made as follows.

At one extreme, one has the structural models of perfect crystals, which have long-range positional order for all the atoms (apart thermal motion). A diffraction experiment on a set of such crystals oriented in one direction (corresponding, in most real cases of polymeric materials, to an oriented fiber) would result in a pattern of sharp reflections organized in layer lines.

Within the class of polymer crystals having, ideally, long-range positional order for all the atoms, different crystalline forms (polymorphs) may arise as a result of having different almost isoenergetic macromolecular conformations (of the main chain, in most known cases) or as a result of different, almost isoenergetic modes of packing of macromolecules with identical conformations [1–3].

The thermodynamic transition between different forms as the above described is formally discontinuous. The difference between polymorphs is shown in general also by a different metrical description of the corresponding lattices.

Disorder in crystals is fairly common in polymeric materials, more common probably than in corresponding materials of low molecular mass [1–5]. Disorder may occur in the packing of the macromolecules, while long-range *positional* order of some structural feature is maintained. According to the different outcomes of an idealized diffraction experiment on a fiber, it is useful to distinguish three main categories in the manner in which positional disorder is introduced:

i) Not all, but only some characterizing points of the structure maintain long-range periodicity in three dimensions and hence a well-defined three-dimensional lattice.

ii) Long-range positional order in three dimensional is maintained only for structural features which are not point-centered (e.g., for the chain axes, for which two periodicities only are sufficient to define a three-dimensional repetition)

iii) Long-range positional order of some feature is maintained only in two or in one dimension (e.g., only along each chain axis).

In the first case, there is only partial instead of complete long-range three-dimensional order. Fiber spectrum features are diffuse haloes (besides sharp reflections) on the layer lines.

As an example of the second case, we may have conformationally disordered chains, but long-range order in the positions of the chain axes (*condis crystals* [5]). Fiber spectrum features are the occurrence of sharp reflections on the equator only and diffuse reflections on the other layer lines.

As an example of the third case, we may have conformationally ordered chains, parallel among themselves, with short-range order in the lateral packing. Fiber spectrum features are well-defined layer lines, with diffuse reflections only.

All solid forms lacking order to a substantial degree and ordered liquids (smectic, nematic) are called at times *mesomorphic forms*.

It is important to note that, for important sub-cases of case *i*), which will be discussed in more detail in Sect. 2.4, there is a low extent of disorder: entropy effects, if any, are small and changes of the lattice dimensions are absent or small. These particular disordered forms are not considered as mesomorphic. In such cases, the limiting models which are fully ordered or fully disordered may be designated respectively as *ordered* or *disordered crystalline modifications*, if their consideration is useful for the structural description of a polymeric material. Note that in real cases the model structures are, in general, intermediate between fully ordered and fully disordered modifications.

The various kinds of long-range positional order (periodicity) of the equilibrium positions of the structural elements, which characterize crystalline matter, are generally lost in the case of polymers after a not too big number of repetitions. The corresponding lattice distortions (different from the thermal ones) have been called distorsions of the second kind. According to Hosemann [6], we may speak in this case of *paracrystals* rather than crystals; and he has given criteria to recognize the effects of distorsions of the second kind on the broadening of X-ray reflections.

Finally, whenever in matter there is no long range *positional* but still long-range *orientational* order, we have ordered liquids, instead of solids. The X-ray spectral features of an ordered liquid with a *smectic* structure is the occurrence of one or a few meridional sharp reflections in the fiber spectrum, plus polarized halos; for an ordered liquid with *nematic* structure the occurrence of polarized haloes, only.

The spectrum features of a completely disordered liquid consist of haloes only.

In this review, phenomena connected with the paracrystallinity as well as ordered liquids are not considered.

2.2 Packing of Chains Having Different Conformations

Polymorphic forms characterized by *widely different conformations* are observed in several cases.

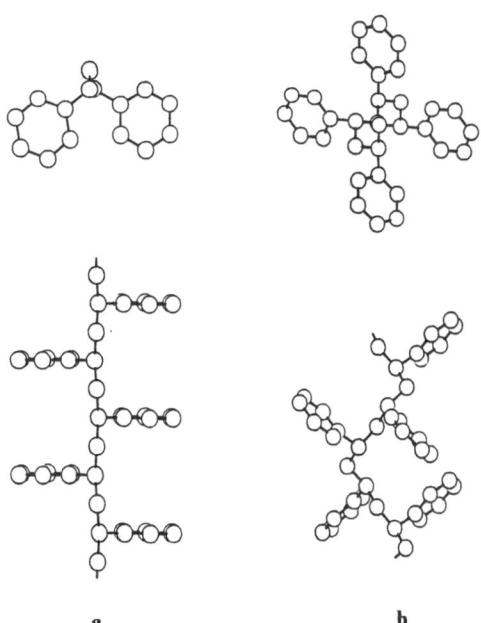

Fig. 1 a, b. Projection along the chain axis and side view of models of syndiotactic polystyrene in the: a) *trans*-planar conformation; b) s(2/1)2 helical conformation

a   b

It is typical, for instance, of syndiotactic polystyrene (s-PS) [7–9] and syndiotactic poly-*p*-methylstyrene (s-PPMS) [10] to present crystalline forms with a *trans*-planar conformation of the chains (shown for s-PS in Fig. 1) as well as crystalline forms with sequences of dihedral angles of the kind TTG^+G^+ (or the equivalent G^-G^-TT), corresponding to a s(2/1)2 helical symmetry of the chains (shown for s-PS in Fig. 1).

In particular, for s-PS, *trans*-planar chains are present in two crystalline forms, named α and β [8], while chains with a s(2/1)2 symmetry are present in a third crystalline form [7–9], named γ [9].

A polymorphic behavior involving packing of chains having completely different conformations has been found also for isotactic polymers. For instance, isotactic polystyrene, under suitable experimental conditions, can produce crystalline gels in which the chains assume a nearly fully extended conformation [11, 12], very close to a *trans*-planar, rather than the classical conformation of three-fold helix [13]. The two possible conformations proposed for the two crystalline forms of i-PS are shown in Fig. 2.

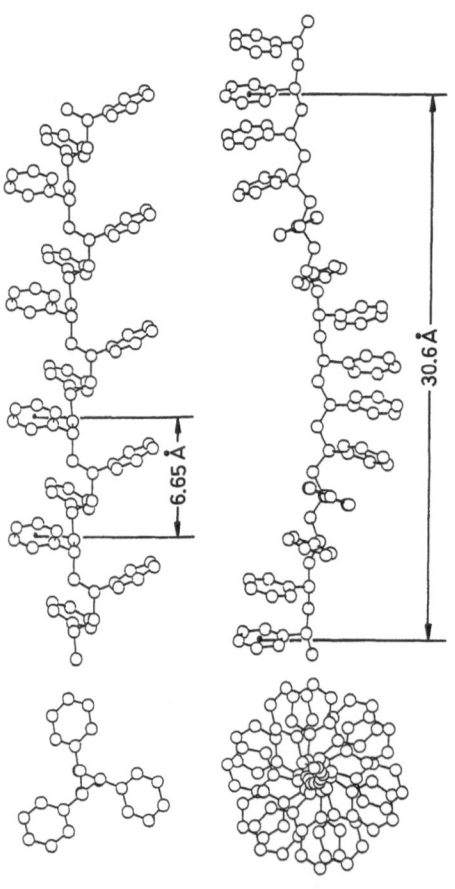

Fig. 2 a, b. Side view and projection along the chain axis of models of isotactic polystyrene in the: **a)** s(3/1) helical conformation; **b)** nearly *trans*-planar conformation, proposed for the crystalline gels [12]

Three different conformations have been observed for the case of the very complex polymorphic behavior of poly (vinylidene fluoride) (PVDF) [14–15]. In fact, sequences of dihedral angles of the kind TG^+TG^- with $T \simeq 180°$ and $G \simeq 45°$, that is a conformation with glide symmetry (tc), are present in the so-called α form (or form II) [16]. A *trans*-planar conformation characterizes the β form (or form I) [17, 18], which is the crystalline form with piezoelectric properties. A conformation of the kind $TTTG^+TTTG^-$, which again corresponds to a glide symmetry, but with a conformational repeating unit corresponding to two monomeric units, has been instead observed for the γ form (form III) [19, 20].

In other cases, polymorphic forms are characterized by *slightly different conformations*. In other words, while the chain conformations packed in the different polymorphs are different, they correspond, however, to small variations in the sequences of the dihedral angles along the main chain.

It is well known, for instance, the case of isotactic polybutene (i-PB), in which the three known crystalline forms (referred to as I, II, and III) contain helices with a number of constitutional repeating units per turn in the range 3–4 (3/1, 11/3, 4/1 helices, respectively) all corresponding to regular sequences of nearly *trans* and nearly *gauche* dihedral angles [3, 21].

A similar behavior has been recently found also for syndiotactic polybutene (s-PB). In fact, two different crystalline forms have been found, which present chains with s(2/1)2 and s(5/3)2 symmetries (shown in Fig. 3a and 3b, respectively).

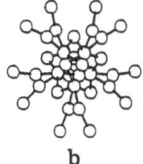

Fig. 3 a, b. Side view and projection along the chain axis of models of syndiotactic polybutene in the: **a)** s(2/1)2 conformation, proposed for form I; **b)** s(5/3)2 conformation, proposed for form II [22]

a   b

Both conformations roughly correspond to sequences of dihedral angles along the chain of the kind TTGG [22].

The different chain conformations observed in different polymorphic forms of a polymer are generally associated to nearly equivalent minima in the conformational energy maps, calculated for isolated chain models [2, 3].

As an example we report in this paper the conformational energy maps of two already cited stereoregular polymers, which have been obtained very recently, syndiotactic polystyrene s-PS and syndiotactic polybutene s-PB (Fig. 4 and 5, respectively). In fact, the energy map calculated for s-PS shows

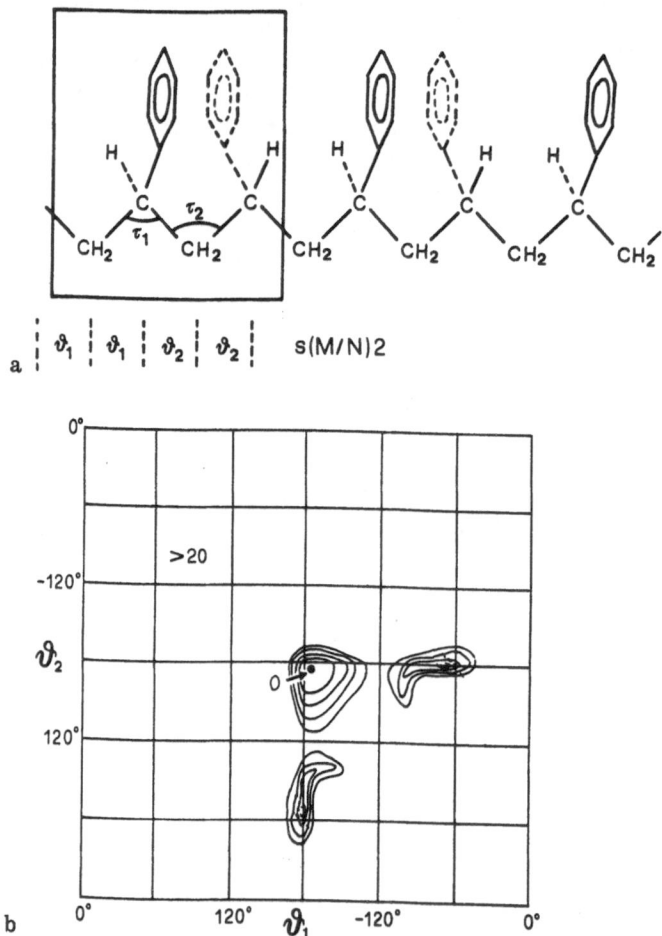

Fig. 4 a. Portion of the chain of s-PS considered for the conformational energy map and used symbols: **b)** Map of the conformational energy of s-PS as a function of θ_1 and θ_2, minimized with respect to θ_3, in the s(M/N)2 line repetition group, for $\tau_1 = 111°$ and $\tau_2 = 113°$. The curves are reported at intervals of 5 kJ/mol of monomeric units, with respect to the absolute minimum of the map assumed as zero [23]

similar energy values for the minima corresponding to the TTTT and TTG$^+$G$^+$ (or G$^-$G$^-$TT) conformations [23] (Fig. 4), observed in the different polymorphs of s-PS. The analogous map for S-PB shows that in this case the *trans*-planar conformation is higher in energy, while the TTGG minima are splitted into two sub-minima, which are located close to the values of the dihedral angles corresponding to the (2/1)2 and (5/3)2 helices suggested on the basis of the X-ray diffraction for the two known crystalline forms [22] (Fig. 5).

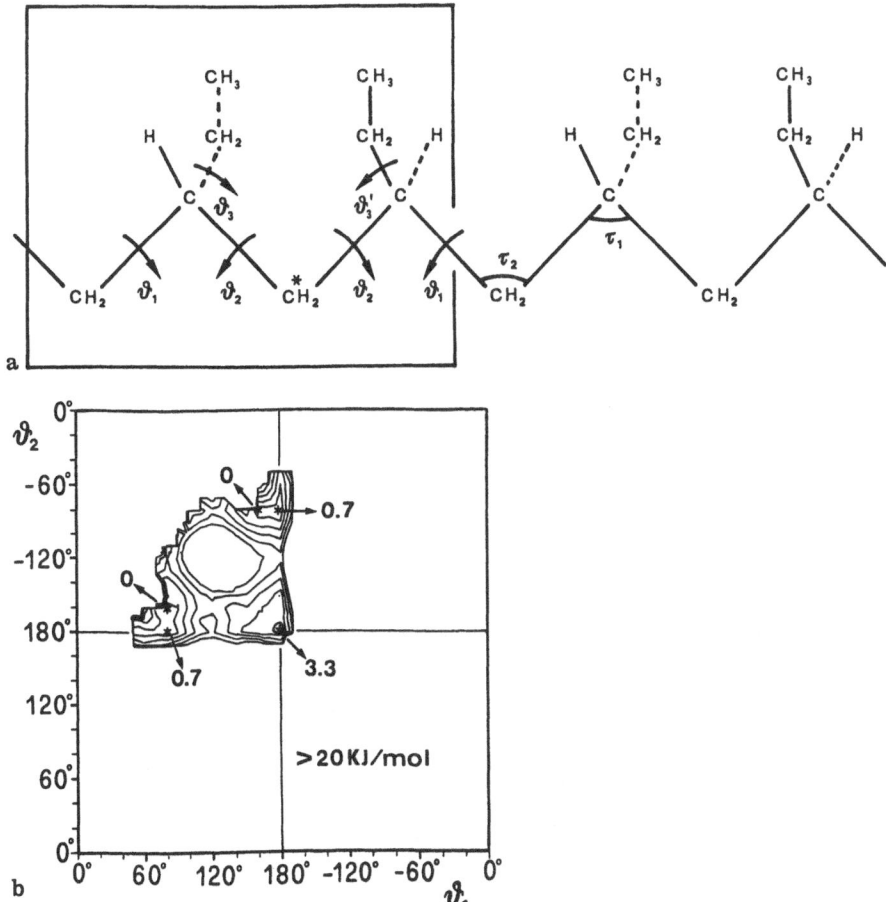

Fig. 5 a. Portion of the chain of s-PB considered for the conformational energy map and used symbols: **b)** Map of the conformational energy of s-PB as a function of θ_1 and θ_2, minimized with respect to θ_3, in the s(M/N)2 line repetition group, for $\tau_1 = 111°$ and $\tau_2 = 113°$. The curves are reported at intervals of 4 kJ/mol of monomeric units, with respect to the absolute minimum of the map assumed as zero [22]

2.3 Modes of Packing of Chains Having Identical Conformations

Chains having identical conformations can be packed in different modes corresponding to different unit cells.

For instance, in the three crystalline forms (α, β, γ) of i-PP the chains are always in the conformation of threefold helix (s(3/1)1 symmetry) but are packed in different ways in monoclinic [24], hexagonal [25], and orthorhombic [26] unit cells, respectively. The X-ray diffraction spectra of unoriented samples in the crystalline forms α, β, γ are reported in Fig. 6.

Different modes of packing in different unit cells exist also for the *trans*-planar chain of s-PS in the so called α and β forms, for which hexagonal [27, 28] and

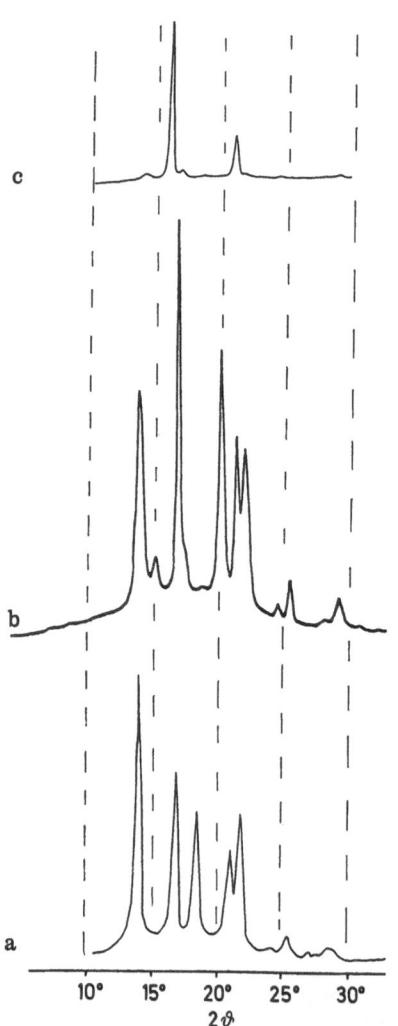

Fig. 6 a–c. X-ray diffraction spectra (CuKα) of unoriented samples of: **a)** α form; **b)** γ form; **c)** β form containing only minor α form impurities [41]

orthorhombic [7, 28, 29] unit cells have been described, respectively. The packing of the chains in the fully ordered hexagonal and orthorhombic modifications of s-PS are shown in Fig. 7 a and 7 b, respectively.

In some cases chains having identical conformations are packed in different modes but in cells which are also substantially identical.

This is, for instance, the case of the α and δ forms of PVDF (also called form II and form IIp, that is form II polarized), both contain chains with a TG^+TG^- conformation and have an orthorhombic unit cell with substantially identical dimensions [30, 31]. Detailed X-ray diffraction analyses have shown that the two chains in the unit cells are with the dipole vectors pointing in opposite directions in the α form, while they are in the same direction in the δ form [32].

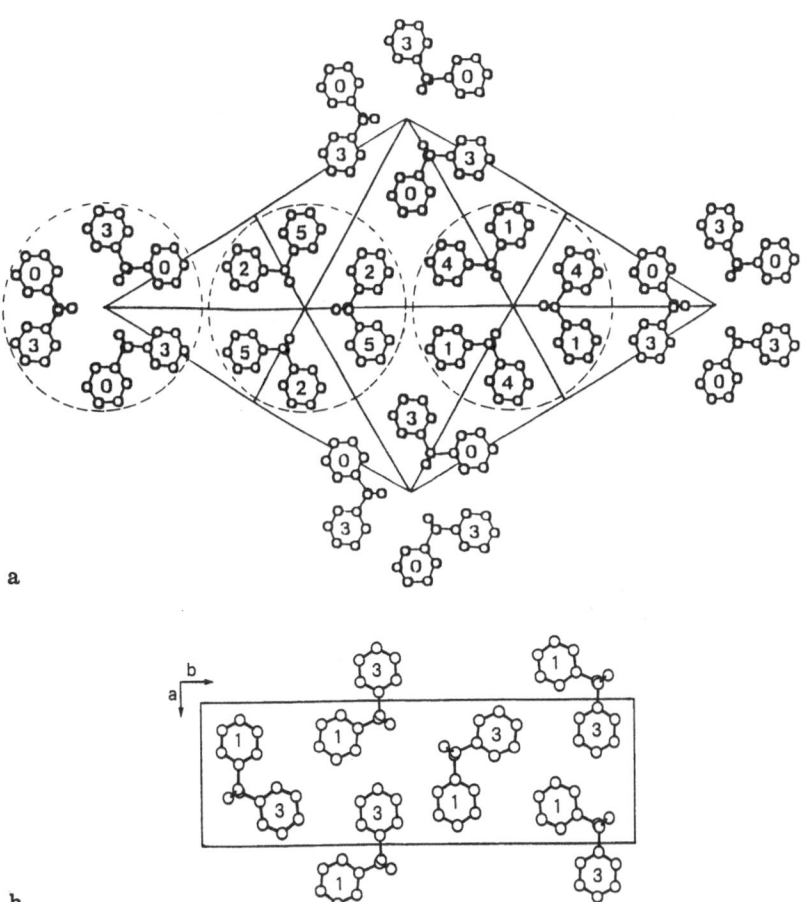

Fig. 7 a, b. Possible models of packing, for *trans*-planar chains of s-PS, for the limiting ordered modifications: **a)** hexagonal (α″); the relative heights of the centers of the phenyl rings are indicated in $c/6$ units [28]; **b)** orthorhombic (β″); the relative heights of the centers of the phenyl rings are indicated in $c/4$ units [7, 29]

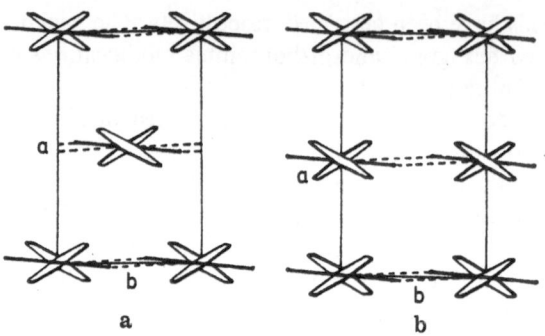

Fig. 8 a, b. Projection along the c axis of the packing of the chains proposed for the two forms of poly-p-phenylene therephtalamide: **a)** for the more common polymorph; the chain axes are localized in $(0, 0, z)$ and $(1/2, 1/2, z)$; **b)** for the other polymorph; the chain axes are localized in $(0, 0, z)$ and $(1/2, 0, z)$ [36]

An analogous case, of identical chain conformations as well as of similar unit cell dimensions, have been described for the two crystalline forms of poly-p-phenylene terephtalamide [33–36] (better known with the trade name of Kevlar). The projections along the c axis of the packing of the chains proposed for the two forms [36] has been sketched in Fig. 8, corresponding to the localization of the chain axes in $(0, 0, z)$ and $(1/2, 1/2, z)$ for the more common polymorph, in $(0, 0, z)$ and $(1/2, 0, z)$ for the other polymorph.

A case of polymorphism which is strictly analogous to that of Kevlar seems to be present in syndiotactic polypropylene (s-PP). In fact, depending on the

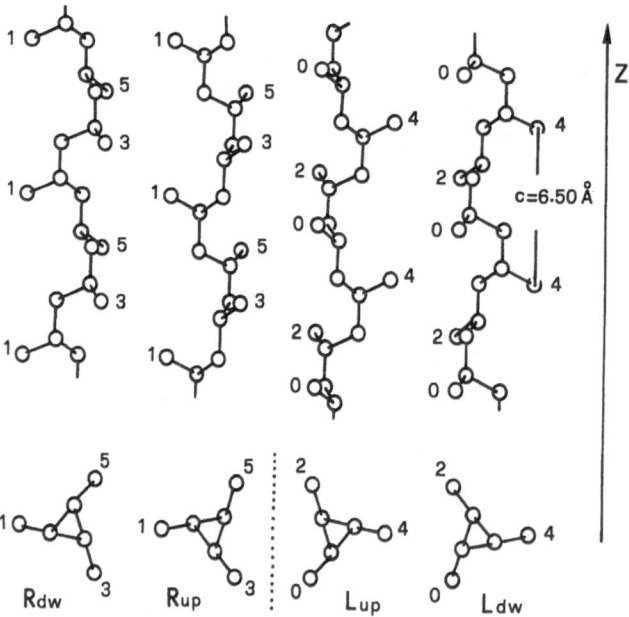

Fig. 9. The right-handed (R) and left-handed (L) three-fold helices of i-PP. For each handedness, the two different orientations (*up* or *down*) with respect to the reference axis are shown. The heights of the methyl groups are expressed ic $c/6$ units

experimental crystallization conditions, a localization of the chain axes as proposed by Corradini et al. [37] or as proposed more recently by Lotz et al. [38] may be more suitable for the interpretation of the X-ray or electron diffraction patterns.

2.4 Ordered and Disordered Modifications of Crystalline Forms

Let us consider a structural limiting model, in which the polymer molecules, presenting a periodic conformation, are packed in a crystal lattice with a perfect three-dimensional order. Besides this limiting ordered model, it is possible to consider models of disordered structures having a substantially identical lattice geometry.

Because the single macromolecules in the considered structures have a same periodical conformation, for a length of the chains which in the ideal case tends to infinity, the difference between the entropies of an ideally ordered structure and a disordered one is small.

Crystalline forms corresponding to limiting ordered or disordered models, with equal lattice geometry, can be obtained with different procedures and can present dif-

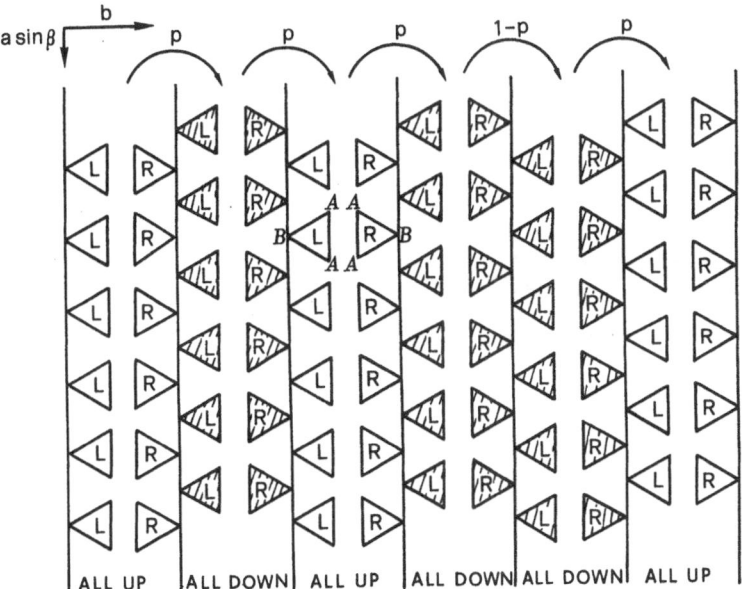

Fig. 10. Mode of packing of right- (R) and left-handed (L) helices in the α form of i-PP, viewed along the c axis. The triangles are schematic representations of the three-fold helices, with the methyl groups projecting at the vertices.

All the modifications of the α form would present, perpendicular to the b axis, macromolecular bilayers including all isoclined chains. A regular succession of bilayers with anticlined helices would correspond to the limiting ordered modification (α_2), while a statistical succession of bilayers would correspond to the limiting disordered modification (α_1) [40].

The letters A and B in a pair of chains in the middle of the figure refer to NMR non-equivalent methyl groups (Sect. 5.2)

ferent physical properties. Therefore, these forms are considered in this article and denoted, as previously indicated (Sect. 2.1), as ordered or disordered crystalline modifications.

These modifications constitute important sub-cases of the case of positional disorder for which only some characterizing points of the structure maintain long-range three-dimensional periodicity (indicated as case *i* in Sect. 2.1).

A condition to have, together with a fully ordered modification, more or less disordered modifications corresponding to a same unit cell is a substantial equality of steric hindrances in the space regions, where the statistical diadochy is achieved, as will be shown in the following.

An important sub-case of this kind corresponds to the occurrence of long-range positional order of all the atoms in two dimensions within layers of macromolecules (which may be single layers or bilayers, etc.) and disorder in the stacking of such layers, whereas some characterizing points of the layers maintain long-range periodicity and a well defined 3-D lattice.

As an example, let us recall the case of the monoclinic (α) form of i-PP. Detailed structural studies have shown that the crystalline form can present different degrees of disorder in the *up* and *down* positioning of the chains [39–41]. We recall that chains of a given handedness (e.g., right-handed helices) can present the lateral methyl groups *up* or *down* with respect to the chain atoms to which they are bonded [42] (Fig. 9). In the framework of a same ordered arrangement of right and left helices, it has been supposed the existence of two limiting modifications, with respect to the *up* and *down* positioning of the chains: an ordered modification (α_1) and a

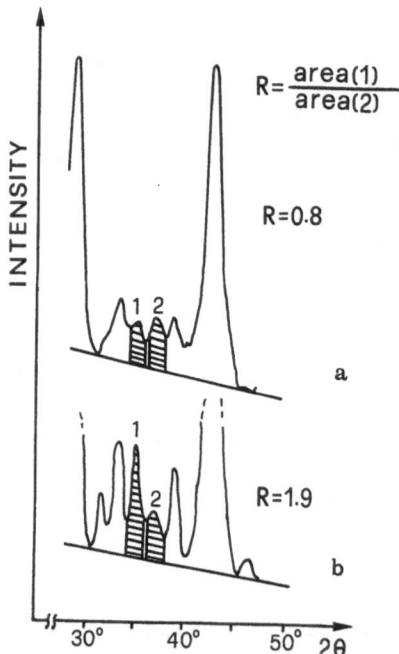

Fig. 11 a, b. The definition of the *up/down* degree of order, as used for i-PP in Refs. [43, 44]. Two examples of evaluation of R, for: **a)** an unannealed sample; **b)** a sample annealed at 167 °C

disordered modification (α_2). In particular, it has been suggested that in the disordered modifications centers near to the methyl groups maintain a long-range 3-D periodicity while a long-range positional order of all the atoms would be maintained within bilayers of macromolecules [40] (Fig. 10). A continuum of crystalline modifications, between the two limiting ones, can be obtained, depending on the crystallization and annealing conditions. The degree of order can be evaluated from the relative intensities of some specific reflections in the X-ray diffraction patterns, which are in other respects very similar for all the modifications of the α form [39, 40, 43, 44] (Fig. 11).

In the α modifications of i-PP, bilayers of macromolecules are stacked one on the top of the other in such a way that the top layer on one bilayer and the bottom layer of the bilayer in contact are made up of helices which are enantiomorphous. such helices are regularly anticlined for the ordered α_2 modification, more or less at chance isoclined or anticlined for the disordered α_1 modifications.

It may be interesting to note that in the case of the γ form, the structure is built up of bilayers of the same kind as those of Fig. 10. However, the top layer of one bilayer and the bottom layer of the bilayer in contact are made up of helices which are isomorphous. This is possible, with a good interlocking of the bulges and hollows (mainly provided by the methyl groups, Fig. 12), if each bilayer is inclined at an angle of 81 with respect to the neighboring one [26, 41]. This gives rise to a structure in which the chains are not parallel among themselves, and yet periodic, since they repeat along the equivalent ($\bar{a} + \bar{b}$) and ($\bar{a} - \bar{b}$) axes of a face centered orthorhombic unit cell. The axes ($\bar{a} + \bar{b}$) and ($\bar{a} - \bar{b}$) have a length (13.09 Å), which is twice the identity period of the three-fold helix of i-PP (c = 6.50 Å for the α form).

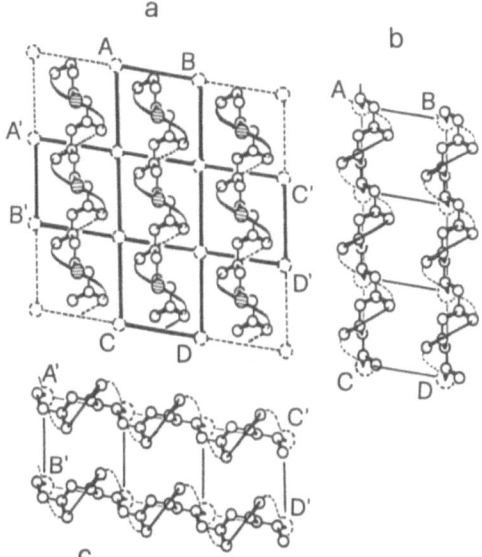

Fig. 12 a–c. Possible chain axes orientations in the crystal packing of i-PP. (a) A layer of left-handed helices is reported reproducing the (0 1 0) face of the α form. The methyl groups which project out of the mean plane toward the reader are shaded, while *broken line circles* indicate the available sites for the methyl groups of the next layer. (b) Next coming layer built up with helices of opposite chirality (right-handed): a parallel arrangement of the chain axes allows for a favorable packing. (c) Next coming layer built up with helices of the same chirality (left-handed): a tilting of the chain axes of 81° is required for a favorable interdigitation of methyl groups to occur

198   P. Corradini and G. Guerra

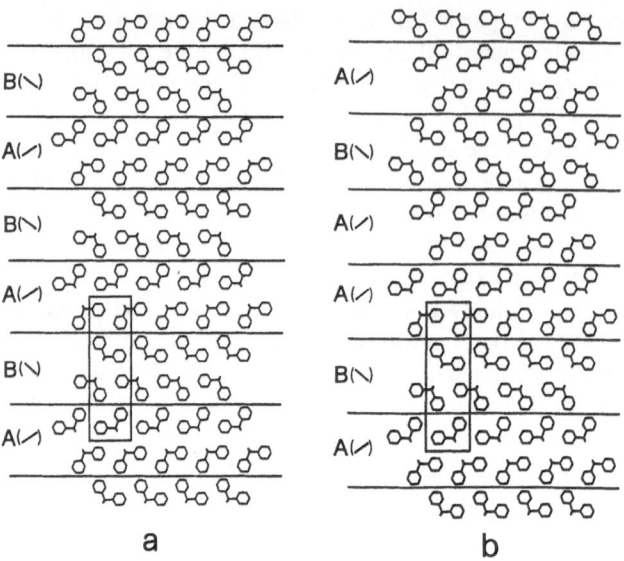

Fig. 13 a, b. Possible stacking of macromolecular bilayers in the β form crystals of s-PS. The regular succession of bilayers ABAB ... gives rise to the ordered β″ modification (**a**); defects, corresponding to pairs of bilayers of the kind AA or BB, would characterize the disordered β′ modification; an AA defect is reported in (**b**). The symbols (/) and (\) indicate the orientation of the lines connecting the adjacent phenyl rings of each chain inside the macromolecular bilayers A and B, respectively [29]

A disorder of the same kind has been suggested also for the orthorhombic form of s-PS. In fact, in the so-called disordered β′ modification the disorder would be in the stacking of ordered macromolecular bilayers and a 3-D long range periodicity would characterize only the mean position and orientation of the phenyl rings [28, 29] (Fig. 13).

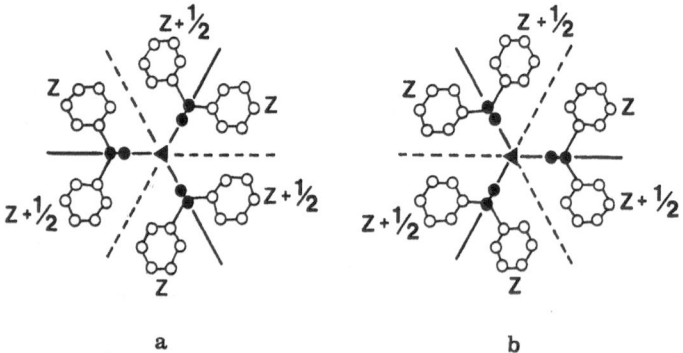

Fig. 14 a, b. Two different and isosteric orientations of triplets of s-PS chains; *dark circles* indicate the main-chain carbon atoms. The presence of a statistical disorder of these orientations, in the positions indicated by *large dashed circles* in Fig. 7a, characterize the disordered α′ modification [27, 9, 28]

Another important sub-case, of disorder in macromolecular crystals, corresponds to the statistical occurrence of two specific orientations only, at well defined positions in a 3-D lattice, of a group of macromolecules or of each single macromolecule.

This kind of disorder is present, for instance in the hexagonal form of s-PS. The disorder present in the so-called α' modification [27, 28], would correspond, in fact, to the statistical occurrence of two specific orientations of triplets of *trans*-planar chains, which leave unaltered the positions of the barycenters of the phenyl rings (Fig. 14) at well-defined positions in the 3-D lattice.

2.5 Solid Mesomorphic Forms

In this subsection some examples are presented of polymorphic polymers, for which one or more forms are solid mesomorphic, in the sense described in the Sect. 2.1. Let us recall that in this review we will not deal with liquid mesomorphic forms.

Mesomorphic can be considered, for instance, both "crystalline" forms of the alternated ethylene-tetrafluoroethylene copolymer (ETFE). In fact, in the low-temperature orthorhombic phase, there is, besides an intramolecular order in the *trans*-planar chains, a nearly perfect intermolecular order in the *ab* projection but a nearly complete absence of intermolecular order (corresponding to random relative displacements of neighbouring chains) along the *c* axis [45, 46]. This is clearly suggested by the fiber spectra showing sharp reflections on the equator and only diffuse reflections on the other layer lines [45, 46]. In the high-temperature hexagonal phase of ETFE, besides a translational disorder also a rotational disorder of the chain is present. A long-range 3-D order is possibly maintained only inside each chain (remaining *trans*-planar) and in the pseudo-hexagonal arrangement of the chain axes [47, 48].

A conformationally disordered mesomorphic form is present, for instance, in the high-temperature phase I of PTFE. In this form, a long-range 3-D order is present only in the periodic pseudohexagonal placement of the chain axes [49]. In fact intramolecular helix reversals would produce the conformational disorder [50–52] and a complete intermolecular rotational disorder would be also present [49, 52, 53].

Another well-known case of polymorphism which involves a conformationally disordered mesomorphic form is the 1–4, *trans*-polybutadiene [54–58]. In this case, there is a transition for which the enthalpy and entropy changes (at $\simeq 350$ K) are consistently higher than those at the melting temperature (at $\simeq 420$ K). In the high temperature modification the macromolecules are packed in a pseudo-hexagonal array and show a statistical identity period c = 4.65 Å, shortened with respect to that of the low temperature modification (c = 4.85 Å). Correspondingly there is a volume expansion of 9%. The large entropy change must be attributed to the conformational freedom gained by the chains at the transition; it may be indeed shown [55, 58] that a surprisingly high number of conformations are available to the chains of 1–4, *trans*-polybutadiene, while practically full-chain extension is maintained and the position of the chain axes maintain three-dimensional long-range order.

One modification of polyethylene (PE), which appears to be stable at high temperatures and pressures (see Sect. 3.2), also presents a pseudohexagonal packing and a disorderd chain conformation [59].

Differently from the cases of ETFE, PTFE, 1,4-*trans*-polybutadiene, and PE, in the mesomorphic form of i-PP it is believed that there is a nearly complete conformational order, in the chains packed with parallel axes, (intrachain long range 3-D order) and only short-range lateral order in the positioning of the chain axes [60, 61].

2.6 Clathrates

A special case of polymorphic forms can be considered the clathrates, that is forms in which polymer molecules interact with solvents in the crystalline state and form inclusion compounds.

For instance, one of the various crystalline forms of polyoxacyclobutane is a hydrate [62]. Syndiotactic polymethylmethacrylate also forms nonstoichiometric inclusion compounds with a variety of solvents [63, 64].

Clathrate structures have been recently obtained also for s-PS [8] and s-PPMS [10]. In particular for s-PS, the treatment of amorphous samples, as well as of crystalline samples in the α or in the γ form, produces clathrate structures including helices having s(2/1)2 symmetries, which present similar diffraction patterns, independently of the considered solvent. The treatment of samples of s-PPMS with suitable solvents also produces clathrate structures including s(2/1)2 helices; however, the large differences in the X-ray diffraction patterns suggest different modes of packing, depending on the included solvent.

3 Crystallizations and Interconversions of Polymorphic Forms

3.1 Thermodynamic and Kinetic Considerations

The obtainment of a particular polymorph, under given crystallization conditions, corresponds in most cases to the one which is thermodynamically more stable. In some cases, however, (when more than one minimum of free energy is available) the form which is obained is simply that one produced more rapidly (indicated as kinetically favored).

Let us consider, for instance, the case of i-PB. By crystallization on cooling from the melt, the metastable form II is obtained, which spontaneously and gradually in a few days at room temperature is transformed into the crystalline form I, which is the most thermodynamically stable [65].

Another interesting example is the melt crystallization of s-PS. For the case of rapid cooling from the melt, the hexagonal α form is obtained [7–9], while for low cooling rates or for isothermal crystallizations, the crystalline form which is obtained (α, β, or mixed) is depending on the crystalline structure of the starting material [9]. In particular, if the starting material is in the β form, always β form crystals are obtained; if the starting material is in the α or γ forms, the produced material is in the α or β form depending on the maximum temperature reached in the melt and the residence time at that temperature [9] (Fig. 15). These results can be interpreted by considering the α form as kinetically favored, particularly in the presence of a "memory" of the hexagonal crystals in the melt. When this memory in the melt is absent or deleted at sufficiently high temperature, the thermodynamically favored orthorhombic β form is obtained. It is interesting to note that in this case the relative stabilities of the two forms is probably very similar. The similarity is indicated by the fact that, the melting temperature for both forms is close to 270 °C and that both forms are not transformed into the other one up to the melting [9].

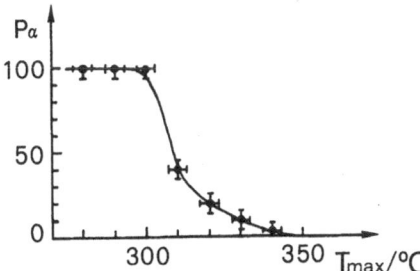

Fig. 15. Percent content of the α form in the crystalline phase (P_α) for compression moulded samples from γ form powders of s-PS, versus the maximum temperature of heating of the melt (T_{max}) [9]. The percent content of the β form is (100 − P_α)

For PVDF, several experiments have shown that at room temperature the β form is thermodynamically the most stable, while the α form is kinetically the most advantageous. For instance, by solution crystallization (casting from polar hexamethylphosphamide solutions) β form, γ form and α form crystals are obtained for low, intermediate and high evaporation rates, respectively [15, 66].

3.2 Phase Transitions Induced by Thermal Treatments

For several polymers, transitions between different polymorphs are induced by thermal treatments.

In some cases the transition occurs reversibly between two forms, which are thermodynamically favored belove or above a given temperature.

This is, for instance, the case of PTFE, which at atmospheric pressure presents two reversible first-order transitions at 19 °C and 30 °C [67]. In the transition at 19 °C the molecular conformation changes slightly, from a 13/6 to a 15/7 helix and the molecular packing changes from an ordered structure with a triclinic unit cell (corresponding to a positioning of the chain axes nearly hexagonal) toward a partially disordered structure (partial intermolecular rotational disorder) with a hexagonal positioning of the chain axes [49, 68, 69]. Above 30 °C the conformationally disordered mesomorphic form described in the Sect. 2.5, is reversibly obtained.

An interesting case of thermally induced reversible transition occurs between the two mesomorphic forms of the alternated copolymer ethylene-tetrafluoroethylene (ETFE), from the more ordered orthorhombic toward the less ordered hexagonal one. In this case, the transition occurs in a very broad temperature range ($\simeq 100$ °C) [47, 48]. The X-ray diffraction data have suggested, in this case, the occurrence of a transformation which is topotactic and hence not involving substantial relative movements of the macromolecules. In the transformation each hexagonal disordered crystal gives rise on cooling to several more ordered, but smaller, orthorhombic crystals [48].

Also for polyethylene, by increasing the temperature, a first-order reversible transition can occur from the usual orthorhombic form toward a hexagonal disordered form, but only for sufficiently high values of the pressure (above 4 kbar) [70–73, 5].

In other cases, the transition induced by thermal treatments are from a metastable to a stable form and are therefore irreversible. This is observed, for instance, for the transition from the γ form of s-PS (with helical conformation of the chain) toward the α form (having a *trans*-planar conformation of the chain) which has been shown to occur, in reasonable times, only above T $\simeq 180$ °C [8, 9].

Irreversible transformations, from more disordered toward more ordered modifications of a given form have been, for instance, observed for the α form of i-PP [39, 40, 43, 44] (see Sect. 2.4), as well as for the most common form of i-PS [74]. In these cases the transitions occur by recrystallization processes in the respective melting regions.

3.3 Transitions Induced by External Field

Very frequent are the cases of stress-induced crystallizations. A typical case is that of slightly vulcanized natural rubber (1,4-*cis*-polyisoprene) which, under tension producing a sufficient chain orientation, is able to crystallize, while it reverts to its original amorphous phase by relaxation [75].

Also common is the occurrence of transitions between different crystalline forms under tensile stresses. We recall here, for instance, the solid-to-solid transitions under stress of nylon 6 [76], PVDF [77, 78], and polybutylene terephtalate (PBT) [79–83].

The transition obtained under stress can be in some cases reversible, as found, for instance, for PBT. In that case, careful studies of the stress and strain dependence of the molar fractions of the two forms have been reported [83]. The observed stress-strain curves (Fig. 16) have been interpreted as due to the elastic deformation of the α form, followed by a plateau region corresponding to the α toward β transition and then followed by the elastic deformation of the β form. On the basis of the changes with the temperature of the critical stresses (associated to the plateau region) also the enthalpy and the entropy of the transition have been evaluated [83].

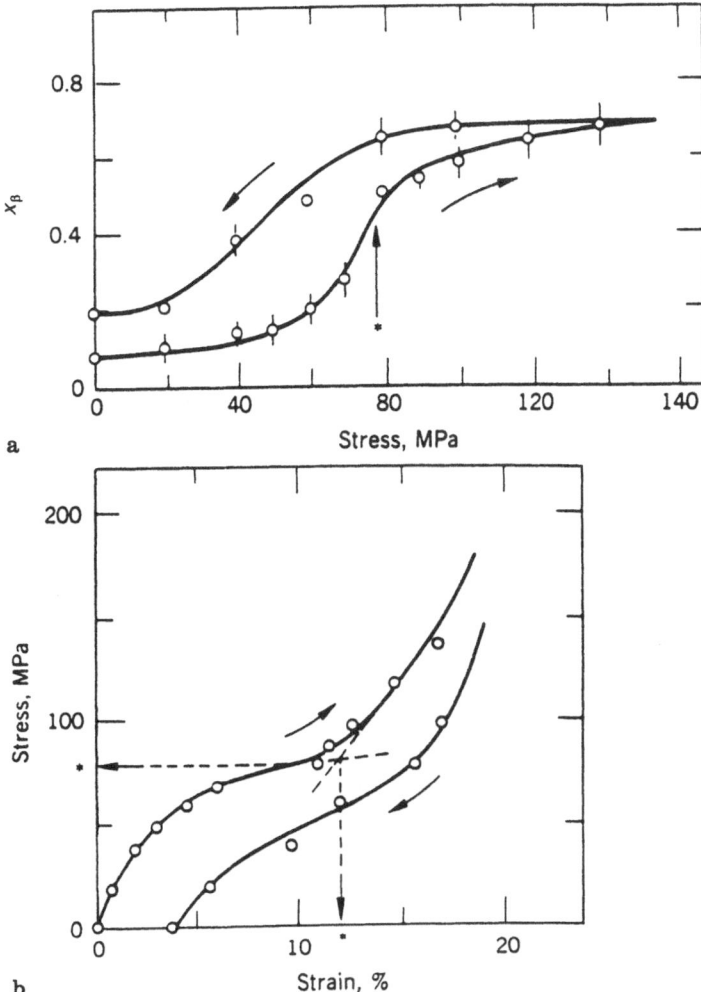

Fig. 16 a. Stress dependence of the molar fraction of the β form ($x_β$) and **b)** bulk stress-strain curve measured for uniaxially oriented polybutyleneterephtalate [83]

In several cases the transitions under tension involve the transformation of *gauche*-bonds to *trans*-bonds: for instance, for **PBT** the transition is from a form with dihedral angles, relative to the bonds of the tetramethylene group, of the kind GGTḠḠ toward a form with the same dihedral angles of the kind TTTTT. Analogously for **PVDF**, under tension, there is a transition from the α form (TGTḠ chains) toward the β form (TTTT chains).

In other cases the conformational changes are small, but in the sense of lengthening of the unit height in the chain conformation. This has been for instance, recently observed for s-PB. The transition from form I (s(2/1)2 helices) toward form II (s(5/3)2 helices (see Sect. 2.2) occurs under tensile stress and involves only a small unrolling of the helices with the unit height increasing from 3.86 to 4.0 Å, corresponding to small variations of the dihedral angles along the chain [22] (Fig. 3).

A phase transition can be induced also by a high electric field. This is, for instance, the case of the conversion of the non polar α form (form II) of PVDF to the polar δ form (form IIp) [31] (see Sect. 2.3).

4 Changes in the Polymorphic Behavior of Polymers

4.1 Polymorphism and Comonomeric Units

The presence of comonomeric units can alter the polymorphic behavior of a polymer, favoring in particular conditions the obtainment of a given crystalline form with respect to the other ones.

As usual, this can be due both to thermodynamic and kinetic reasons. In fact, the presence of comonomeric units increases, in general, the energy content of all the crystalline forms, but, since the extent of increase may be different, it may destabilize some chain conformation or some kind of packing more than other ones. On the other hand, the influence of the comonomeric units on the polymorphic behavior of a polymer can be due to a change in the crystallization rates of the various forms.

It is known the case of i-PP, for which the copolymerization with small amounts of ethylene tends to stabilize the γ form [84]: for instance, by melt crystallization of a copolymer with 6% by mol of ethylene more than 80% of the crystalline phase is in the γ form [85]. It is also known that the obtainment of the γ form by melt crystallization, is also favored for samples of low molecular mass [86, 87] and for "stereoblock" fractions [88]. This seems to suggest that, whenever the preferential crystallization of the γ-form is observed, there is the concomitant occurrence of a reduction in the polymer of the length of the chain stretches with polypropylene head to tail constitution and isotactic configuration.

Particularly relevant is the case of some copolymers of PVDF. Already small amounts (5–20% by mol) of a fluorolefinic comonomer (vinyl fluoride (VF) [89–90], trifluoroethylene [91–93], tetrafluorethylene [94, 95]) can force the polymers to a melt crystallization in the piezoelectric β form. (We recall that the homopolymer crystallizes in the non-piezoelectric α form, by melt crystallization).

Just as an example, the X-ray diffraction patterns of compression moulded samples of PVDF, poly(vinylfluoride), and of some VDF-VF copolymers of different compositions are shown in Fig. 17 [90]. The degrees of crystallinity of the copolymer samples (40–50%) are high and analogous to those of the homopolymer samples. This indicates a nearly perfect isomorphism between the VF and VDF monomeric units [90, 96]. The diffraction patterns and the crystal structures of the copolymers are similar to those of PVF, which are in turn similar to the X-ray pattern and crystalline structure of the β form of PVDF. On the contrary, the X-ray pattern of a PVDF sample crystallized under the same conditions (Fig. 17a) is completely different, that is typical of the non-piezoelectric α form [90].

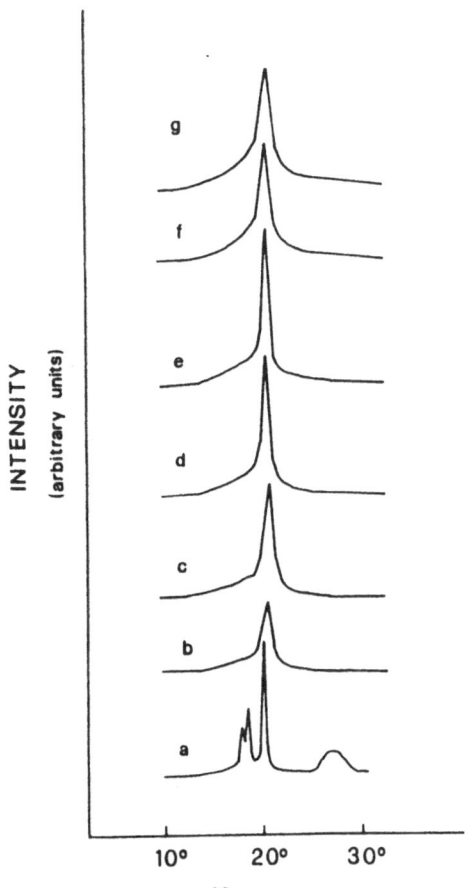

Fig. 17 a–g. X-ray diffraction patterns of compression moulded samples of: (a) PVDF, (b) 88/12 VDF/VF copolymer, (c) 84/16 VDF/VF copolymer, (d) 51/49 VDF/VF copolymer, (e) 43/57 VDF/VF copolymer, (f) 3/97 VDF/VF copolymer, (g)PVF. (All copolymer compositions in mol%) [90]

For the case of polymers which present reversible solid-solid transitions, producing more disordered forms, with increasing the temperature, (e.g. PTFE, ETFE, 1,4-*trans*-polybutadiene (see Sects. 2.5 and 3.2)) the introduction in the chains of comonomeric units, as well as of other constitutional defects, tends to stabilize the more disordered structure with respect to the more ordered one, and hence to lower the transition temperatures.

For instance, the temperature of the solid-solid transition at room temperature of PTFE is lowered of nearly 40 °C for the random copolymerization of 5% by mol of hexafluoropropylene with tetrafluoroethylene [97, 98].

4.2 Polymorphism and Miscible Polymer Blends

First of all it is worth to remind that there are not well documented cases of cocrystallizations between different components of a polymer blend. Hence, also in the cases of miscible blends of semicrystalline polymers the miscibility refers only to the amorphous phase.

For instance, also the homopolymers PVF and PVDF have been described to crystallize in separate crystals in their blends [99] (Though constituted by isomorphous monomeric units which can cocrystallize in the copolymers in the whole range of composition, as seen in Sect. 4.1). Moreover, at least for the studied conditions, the polymorphic behavior of PVDF is not altered by the presence of PVF [99].

However, a few cases of miscible blends have been described recently in the literature in which, in particular conditions, the polymorphic behavior of a polymer appears to be altered.

Since interactions at the molecular level between polymer components in the blends occur only in the amorphous phase, it is reasonable to assume that these effects are due to kinetic factors and, in particular, to the influence of a polymer component on the nucleation or crystallization kinetics of the other one.

In particular, blends of PVDF with a series of different polymers (polymethyl-methacrylate [100–102], polyethylmethacrylate [101], polyvinyl acetate [101]), for suitable compositions, if quenched from the melt and then annealed above the glass transition temperature, yield the piezoelectric β form, rather than the normally obtained α form. The change in the location of the glass transition temperature due to the blending, which would produce changes in the nucleation rates, has been suggested as responsible for this behavior. A second factor which was identified as controlling this behavior is the increase of local *trans*-planar conformations in the mixed amorphous phase, due to specific interactions between the polymers [102].

The polymorphic behavior of PVDF has been found to be changed also in the presence of functionalized ethylene-propylene copolymers, for the case of samples isothermally crstallized from the melt [103].

Also the polymorphic behavior of s-PS can be altered by blending, in particular with poly-2,6-dimethyl-1,4-phenylene oxide (PPO), both for the case of crystalliza-tion from the melt [104] and for the case of crystallization from the quenched amorphous phase [105].

For the case of melt crystallization, the blending with PPO favors the obtainment of the β form. On the basis of several experimental results, it has been suggested that this behavior would be due to the interactions between PS and PPO chains in the melt, which would produce a more rapid disappearence of a "memory" (that is of possible nuclei) of the α form in the melt [104].

For the case of the crystallization from the amorphous phase, the blending with PPO for lower contents (less than 30 wt%) favours the obtainment of the α'' ordered modification with respect to the α' disordered modification, which is obtained for the unblended polymer. For higher PPO contents the obtainment of the β form is favored [105]. This behavior would be simply due to the increases of the glass transition temperature, and hence of the crystallization temperature on heating, which correspond to increased PPO contents in the blends [105].

5 Subsidiary Characterization Techniques

Although the diffraction techniques are unique in providing detailed information on the structural organization at the molecular level in the different crystalline forms, there are other characterization techniques which are sensitive to the chain conformation and in some cases to the chain packing, which can be used advantageously (and in some case more efficiently than diffraction techniques) in the recognition and quantification of the different polymorphs in polymeric materials.

In the following, some examples of applications of Fourier transform infrared (FTIR) Spectroscopy and of solid-state nuclear magnetic resonance (NMR) to the study of polymorphism in polymers are described.

5.1 FTIR

Several bands of the FTIR spectra are conformationally sensitive. For this reason by this technique it is, in general, easy to distinguish between polymorphic forms presenting different chain conformations.

This is, for instance, shown for the case of the α and γ forms of s-PS, having *trans*-planar and s(2/1)2 helical conformations, respectively, for which different FTIR spectra are obtained [106–110] (Fig. 18). Note, for instance, the strong bands at 1222 and 540 cm^{-1} typical of the *trans*-planar chains, and the bands at 943, 934, 572, 548, and 502 cm^{-1} typical of the s(2/1)2 helices of s-PS.

Although in the frequency region of the conventionally measured infrared and Raman spectra (400–4000 cm^{-1}) only intramolecular modes appear, some particular bands can be sensitive to intermolecular interactions typical of the different modes of packing of chains with identical conformations.

Well-defined differences exist, for instance, between the spectra of samples in the α and β crystalline forms (although both contain chains of *trans*-planar conformation) as shown by the expanded FTIR spectra, in two different regions, of Fig. 19 [110].

Interesting are, as an example, the bands at 902 and 911 cm^{-1} which characterize α and β forms, respectively, both presenting a shoulder at 906 cm^{-1}, due to the presence of the amorphous phase. These spectral differences appear to be essentially independent of the particular modification obtained (α' or α'', β' or β'') and of the preparative route [110].

In the identification of different polymorphs in polymers the FTIR technique presents, with respect to the diffraction techniques, the advantage of easier and more rapid measurements. In particular, the high speed of the measurements allows to study the polymorphic behavior under dynamic conditions. As an example let us recall the study of the transition from the α toward the β form of PBT induced by tensile stresses, evaluated by quantitative analysis of the infrared spectra [83].

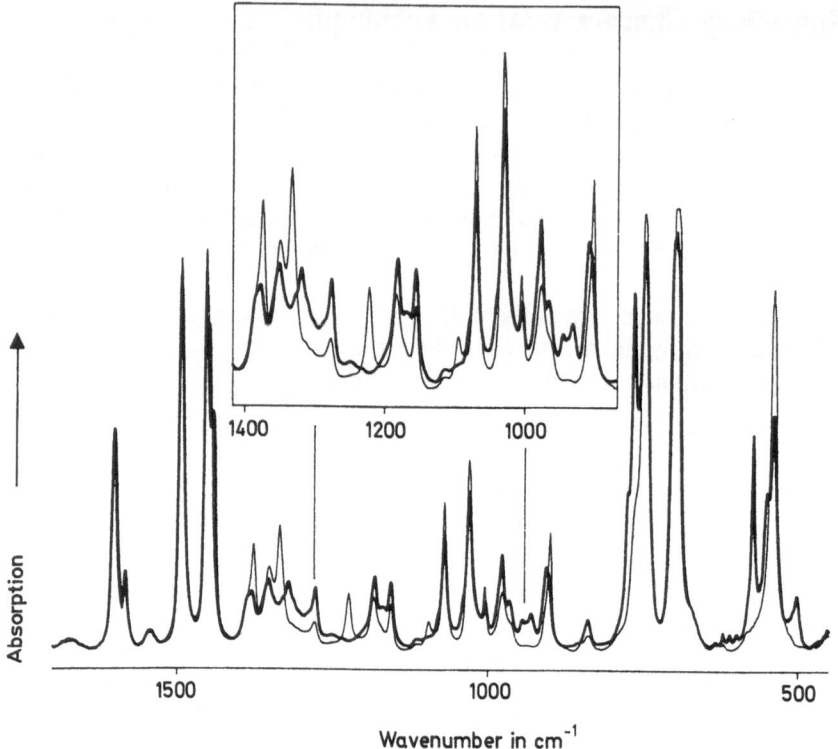

Fig. 18. Infrared spectrum of a s-PS sample in the γ form (*heavy line*) compared to the spectrum of the same sample, after transformation into the α form, by annealing at 200 °C (*light line*) [110]

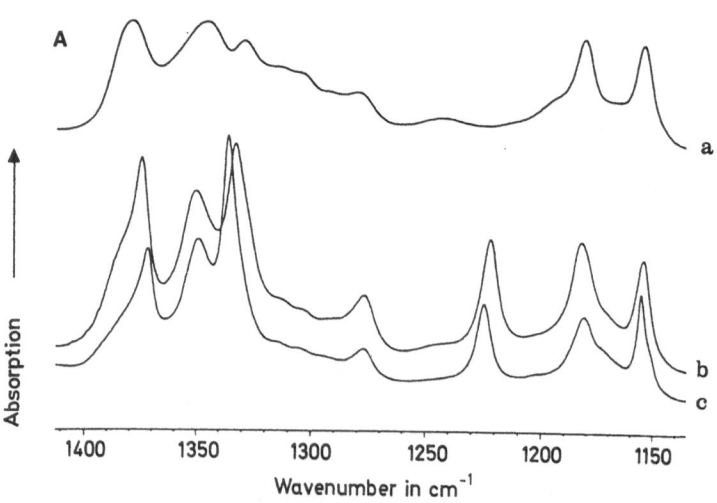

Fig. 19 A

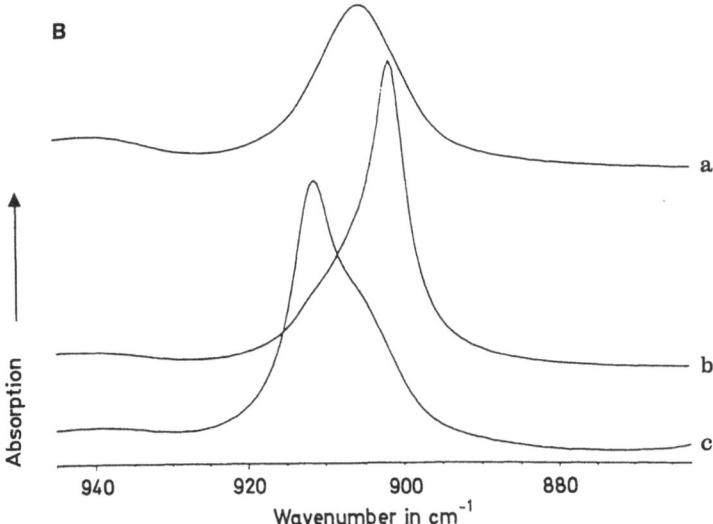

Fig. 19 A, B. Expanded infrared spectra of s-PS samples in three different regions:
A) 1100–1400 cm^{-1}, **B)** 860–940 cm^{-1}; *a*) amorphous; *b*) α form; *c*) β form [110]

The FTIR technique has also the advantage of being easily associated to other techniques. For instance, the FTIR microscopy, which couples the visible image of the samples to the corresponding infrared spectra, allows the selection of microareas of interest, and thus an easy determination of the crystalline forms corresponding to different regions, in samples in which more than one polymorph is present [103].

5.2 Solid-State NMR

As a consequence of restricted internal mobility in molecules in the crystalline state, nuclei in different conformation environments, but identical in other respects, can produce different signals in ^{13}C cross polarization, magic angle spinning (CPMAS) solid-state NMR. This analysis is not necessarily limited to crystalline regions, since signals of different conformations are resolved if the exchange is slow with respect to the time scale of the NMR experiment.

Reviews on the use of the solid-state NMR in the study of polymorphism in polymers has been presented in chapters of two relatively recent books [111, 112]. However, it is worthy to recall some aspects in the present article.

In some crystalline polymers chemical shift differences between crystalline and amorphous phases have been observed and interpreted and for several crystalline forms the signals to be attributed to nuclei in different conformational environments have been identified [111, 112].

Let us recall, for instance, the case of the crystalline form of s-PP [113], or of s-PS [114, 115], which contain s(2/1)2 helices. In this kind of helix (Fig. 2) there are two non-equivalent sets of methylene carbons in the main chain, the first one on the chain axis and the second one far from the chain axis. While for the methylene on the axis there are two γ-carbons (that is carbons separated by three bonds) in G conformation, for the methylene on the periphery of the helix there are two γ-carbons in the T conformation. This generates the so called γ-effect, that is a shift difference between the resonances of the two methylene carbons (8.7 ppm for s-PP, 10 ppm for s-PS).

It is hence easy to detect by this technique different polymorphic forms having different chain conformations. For instance, the α or β forms of s-PS (*trans*-planar chain conformations) present only a single methylene resonance at 48.1 ppm (vs.TMS), while the γ form (helical conformation) presents two methylene resonances at 37.3 and 47.3 ppm (Fig. 20) [114].

The sensitivity of ^{13}C solid-state chemical shifts to small conformational changes is well illustrated by the case of i-PB. Table 1 summarizes some chemical shifts differences for the forms, which have been interpreted in terms of variations of the γ-shielding parameter corrected for the deviations, with respect to the exactly G conformations, in the slightly different nearly *gauche* − nearly *trans* sequences, characterizing the three crystalline forms of i-PB [116].

Differences in the solid-state NMR signals of crystalline forms having identical conformations have been also observed. For instance, well-crystallized α form samples of i-PP show splittings for the methyl (22.6, 22.1 ppm) and methylene resonances (45.2, 44.2 ppm) into two lines with relative intensities 2:1 [117, 118]. These splittings have been interpreted in terms of the known crystalline packing of the α form, which is characterized by pairs of 3/1 helices of opposite handedness at closer distances (Fig. 10). This generates inequivalence between the carbons indicated as A and those indicated as B in Fig. 10 [117, 118].

Single resonances are instead observed for the β form of i-PP, for which these kinds of pairs of helices are not present. However, these splittings of the NMR signals are not particularly useful for the identification of the various polymorphs of i-PP, since they are small (being intermolecular in origin) and since they are not evident in not well-crystallized α form samples [117, 119].

Table 1. ^{13}C solid-state chemical shifts[a] of some resonances observed in the NMR spectra of different forms of isotactic polybutene [116]

State	CH_3	CH
Form I	1.55	20.68
Form II	0.75	22
Form III	3.05	25.21
	2.25	
Amorphous	0.00	23.46

[a] Referred to the CH_3 resonance of the amorphous i-PB sample as zero

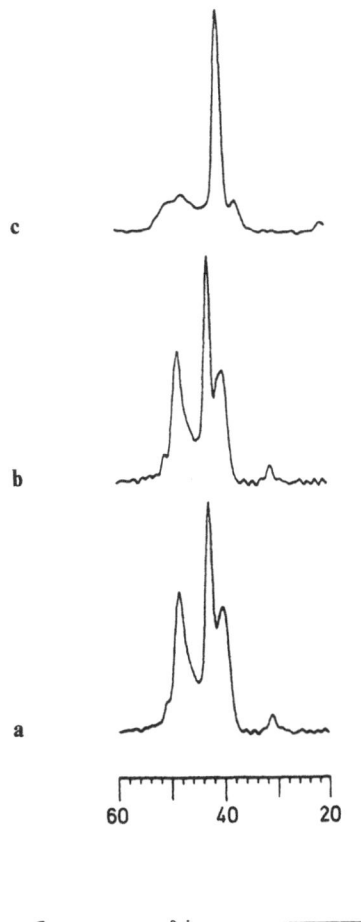

Fig. 20 a–c. Aliphatic region of the ^{13}C-NMR spectra for the different crystalline forms of s-PS: **a)** α form; **b)** β form; **c)** γ form [114]

δ in ppm

6 Polymorphism and Properties

6.1 Possible Inconveniences due to the Polymorphism in Polymeric Materials

In some case the presence of phenomena of polymorphism can be detrimental for the possible applications and/or the processing of a polymeric material.

This occurs, for instance, when a molded manufact tends to have a transition between polymorphic forms in conditions of use. In fact solid-solid transitions can generate problems of dimensional instability of the manufacts.

It is interesting the case of i-PB, crystallizing by rapid cooling from the melt in the metastable form II, which spontaneously and gradually in a few days is transformed into the more stable crystalline form I (Sect. 3.1). This is certainly a main factor in determining the limited commercial relevance of this polymer.

An analogous case is that one of 1,4-*trans*-polybutadiene. The reversible solid-solid transition at nearly 70 °C (Sect. 2.5) could create problems in possible applications as a material, since the volume of the crystals changes by nearly 9% at the transition [54, 56]. Incidentally we note that, for the large enthalpy and volume changes at the transition, this material has been suggested as useful for devices converting thermal energy into mechanical energy [120].

The polymorphism can be inconvenient also for the processing, when small variations of the processing conditions can produce samples in different crystalline forms. This is, for instance, the case of s-PS, for which the crystalline form obtained by cooling from the melt (α and/or β) is dependent not only on the cooling rate but also on the crystalline form of the starting material, on the maximum temperature to which the melt is heated as well as on the time for which the melt is held at this maximum temperature (Sect. 3.1). This requires, in order to get reproducible manufacts, an extremely accurate control of the processing conditions, which is, of course, undesirable in thermoplastic materials.

6.2 Properties of Specific Crystalline Forms Useful in the Applications

A precise knowledge of the physical properties (mechanical, thermal, electrical, solvent resistance, etc.) of the different polymorphic forms of a given polymeric material can be advantageously used for possible applications.

The most remarkable case of commercial use of samples in a particular crystalline forms, not obtained under standard processing conditions, is possibly the piezo-electric (converting mechanical energy into electrical energy) and pyroelectric (converting thermal energy into electrical energy) β form of PVDF. With respect to piezoelectric and pyroelectric materials of low molecular mass, this polymeric material present the great potential to be inexpensively fabricated into large area thin sheets. Typical applications include audio frequency transducers (such as microphones, headphones, and loudspeaker tweeters), ultrasonic transducers for underwater applications and pyroelectric detectors [121].

Another interesting case is the much higher solvent resistance of the β crystalline form of s-PS, with respect to the other ones. In fact, it has been found that the sorption of solvents (which are suitable to produce transformations from the α or the γ form toward clathrate structures) occurs only in the amorphous phase, for the case of β form samples [122–124]. Sorption kinetic curves of liquid methylene chloride in s-PS samples in the α and β form are, for instance, compared in Fig. 21 [124].

It is also well known that different polymorphic forms can present largely different crystallite moduli along the chain axis (both observed and calculated). These differences can be large if large variations in the chain conformations are involved, and can have a significant influence on the bulk properties [15].

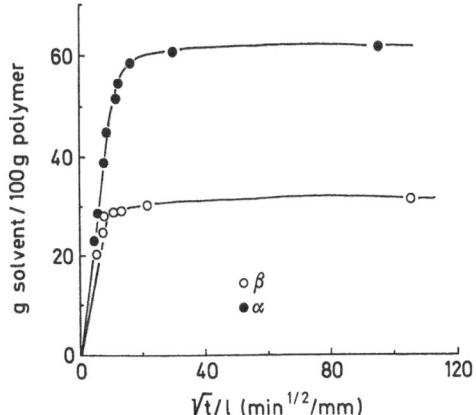

Fig. 21. Sorption kinetics of liquid methylene chloride in semicrystalline samples of s-PS (degree of crystallinity close to 55%) in the α form or in the β form [124]

Let us recall, for instance, the observed crystallite moduli along the chain axis of the β and α forms of PVDF (177 and 59 GPa, respectively) and of the α and γ forms of nylon-6 (165 and 27 GPa, respectively).

Significant differences in physical properties can be found also between different modifications of a given crystalline form. For instance, for the case of the α form of i-PP a nearly linear increase of the melting temperature (up to $\simeq 15\,°C$) with the increase of an order parameter (which allows to differentiate the various modifications) has been observed [43, 44]. This relationship holds for a given i-PP sample independently of its thermal history (cooling rate from the melt, annealing procedures, temperatures of isothermal crystallization). These studies, of course, do not indicate that the melting temperature depends only on the degree of *up-down* order in the crystals (other features are relevant, like, for instance, the size of crystals). However, it has been found that processes which produce increases in the melting temperature produce correspondingly increases of the *up-down* degree of order, and vice versa [43, 44].

6.3 Properties of Specific Forms Useful in the Processing

A detailed knowledge of the polymorphic behavior of polymeric materials allows sometime to define better conditions for their processing.

We have already cited (Sect. 4.1) that a small amount of suitable comonomeric units in PVDF allows to obtain the piezoelectric β form using normal melt processing conditions, without a significant reduction of the degree of crystallinity. This is the main reason for the recent commercial development of the VDF-trifluorethylene copolymers [121].

The advantage of using for the melt processing powder or pellets in the β form, rather than in other forms of s-PS, has been also shown [122]. In this case, in fact, the crystalline form of the manufacts depends only on the cooling rate, being independent of the melting conditions [9].

It is, possibly, also worthy to cite the relevance of some mesomorphic forms in the optimization of the drawing conditions to get highly oriented manufacts; although this is, possibly, mainly related to the peculiar morphological organization of these solid mesomorphic forms.

In particular, it has been shown that the most important factor determining the drawing conditions of fibers and films of i-PP is the structure (α or mesomorphic) which characterize the yarn or the film obtained by extrusion. The drawing from mesomorphic samples requires, indeed, a lower tension and, generally, higher draw ratios are obtained [125, 126].

Analogously, for polyethyleneterephtalate, drawing procedures in multiple steps, with preliminary drawings at room temperature generating a mesomorphic form [127], have been suggested by several authors [128–130], in order to get high modulus and high tenacity fibers.

Acknowledgements: The authors are grateful to Prof. V. Petraccone, Prof. B. Pirozzi, Dr. C. De Rosa, Dr. V. Venditto (University of Naples) for their advice and support. Financial support by the "Ministero dell'Università e della Ricerca Scientifica e Tecnologica" (Italy) is acknowledged.

7 References

1. Corradini P (1968) in: Ketley AD (ed) The stereochemistry of macromolecules, Marcel Dekker, New York, p 1
2. Corradini P, Guerra G, Pirozzi B (1980) in: Lenz RV, Ciardelli F (ed) Preparation and properties of stereoregular polymers, Reidel, Dordrecht, p 317
3. Corradini P (1981) in: Ciardelli F, Giusti P (eds) Structural order in polymers, Pergamon Press, Oxford, p 25
4. Petraccone V (1986) God Jugosl cent kristalogr 21: 105
5. Wunderlich B, Grebowicz J (1984) Adv Polym Sci 60/61: 1
6. Hosemann R, Bagchi SN (1962) Direct analysis of diffraction by matter, North-Holland, Amsterdam
7. Chatani Y, Fujii Y, Shimane Y, Ijitsu T (1988) Polym Prepr, Japan (Eng Ed) 37: E428
8. Immirzi A, De Candia F, Iannelli P, Vittoria V, Zambelli A (1988) Makromol Chem, Rapid Commun 9: 761
9. Guerra G, Vitagliano VM, De Rosa C, Petraccone V, Corradini P (1990) Macromolecules 23: 1539
10. Iuliano M, Guerra G, Petraccone V,. Corradini P, Pellecchia C, New Polym Mater (in press)
11. Atkins AJ, Isaac DH, Keller A, Miyasaka K (1977) J Polym Sci, Polym Phys Ed 15: 211
12. Corradini P, Guerra G, Petraccone V, Pirozzi B (1980) Eur Polym J 16: 1089
13. Natta G, Corradini P (1955) Makromol Chem 16: 77
14. Lovinger AJ (1982) in: Basset DC (ed) Developments in crystallline polymers-1, Appl Sci Publ, London, p 195
15. Tashiro K, Tadokoro H (1989) in: Encyclopedia of polymer science and engineering, Supplement, p 187
16. Hasegawa R, Takahashi Y, Chatani Y, Tadokoro H (1972) Polymer J 3: 600
17. Gal'perin Ye L, Strogalin YV, Mlenik MP (1965) Vysokomol Soed 7: 933
18. Farmer BL, Hopfinger AJ, Lando JB (1972) J Appl Phys 43: 4293
19. Takahashi Y, Tadokoro H (1980) Macromolecules 13: 1317

20. Lovinger AJ (1981) Macromolecules 14: 322
21. Corradini P, Petraccone V, Pirozzi B (1976) Eur Polym J 12: 831
22. DeRosa C, Venditto V, Guerra G, Pirozzi B, Corradini P, (Macromolecules) (in press)
23. Corradini P, Napolitano R, Pirozzi B (1990) Eur Polym J 26: 157
24. Natta G, Corradini P (1960) Nuovo Cimento, Supplemento 15: 40
25. Turner-Jones A, Aizlewood JM, Beckett DR (1964) Makromol Chem 17: 134
26. Brukner S, Meille SV (1989) Nature 340: 455
27. Greis O, Xu Y, Asano T, Peterman J (1989) Polymer 30: 590
28. DeRosa C, Guerra G, Corradini P, Rend Accad Naz Lincei (in press)
29. DeRosa C, Rapacciuolo T, Guerra G, Petraccone V, Corradini P, Polymer (in press)
30. Naegele D, Yoon DH, Broadhurst MG (1978) Macromolecules 11: 1297
31. Davis GT, McKinney JE, Broadhurst MG, Roth SC (1978) J Appl Phys 49: 4998
32. Bachmann M, Gordon WL, Weinhold S, Lando JB (1980) J Appl Phys 51: 5095
33. Tashiro K, Kobayashi M, Tadokoro H (1977) Macromolecules 10: 413
34. Northolt MG, van Aartsen JJ (1973) J Polym Sci, Polym Phys Ed 11: 333
35. Northolt MG (1974) Eur Polym J 10: 799
36. Haraguchi K, Tisato K, Motowo T (1979) J Appl Polym Sci 23: 915
37. Corradini P, Natta G, Ganis P, Temussi PA (1967) J Polym Sci Part C 16: 2477
38. Lotz B, Lovinger AJ, Cais RE (1988) Macromolecules 21: 2375
39. Ikosaka M, Seto T (1973) Polym J 5: 111
40. Corradini P, Giunchi G, Petraccone V, Pirozzi B, Vidal HM (1980) Gazz Chim Ital 110: 413
41. Brukner S, Meille V, Petraccone V, Pirozzi B, Progr Polym Sci (in press)
42. Natta G, Corradini P (1960) Nuovo Cimento, Supplemento 15: 9
43. Guerra G, Petraccone V, Corradini P, DeRosa C, Napolitano R, Pirozzi B, Giunchi G (1984) J Polym Sci, Polym Phys Ed 22: 1024
44. DeRosa C, Guerra G, Napolitano R, Petraccone V, Pirozzi B (1985) J Therm Anal 30: 133
45. Tanigami T, Yamaura K, Matsuzawa S, Ishikawa M, Mizoguchi K, Miyasaka K (1986) Polymer 27: 999
46. Petraccone V, DeRosa C, Guerra G, Iuliano M, Corradini P, Polymer (in press)
47. Tanigami T, Yamaura K, Matsuzawa S, Ishikawa M, Mizoguchi K, Miyasaka K (1986) Polymer 27: 1521
48. Iuliano M, DeRosa C, Guerra G, Petraccone V, Corradini C (1989) Makromol Chem 190: 827
49. Clark ES, Muus LT (1962) Z Kristallogr 117: 119
50. Corradini P, Guerra G (1977) Macromolecules 10: 1410
51. Tanaka H, Takemura T (1980) Polymer J 12: 355
52. Corradini P, DeRosa C, Guerra G, Petraccone V (1987) Macromolecules 20: 3043
53. DeRosa C, Guerra G, Petraccone V, Centore R, Corradini P (1988) Macromolecules 21: 1174
54. Natta G, Corradini P (1959) J Polym Sci 39: 29
55. Corradini P (1969) J Polym Sci, Polym Lett 7: 211
56. Suehiro K, Takayanagi N (1970) J Macromol Sci Phys B4: 39
57. Finter J, Wegner G (1981) Makromol Chem 182: 1859
58. DeRosa C, Napolitano R, Pirozzi B (1985) Polymer 26: 2039
59. Basset DC (1982) in: Basset DC (ed) Developments in crystalline polymers-1, Appl Sci Publ, London, Chap. 3, p 115
60. Natta G, Peraldo M, Corradini P (1959) Rend Accad Naz Lincei 24: 14
61. Corradini P, Petraccone V, DeRosa C, Guerra G (1986) Macromolecules 19: 2699
62. Kakida H, Makino D, Chatani Y, Tadokoro H (1970) Macromolecules 3: 569
63. Kusuyama H, Takase M, Higashihata Y, Tseng HT, Chatani Y, Tadokoro H (1982) Polymer 23: 1256
64. Kusuyama H, Miyamoto N, Chatani Y, Tadokoro H (1983) Polym Commun 24: 119
65. Danusso F, Gianotti G (1963) Makromol Chem 61: 139

66. Tashiro K, Kobayashi M (1987) Rep Progr Polym Phys Jpn 30: 119
67. Bunn CW, Howells ER (1954) Nature 174: 549
68. Sperati CA, Starkweather HW Jr (1961) Fortschr Hochpolym Forsch 2: 465
69. Yamamoto T, Hara T (1986) Polymer 27: 986
70. Wunderlich B, Arakawa T (1964) J Polym Sci Part A 2: 3697
71. Basset DC, Turner B (1972) Nature Phys Sci 240: 146
72. Basset DC, Block S, Pirmarini GJ (1974) J Appl Phys 45: 4146
73. Hikosaka M, Minomura S, Seto T (1980) J Appl Phys 19: 1763
74. Petraccone V, DeRosa C, Tuzi A, Fusco R, Oliva L (1988) Eur Polym J 24: 297
75. Mandelkern L (1964) Crystallization of polymers, McGraw-Hill, New York
76. Miyasaka K, Ishikawa K (1968) J Polym Sci Part A2, 6: 1317
77. Lando JB, Olf HG, Peterlin A (1966) J Polym Sci Part A 4: 941
78. Hasegawa R, Takahashi Y, Chatani Y, Tadokoro H (1972) Polym J 3: 600
79. Jakeways R, Ward IM, Wilding MA, Hall IH, Desborough IJ, Pass MG (1975) J Polym Sci Polym Phys Ed 13: 799
80. Yokouchi M, Sakakibara Y, Chatani Y, Tadokoro H, Tanaka T, Yoda K (1976) Macromolecules 9: 266
81. Jakeways R, Smith T, Ward IM, Wilding MA (1976) J Polym Sci Polym Lett Ed 14: 41
82. Ward IM, Wilding MA (1977) Polymer 18: 327
83. Tashiro K, Nakai J, Kobayashi M, Tadokoro H (1980) Macromolecules 13: 137
84. Turner-Jones A (1971) Polymer 12: 487
85. Busico V, Corradini P, De Rosa C, Di Benedetto E (1985) Eur Polym J 21: 239
86. Kojima M (1968) J Polym Sci Part A2, 6: 1255
87. Lotz B, Graff S, Wittman JC (1986) J Polym Sci Polym Phys Ed 24: 2017
88. Natta G, Mazzanti G, Crespi G, Moraglio G (1957) Chim Ind 39: 275
89. Balik CM, Farmer R, Baer E (1979) J Mater Sci 14: 1511
90. Guerra G, Di Dino G, Centore R, Petraccone V, Obrzut J, Karasz FE, MacKnight WJ (1989) Makromol Chem 190: 2203
91. Yagi T (1979) Polymer J 11: 353
92. Yagi T, Tatemoto M (1979) Polymer J 11: 429
93. Yagi T, Tatemoto M, Sako J (1980) Polymer J 12: 209
94. Lovinger AJ (1983) Macromolecules 16: 1529
95. Lovinger AJ, Johnson GE, Bair HE, Anderson EW (1984) J Appl Phys 56: 2412
96. Natta G, Allegra G, Bassi IW, Sianesi D, Caporiccio G, Torti E (1965) J Polym Sci Part A 3: 4263
97. Weeks JJ, Sanchez IC, Eby RK, Poser CI (1980) Polymer 21: 325
98. Centore R, De Rosa C, Guerra G, Petraccone V, Corradini P, Villani V (1988) Eur Polym J 24: 445
99. Guerra G, Karasz FE, MacKnight WJ (1986) Macromolecules 19: 1935
100. Leonard C, Halary JL, Monnerie L, Broussoux D, Servet B, Micheron F (1983) Polym Commun 24: 110
101. Leonard C, Halary JL, Monnerie L (1987) Proc I Eur Symp on Polymer Blends, Strasbourg, 25–27 May, p 53
102. Leonard C, Halary JL, Monnerie L (1988) Macromolecules 21: 2988
103. Benedetti E, Pracella M, Galleschi F, D'Alessio A, Moggi G, Vergamini P, Aglietto M, Del Fanti N (1990) Proc III European Polymer Federation Symposium, Sorrento 1–5 October, p 302
104. Guerra G, Vitagliano VM, DeRosa C, Petraccone V, Corradini P, (1991) J Polym Sci Polym Phys Ed 29: 265
105. Guerra G, DeRosa C, Vitagliano VM, Petraccone V, Corradini P, Karasz FE, (1991) Polym Commun 32: 30
106. Reynolds NM, Savage JD, Hsu SL (1989) Macromolecules 22: 2867
107. Reynolds NM, Hsu SL (1990) Macromolecules 23: 3463
108. Kobayashi M, Nakaoki T, Ishihara N (1990) Macromolecules 23: 78
109. Vittoria V (1990) Polym Commun 31: 263
110. Guerra G, Musto P, Karasz FE, MacKnight WJ (1990) Makromol Chem 191: 2111

111. Axelson DE (1986) in: Komoroski (ed) High-resolution NMR spectroscopy of synthetic polymers in bulk. VCH, Weinheim, Chaps. 5 and 6
112. Tonelli AE (1989) NMR spectroscopy and polymer microstructure: the conformational connection. VCH, Weinheim
113. Bunn A, Cudby EA, Harris RK, Packer KJ, Say BJ (1981) Chem Commun 15
114. Grassi A, Longo P, Guerra G (1989) Makromol Chem Rapid Commun 10: 687
115. Gomez MA, Tonelli AE (1990) Macromolecules 23: 3385
116. Belfiore LA, Schilling FC, Tonelli AE, Lovinger AJ, Bovey FA (1984) Macromolecules 17: 2561
117. Bunn A, Cudby MEA, Harris R, Packer KJ, Say BJ (1982) Polymer 28: 2227
118. Gomez MA, Tanaka H, Tonelli AE (1987) Polymer 28: 2227
119. Saito S, Mateki Y, Nakagawa M, Horii F, Kitamaru R (1988) Polym Prepr 29: 1
120. Natta G, Pegoraro M, Szilàgyi (1967) Chim Ind 49: 1, 49: 7
121. Wang TT, Herbert JM, Glass AM (1987) The applications of ferroelectric polymers. Blackie, Glasgow
122. Guerra G, Vitagliano VM, Corradini P, Albizzati E (1989) It Pat 19588 (Himont Inc)
123. Vittoria V, Russo R, De Candia F (1989) J Macromol Sci Phys 28: 419
124. Rapacciuolo M, DeRosa C, Guerra G, Mensitieri G, Apicella A, Del Nobile MA, J Mater Sci Lett (in press)
125. Ahmed M (1982) Polypropylene fibers-Science and technology Elsevier, Amsterdam
126. Saraf RF, Porter RS (1988) J Polym Sci Polym Phys Ed 26: 1049
127. Asano T, Seto T (1973) Polymer J 5: 72
128. Ito M, Tanaka K, Kanamoto T (1987) J Polym Sci Polym Phys Ed 25: 2127
129. Ito M, Takahashi K, Kanamoto T (1990) Polymer 31: 58
130. Fakirov S, Evstatiev M (1990) Polymer 31: 431

Editor: P. Corradini
Received January 7, 1991

Polymer Crystallization Theories

K. Armitstead and G. Goldbeck-Wood,
with a Foreword by A. Keller
H. H. Wills Physics Laboratory, University of Bristol, Tyndall Avenue,
Bristol BS8 1TL, UK

Foreword

The growth of polymer crystals has a special position in the fields of Polymer Science and Crystal Growth. While, necessarily, it has its roots in both, the basic facts as they emerged experimentally, could not have been derived from or predicted by either discipline. The theories were set up subsequently in order to rationalize the apriori unsuspected observations in the course of which they kept drawing both on the traditional science of long chain molecules and on the basic principles of crystal growth of simple substances. To what extent of each, varies from one approach to another and is not always apparent from the individual papers. A treatise placing the subject in the wider perspectives of the underlying basic sciences has therefore long been desirable. I believe that the present review goes a long way towards this goal.

Also, over the decades theories kept being amended and variants proposed in the light of new facts which have kept continually emerging, perhaps more so than usual in most branches of science. As a result, following up the history has become a prohibitively arduous task. On the other hand delving midway into it could be a perplexing experience without familiarity with the preliminaries. Extraction of the essentials and their critical juxtaposition on the basis of original papers and partial reviews requires both extensive labour and thorough immersion in the subject. This is beyond what most otherwise interested readers can be expected to do. I believe that this has now been done for them by the present reviewers with the expertise required for such an undertaking.

The origin of this review may deserve telling. The authors, with backgrounds in statistical mechanics and crystal growth of simple substances respectively, came in contact with polymer crystallization research in our Bristol polymer laboratory. It is in preparation for active involvement by themselves that the present literature survey and evaluation was carried out. First, it was for their own information, then for internal purposes of the laboratory. Myself reading earlier versions felt that it was not only suitable for external presentation, but in fact it would be fulfilling a long and widely felt need in several respects, and encouraged preparation of a version appropriate for submission accordingly.

The needs referred to above include availability of an advanced teaching text, a "lead in" for intended researchers, reference text for the active researcher himself and informational reading both for specialists on crystallization outside the

Advances in Polymer Science, Vol. 100
© Springer-Verlag Berlin Heidelberg 1992

polymer field and for polymer scientists in other branches of polymer science. The construction of the review should allow the reader to select the level appropriate to his sphere and depth of interest.

There are two further features of the review which I feel are specially commendable: its intended impartiality, particularly welcome in a subject punctuated by controversies, and most importantly, open endedness, hopefully leaving the reader receptive to new developments yet to come.

Finally, I cannot resist adding two comments of my own as an experimentalist.

First, I wish to delineate the scope of the review within the broader canvas of polymer science. Namely, polymer crystallization is a much wider field than covered by treatises under that title. What usually passes under such a heading (including that of the present review) is confined to perfectly periodic, fully flexible and chemically unspecific chains. This is a physicist's idealized model system to which polyethylene happens to be the closest approximation. In real life, chains may have faults, or may be outright aperiodic, they have varying degrees of stiffness and specific, variable chemical affinities along their lengths. All these, important as they are in their own right, could be regarded as an overlap on the physicist's idealized system, the theoretical aspects of the latter being the subject of the present review.

Secondly, I wish to counteract anticipated despondency which some of the complexities on the present theoretical scene may perhaps provoke. For this purpose, I wish to invoke the decisive simplicity and definiteness of some of the experimental effects observed within the confines of the above, near ideal systems. This, as I often pointed out elsewhere, is unmatched in the field of crystal growth of simple substances. Complicated as polymers may seem, and subtle as some of the currently relevant theoretical issues, this should not obscure the essential simplicity and reproducibility of the core material. To be specific, the appropriate chains seem to "want" to fold and "know" when and how, and it is hardly possible to deflect them from it. Clearly, such "purposeful" drive towards a predetermined end state should continue to give encouragement to theorists for finding out why?! Those who are resolved to persevere or those who are newly setting out should find the present review a most welcome source and companion.

Bristol   A. Keller

This article reviews the various theories of polymer crystallization which have been proposed over the last thirty years. Their background in thermodynamics and crystal growth theories of simple substances is laid out in some detail. This provides the basis for a discussion of the application of surface nucleation and surface roughening concepts to polymer crystallization. The widely used nucleation model of Lauritzen and Hoffman is examined in depth with emphasis on its consistency with experimental evidence and including its recent modifications and extensions. Alternative approaches are also analyzed, in particular Point's 'multipath' model and the 'roughness/pinning' model by Sadler and Gilmer, but also 'molecular nucleation' and 'chain sliding'. The aim of this article is to lead to a better understanding of the basic hypotheses of the various models and to give guidance to the theorist in improving the models as well as to the experimentalist in interpreting his/her data. It will hopefully make the subject more accessible and stimulate discussion and further research directed at resolving the controversial issues.

List of Symbols

l	Lamellar thickness
l_{min}	Minimum stable thickness
δl	Thickness deviation $l - l_{min}$
A	Surface area of the fold surface
a	Width of a stem
b	Thickness of a stem
σ_e	Fold surface free energy
σ	Lateral surface free energy
ΔF	Bulk free energy of crystallization per unit volume
ΔG	Free energy difference between crystal and liquid per unit volume
ΔH	Bulk enthalpy of melting per unit volume
ΔS	Bulk entropy of melting per unit volume
T	Temperature
$T_m(\infty)$, T_m^0	Equilibrium melting point
ΔT	Supercooling $T_m(\infty) - T$
T_R	Roughening temperature
p	Number of monomer units in a chain
i	Nucleation rate per unit length
g	Step spreading rate
N^*	Critical nucleus size
N^s	Stable nucleus size
ΔF^*	Critical free energy difference
G	Growth rate
G_M	Growth rate in the mononucleation regime
$G_{I, II, III}$	Growth rate in regime I, II, III, respectively
L	Substrate length
L_p	Persistence length
L_k	Kinetic length
$r(x)$	Concentration of right moving steps at position x
$l(x)$	Concentration of left moving steps at position x
z	Lauritzen number $= L^2 i / 4g$
C	Concentration of polymer in solution
γ	Concentration exponent
i'	Addition rate per unit length of solute molecules at immobile steps
i''	Nucleation rate per unit length due to cilia
j	Attachment rate at a reentrant corner
A_0	Rate constant for adding a stem to a flat substrate
B_0	Rate constant for removal of an isolated stem
A_1	Rate constant for adding a stem next to an existing stem of the same length
B_1	Rate constant for removal of a stem next to an existing stem of the same length
A_2	Rate constant for adding a stem to a cavity
B_2	Rate constant for creating a cavity

S	Flux
β	'Transport term', i.e. the rate at which molecules arrive at the surface
ψ	'Apportioning factor': proportion of the bulk free energy released during stem deposition
h	Lateral growth rate of a sector
σ_s	Strain surface free energy
C_l	Net forward rate for folding at lamellar thickness l
ε	Pairwise nearest neighbour attractive energy
k^+	Rate constant for addition of a unit
$k^-(\alpha)$	Rate constant for removal of a unit from configuration α
$C_n(i)$	Concentration of stems of length i at position n
$P_n(i, j)$	Probability of configurations with thickness i at n and j at $n + 1$

1 Introduction

The subject of polymer crystallization has received great interest over several decades, and still provides a fascinating and fruitful area of research. The long chain nature of the molecules gives rise to highly complex and diverse growth forms, whose analysis and characterization provides a very challenging task to the experimentalist. The modern era of polymer crystallization began with the discovery [1–3] that solution grown single crystals are thin platelets, or lamellae, with the molecular backbone oriented along the thin dimension of the crystal. This led to the hypothesis by Keller [1] that the molecules are folded many times upon themselves in a regular fashion. The first theory aiming to explain the growth of these crystals was published three years later [4]. Since that time a vast literature has accumulated and many diverse opinions have been expressed. Many points of dispute remain unresolved and the discussion continues to this day.

The purpose of this article is to provide an overview of the current state of the sometimes controversial crystallization theories of polymer single crystals. It aims to give an understanding of the basic hypotheses of each theory, and therefore the statistical mechanical groundwork is often laid out in some detail. In assessing the validity of the various theories these underlying assumptions must be fully appreciated, as all the theories lead to some degree of agreement with experiment. It is hoped that this type of summary will assist a direct comparison of the different approaches in a logical and unbiased fashion and be a basis for further research in this field which will hopefully lead to wider agreement as to what governs the growth of polymer crystals.

We have attempted to cover all aspects of the field, trying to display its richness of thoughts and models, as well as its still considerable shortcomings. However, it is not possible to detail all developments and we apologize for any omissions which others may have included.

The layout of this article is as follows: Section 2 considers the equilibrium aspects of the crystals whilst Sects. 3 and 4 explain the growth theories, divided into 'nucleation' and 'non-nucleation' theories, respectively. Finally, Sect. 5 provides an overview and suggests future lines of investigation.

2 Foundations of Polymer Crystallization Theory

2.1 General Outline

This chapter serves as a basis for all the growth theories which will be described in subsequent chapters. Although the theories make varying assumptions about the nature of the growth process, they also have several common features and necessarily share many problems in the determination of the physical parameters. We aim to emphasize the simplifications and uncertainties underlying all the approaches which are often passed over, apparently unnoticed. However, if their accumulated errors are sufficiently large, the later discrepancies arising from the differences between the growth theories become somewhat academic.

In the next section we describe a very simple model, which we shall term 'the crystalline model', which is taken to represent the real, complicated crystal. Some additional, more physical, properties are included in the later calculations of the well-established theories (see Sect. 3.6 and 3.7.2), however, they are treated as perturbations about this basic model, and depend upon its being a good first approximation. Then, Sect. 2.1 deals with the information which one would hope to obtain from equilibrium crystals – this includes bulk and surface properties and their relationship to a crystal's melting temperature. Even here, using only thermodynamic arguments, there is no common line of approach to the interpretation of the data, yet this fundamental problem does not appear to have received the attention it warrants. The concluding section of this chapter summarizes and contrasts some further assumptions made about the model, which then lead to the various growth theories. The details of the way in which these assumptions are applied will be dealt with in Sects. 3 and 4.

2.2 The Crystalline Model

The morphology of polymer crystals has been studied extensively by many experimentalists, and an enormous variety and richness has been unveiled (see e.g. Ref. [5]). It is beyond the scope of this review to lay out the evidence in detail. We can only concentrate on some selected points and refer the interested reader to the literature, in particular the review articles by Keller [6], Bassett [7], Dosière [8] and most recently Phillips [9].

In general, polymer single crystals take the form of thin lamellae, which are large in two dimensions but are bounded in the third by the folds which constitute the basal plane. The crystals may occur either in isolation or as an aggregate with other, similar, crystals leading to the formation of 'mats' or 'bundles'. Although the general structure of a single lamella may be easily described the resulting morphology is far from simple and includes rhombic and hexagonal shaped crystals, laths and ribbons with many variations in-between – the variations obtained depending both on the particular polymer and on the choice of growth conditions. The appearance of the aggregates then also depends upon the interactions between the lamellae due to cross-links, entanglements, etc. However, the fact that growth rate and thickness measurements can in general both be described in terms of functions of temperature and supercooling which are common to many crystals, implies that the controlling factors are the same in all cases, and that the morphological aspects are not rate determining in general. In some samples, however, different growth rates are observed for different shapes under the same crystallization conditions [10]. A model which is thought to incorporate all the necessary features is shown in Fig. 2.1 – we call this the *crystalline model*, and it is the starting point for any microscopic theory.

Obviously this model is very simplified, compared with the real crystal which contains many defects, dislocations and entanglements. In particular, it neglects many aspects of the true three-dimensional nature of the lamella which one may have thought to be important: the influence of the stacking of folds, which is

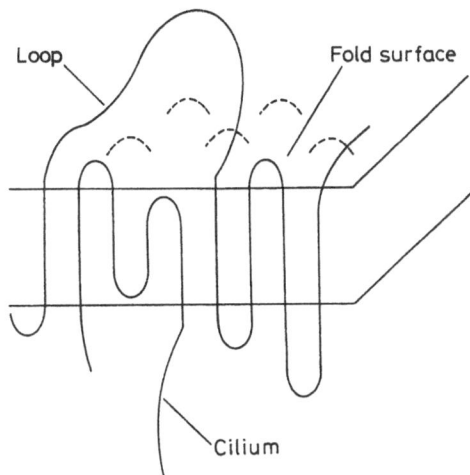

Loop   Fold surface

Cilium

Fig. 2.1. The 'crystalline model': a single crystal in which molecules traverse the lamella perpendicular to the fold surface. Cilia are formed at the end of the molecules outside the crystalline core. The folds are predominantly adjacent and the loop sizes may vary

commonly believed to lead to the pyramidal shapes [6], is lost and the molecules are assumed to traverse the lamella perpendicular to the fold surface. These factors may influence any quantitative predictions through the physical parameters, however they must not be the fundamental reason for growth via chain folding in order for the crystalline model to be valid.

Another important assumption made in this model is that the predominant type of folding involves reentry of an emergent molecule into an adjacent position within the crystal. If the effects of 'loops' or 'cilia', as indicated in Fig. 2.1, are considered at all it is as a correction factor to the predominantly adjacent re-entry growth. This issue was first addressed by Frank [11, 12]. Hoffman [13] described the recent evidence supporting this view, arguing that random re-entry (the 'switchboard model', [14]) necessarily leads to an unphysical density increase outside the crystalline core. A Monte Carlo study of chain folding [15] showed that adjacent and next-nearest neighbour re-entry always dominate. Neutron scattering [16–18] and infrared [19] studies came to a similar conclusion. We shall continue to assume that the evidence is strongly in favour of predominantly adjacent re-entry growth; however, the amount of non-adjacent re-entry and the possibility of a random model should not be forgotten (Disc. Faraday Soc. 68, 1979).

The crystalline model makes no presuppositions about the direction of the folds with respect to either the crystal lattice or to the morphology. Clearly, knowledge of any general pattern in the type of folding, either random or in a preferred direction, could be incorporated into the basic model and used within a theory. If no such pattern exists then, in order to be generally applicable, a theory should be compatible with any type of folding. Experiments on facetted crystals indicate that the folds lie parallel to the growth face [6, 20] and a resulting diamond-shaped polyethylene crystal which has a (110) growth face is shown schematically in Fig. 2.2. Recent experiments by Wittman and Lotz [21] using decoration techniques also demonstrate that there is frequently a tendency for the folds to lie

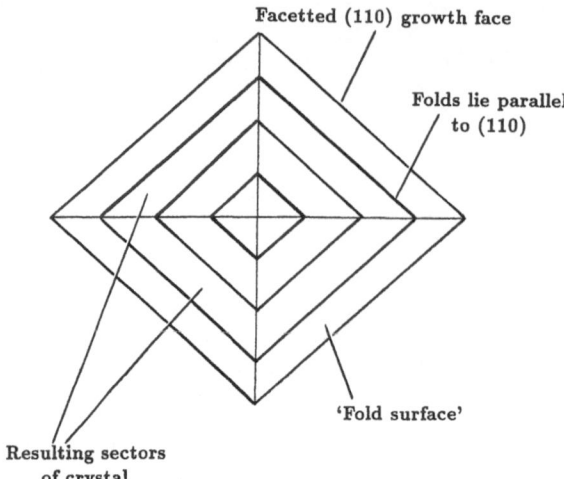

Fig. 2.2. A diamond-shaped crystal showing four sectors and the folds lying parallel to the (110) growth faces

approximately parallel to the local growth front for curved faces of polyethylene and the straight faces of lath-shaped polypropylene. However, there is also evidence of folding perpendicular to the lamellar edge [22], and in general the fold direction in more complicated crystals has not yet been determined. Therefore, as the present evidence is insufficient to support a common type of folding, there should be no unnecessary restrictions in the theories.

Having briefly discussed the way in which the important features of a single crystal are incorporated into a simple model we next consider what information can be obtained on this single crystal from equilibrium thermodynamics.

2.3 Equilibrium Thermodynamics

A basic requirement of all the theories is the evaluation of the difference in Gibbs free energy between a lamella of average thickness l and the uncrystallized polymer, either in the melt or in solution. This is related to the driving force for crystallization and establishes relationships between the physical parameters, such as temperature and thickness, which must be obeyed if a lamella is to be stable compared with the liquid state. Bulk thermodynamic potentials, such as the Gibbs free energy, are only well defined in the thermodynamic limit, that is, for an infinite system so that there are no surface effects. This poses problems when dealing with polymers where finite size immediately enters the problem via the lamellar thickness and the length of the molecules. Unfortunately, a consistent approach has not been developed and most authors work within one framework, apparently unaware that discrepancies may exist. This problem has been recognized by Buckley and Kovacs [23], who discuss the subject in some depth. It would be interesting to analyse experimental data in several ways as an estimate of the errors which may be incurred. Our objective here is to contrast the different 'definitions' which have been used, and to encourage due discussion of the subject. It will also serve to define notation which will be used later in this work and which has been used somewhat ambiguously in the literature.

The Gibbs free energy for a crystal is usually divided into a 'bulk' and a 'surface' contribution. The former is the free energy the same volume of crystal would have in the absence of any surfaces, that is, if it were part of an infinite crystal. The latter may be considered as the increase in free energy over the stable bulk phase due to the presence of the surface — a rigorous definition of these quantities was first expressed by Gibbs [24]. The surface tension, defined as the work required to increase the surface by unit area, is equivalent to the excess surface free energy for a one-component system with no foreign adsorption (see e.g. Ref. [25]), which is our main concern. The total surface free energy of an arbitrarily shaped crystal may be found by integrating the directionally dependent surface tension over the surface area. Let the surface free energy per unit area of the fold surface of a lamellar crystal be σ_e, then the increase in Gibbs free energy due to a single lamella, of average thickness l, immersed in the surrounding liquid may be written:

$$\Delta G = \Sigma_{lat} + 2A\sigma_e - \Delta F \, Al, \tag{2.1}$$

where Σ_{lat} represents the contribution to the free energy from both the thin, lateral surfaces and any edges bounding the crystal, A is the surface area of the fold surface ($2A$ is the total area of both fold planes), and ΔF is the bulk free energy of crystallization per unit volume.

The most stable state minimizes the free energy, that is G is most negative there. If the surface area of the fold plane is very large we may neglect the first term of Eq. (2.1) compared with the remaining terms leading to:

$$\Delta G = 2A\sigma_e - \Delta F \, Al. \tag{2.2}$$

At a temperature such that the solid phase is stable ΔF is necessarily positive and the stable crystal has infinite thickness. However, any crystal which has $\Delta G < 0$ will be stable compared with the liquid, so that a crystal of finite thickness may be metastable if there is a free energy barrier to the formation of an infinite crystal. $\Delta G < 0$ if:

$$l > 2\sigma_e/\Delta F . \tag{2.3}$$

The lower limit of this inequality is the minimum possible thickness l_{min} of a lamella for a given ΔF. Equivalently, a crystal of thickness l will be at its melting point when Eq. (2.3) is satisfied as an equality. It is common to express the average thickness as:

$$l = l_{min} + \delta l . \tag{2.4}$$

If $\delta l/l$ is small and ΔF is a known function of temperature, Eq. (2.3) can be used to determine the melting point and surface free energy of a lamella from experimental data. For long chain molecules there are several difficulties in choosing the relevant form for ΔF. Therefore we first consider the limiting case of infinite molecular weight, where the effects of chain ends may be neglected. For infinite molecular weight:

$$\Delta F[T] = \Delta H[T] - T \, \Delta S[T] \,, \tag{2.5}$$

where $\Delta H[T]$ and $\Delta S[T]$ are the bulk enthalpy and entropy of melting per unit volume, with the temperature dependence explicitly displayed. These bulk values are defined for an infinitely large crystal in which all surface effects are negligible. At the equilibrium melting point, $T_m(\infty)$:

$$\Delta F[T_m(\infty)] = \Delta H[T_m(\infty)] - T_m(\infty) \, \Delta S[T_m(\infty)] = 0 \,. \tag{2.6}$$

Therefore

$$\Delta S[T_m(\infty)] = \Delta H[T_m(\infty)]/T_m(\infty) \,. \tag{2.7}$$

As a first approximation assume that $\Delta H[T]$ and $\Delta S[T]$ do not vary in the vicinity of $T_m(\infty)$, − better approximations may be used [26, 27] − so that for $T \simeq T_m(\infty)$:

$$\Delta F[T] = \Delta H[T_m(\infty)] - T \, \Delta S[T_m(\infty)] \tag{2.8}$$

$$= \Delta H[T_m(\infty)] \, (T_m(\infty) - T)/T_m(\infty) \tag{2.9}$$

$$= \Delta H[T_m(\infty)] \, \Delta T/T_m(\infty) \,, \tag{2.10}$$

where $\Delta T = (T_m(\infty) - T)$ is known as the supercooling. Substitution of Eq. (2.10) into Eq. (2.3) gives:

$$l > \frac{2\sigma_e T_m(\infty)}{\Delta H[T_m(\infty)] \, \Delta T} \tag{2.11}$$

which shows that the minimum thickness of a lamella varies inversely with the supercooling. Equivalently, we may write Eq. (2.11) in terms of a condition which must be obeyed by the temperature, T, given a lamella of thickness l:

$$T < T_m(\infty) \left\{ 1 - \frac{2\sigma_e}{l \, \Delta H[T_m(\infty)]} \right\} \,. \tag{2.12}$$

The upper limit of this inequality gives the melting point of the lamella, $T_m(l)$. Equations (2.11) and (2.12) are displayed graphically in Fig. 2.3.

Experimentally it is found that finite molecular weight effects are negligible at medium to high molecular weights, that is, the melting point of a lamella depends only upon its thickness and not upon the molecular weight of the sample. In these cases Eq. (2.12) may be used to obtain the equilibrium melting point (that is, the melting point for an infinite crystal) from the intercept of a plot of $T_m(l)$ against $1/l$, provided that $\delta l/l$ is small. The gradient of this plot may then be used to obtain σ_e if $\Delta H[T_m(\infty)]$ is known. There are many difficulties in achieving such a plot experimentally, which lie outside the scope of this work (see e.g. Refs. [28, 9]).

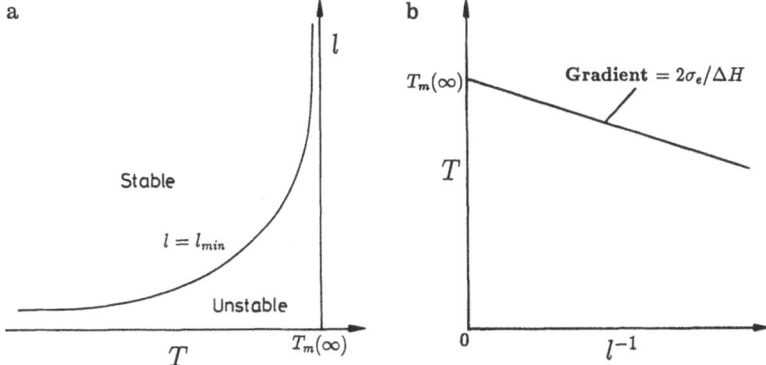

Fig. 2.3a, b. Graphical representation of **a** the stable thickness of a lamella as a function of temperature T; **b** The temperature above which a lamella of a given thickness would be unstable. Other notation in the Figure is defined in the text

A good example of the precautions which need to be taken is given by Leung et al. [29, 30].

We now discuss the effects of finite chain length. The difficulties arise from the definition of a 'bulk' free energy term, when the very nature of the chains constrains the crystal thickness to be finite. There are two different approaches to this problem: the first to be considered is due to Hoffman et al. [31] and is a simple modification of the infinite chain case, but is somewhat lacking in theoretical justification; the second, due to Buckley and Kovacs [23], aims to correct this deficiency and suggests that the interpretation of experimental data given by Hoffman's approach is misleading.

For a chain consisting of p monomer units the most stable crystal, as argued after Eq. (2.1), is that which maximizes the thickness, that is, all chains lie in an extended fashion side-by-side with no chain folds and no non-crystalline component. The melting point of such a lamella will be denoted $T_m(0, p)$, where the '0' indicates that there are no folds at the surface of the crystal. It is less than $T_m(\infty)$ because of the increase of free energy associated with the surface. The total free energy difference between the crystal and its surroundings must be zero at its melting point, giving:

$$\Delta H[T_m(0, p)] - T_m(0, p)\,\Delta S[T_m(0, p)] = 0 \,. \tag{2.13}$$

This is an expression for the overall enthalpy and entropy divided by the volume of the complete lamella and is strictly correct. However, because the total free energy difference is calculated, the effects of the unfolded chain ends lying at the surface are implicitly included in $\Delta H[T_m(0, p)]$ and $\Delta S[T_m(0, p)]$ and it is therefore misleading to consider these as bulk 'per volume' quantities. The proportion due to the contribution from the surfaces will be greatest for thinner crystals, that is for lower molecular weight.

Hoffman takes Eq. (2.13) as the definition of bulk free energy per unit volume at the melting point, neglecting the fact that a surface contribution is also included.

Therefore assuming as before that ΔH and ΔS do not vary significantly in the vicinity of the melting point he obtains:

$$\Delta F_H = \Delta H[T_m(0, p)] - T \,\Delta S[T_m(0, p)] \tag{2.14}$$

$$= \Delta H[T_m(0, p)] \,(T_m(0, p) - T)/T_m(0, p) \,. \tag{2.15}$$

This is the same as Eq. (2.9) with $T_m(\infty)$ replaced by $T_m(0, p)$. This expression is than substituted into Eq. (2.3), to give:

$$l > \frac{2\sigma'_e T_m(0, p)}{\Delta H[T_m(0, p)] \,(T_m(0, p) - T)}, \tag{2.16}$$

$$T < T_m(0, p)\left\{1 - \frac{2\sigma'_e}{l \,\Delta H[T_m(0, p)]}\right\}. \tag{2.17}$$

Hoffman assumes that σ'_e has the same interpretation as for infinite chain length, that is the surface tension of the fold surface. However, as pointed out above, effects of a non-folded surface are already included in $\Delta H[T_m(0, p)]$ and $\Delta S[T_m(0, p)]$, and at best σ'_e could be regarded as the contribution to the surface tension from just the folds, but more realistically as a parameter which is related to the surface tension but which also varies with the thickness of the lamella, that is as the proportion of the number of folds to free ends in the surface changes.

$T_m(0, p)$ may be estimated using the Broadhurst or Flory-Vrij equation [31], by directly measuring the melting point of extended chain crystals [29], or by measuring the intercept of $T_m(l)$ against $1/l$ using Eq. (2.17). These methods give consistent results, however it should be noted that there is no sound theoretical justification for the last method due to the variation of σ_e with thickness, although a straight line may be obtained if the number of folds does not vary appreciably with l. The interpretation of the gradient of such a line, however, would be very suspect.

A different approach [23, 32] considers ΔF in Eq. (2.2) to consist of a part which must be defined with respect to infinite chain length plus an entropy of localization [33] due to the pairing of chain ends which becomes important in the case of closely stacked lamellar crystals [34]. It amounts to $-K \ln (Cp)$ per molecule, where C is a constant related to the flexibility of the chains in the melt, and which arises due to the conformations of the finite chain. Hence:

$$\Delta F(T) = \Delta H[T_m(\infty)] - T \,\Delta S[T_m(\infty)] - kT \,\frac{\ln (Cp)}{pv}, \tag{2.18}$$

where v is the volume of a monomer unit. Note, however, that the applicability of the entropy of localization is still an open question [35]. It leads, for example, to an overestimation of $T_m(\infty)$ for polyethylene [28]. As enthalpy and entropy are defined for an infinite crystal, Eq. (2.7) may be used to eliminate $\Delta S[T_m(\infty)]$ in terms of $\Delta H[T_m(\infty)]$. Equation (2.18) does not include any effects from surface tension, therefore Eqs. (2.2) and (2.3) may be used with σ_e as the total surface tension from the fold surface, although again this must still vary as the relative number of folds and free ends changes. The resulting expression for T in terms of l is cumbersome, but may be simplified for small ln (Cp) to give [23]:

$$T \le T_m(\infty) \left\{ \left(1 - \frac{2\sigma_e}{l\,\Delta H[T_m(\infty)]} \right) - \frac{RT_m(\infty)\ln (p)}{p\,\Delta H[T_m(\infty)]} \right\}. \qquad (2.19)$$

A plot of $T_m(l)$ against $1/l$ would give a straight line with an intercept $T_m(0, p)$ given by the Flory-Vrij equation and a gradient having a different interpretation than that from Eq. (2.17).

This discrepancy in definition of the melting point has not often been high-lighted, yet it is of considerable importance bearing in mind the accuracy claimed for the deduction of σ_e.

2.4 Growth Theories

2.4.1 Equilibrium Theory

Having discussed some equilibrium properties of a crystal, we now outline and contrast the bases of the growth theories which will be dealt with in more detail later. The theories may be broadly split into two categories: equilibrium and kinetic. The former [36–42] explain some features of the lamellar thickness, however the intrinsic folding habit is not accounted for. Therefore, at best, the theory must be considered to be incomplete, and today is usually completely ignored. We give a brief summary of the approach and refer the interested reader to the original articles. The kinetic theories will be the topic of the remainder of the review.

The equilibrium theory calculates the free energy of a crystal as a function of segment length and assumes that the crystal will be in a minimum free energy state. The chain undergoes linear vibrations and torsional oscillations which increase in amplitude with the temperature and with the chain length. If the molecular motions are incoherent there will be a significant contribution to the free energy, in contrast to low molecular weight material which does not exhibit this chain-like behaviour. They predict, as the result of calculations, that below a certain critical temperature there are two minima in the free energy — one at a finite thickness and one at infinite thickness. The segment length corresponding to finite thickness increases with temperature up to the critical temperature, above which only an infinite crystal is stable. Hence, below the critical temperature the crystal is predicted to reach the smaller minimum, whilst at higher temperatures it would thicken without limit. This certainly explains annealing experiments and crystal thickening, however the existence of the folds is not explicitly included and the strong correlation between thickness and supercooling is not explained.

At this point a third intermediate approach deserves mentioning. It is due to Allegra [43] who proposed that polymer crystallization is controlled by a metastable equilibrium distribution of intramolecular clusters, the so-called 'bundles', forming in the liquid phase. These subsequently aggregate to the side surfaces of the crystals, driven by van der Waals interactions. The lamellar thickness is determined by the average contour length of the loops within the bundles. Although the model can claim some reasonable agreement with observed thickness data and offers an explanation for the segregation of short chains (see also Sect. 3.8.1), there is no strong evidence for the existence of 'bundles', and it admits to be incomplete because growth kinetics are not addressed.

In conclusion, it seems more likely that the folds go into the crystal sequentially as it is being formed, rather than forming through equilibrium considerations — this idea is used as a basis for kinetic theories.

2.4.2 Kinetic Theory

There is now wide agreement that the morphology of polymer crystals is determined by kinetic rather than equilibrium factors. Although in principle a crystal would equilibrate given enough time, the free energy barriers are so great that the time required is effectively infinite. The growth faces which are observed are in the slowest growing directions — the fast growing fronts will initially grow out until the crystal is bounded by the slowest growing fronts. The kinetic theories assume that the observed growth faces would have a range of possible thicknesses, each having a growth rate which depends on thickness. The preferred thickness for the crystal is that which maximizes the growth rate — hence the fundamentally kinetic origin of the model.

In any theory there must be some 'driving force' which is causing crystallization and is related to the supercooling: the greater the supercooling the more the supercooled liquid wants to crystallize. If we imagine a crystal of specific thickness, l, then the driving force will also depend on that thickness. For $l < l_{min}$ there will be no driving force forward as by definition this would lead to an overall free energy increase, but for $l > l_{min}$ the driving force will be positive and increases with l. At first it increases rapidly as the bulk free energy gain grows compared with the effect of the surfaces. At higher l the driving force flattens out as the effect of the surface energy as a proportion of the total becomes smaller. The expected general form of the driving force is shown in Fig. 2.4.

If this were the only force acting on the crystallization process then the fastest growth rate would correspond to a crystal having infinite thickness. This is inhibited by a 'barrier' term which disfavours a thick crystal. The origin of this barrier is the starting point for different kinetic theories. The so-called 'nucleation' theories invoke an energy barrier, which must be overcome via a random fluctuation as a molecule attaches itself to the crystal, and which increases with thickness. The more recent 'entropic' theories [44] argue in favour of a configurational degeneracy: a molecule explores many possible paths during crystallization, only a few of which are favourable to further growth. The crystal growth may be prevented from continuing due to an incorrectly placed molecule — that is a 'wrong' configuration. The number of wrong configurations increases with the thickness of the crystal, and hence the barrier also increases. Whatever the cause of the barrier its general form is shown in Fig. 2.4.

The driving force and barrier terms together give a growth rate which is shown by the dashed curve in Fig. 2.4. The thickness corresponding to the maximum growth rate is slightly above l_{min}, so it is the underlying dependence of l_{min} on ΔT

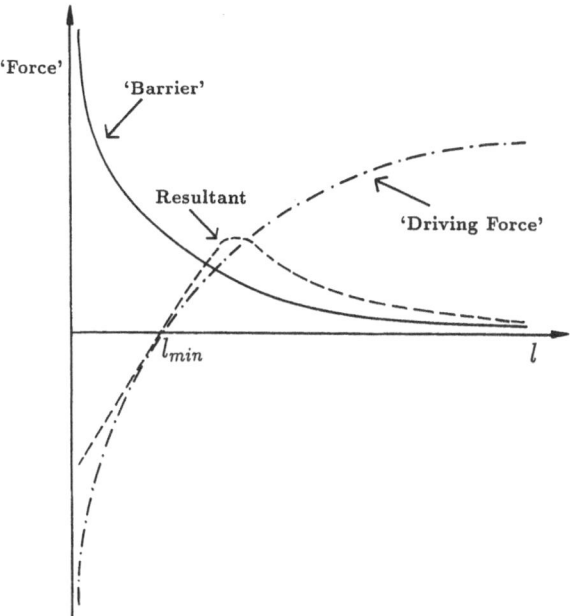

Fig. 2.4. The resultant growth rate which is due to a barrier and a driving force term. The driving force increases with l, but at a slower rate than the decrease of the force due to the barrier term. Hence the overall growth rate is positive for $l > l_{min}$ but decreases to zero at large l, with a maximum in between

which leads to the thickness dependence on ΔT. The absolute value of the growth rate depends upon the details of the theory. However, if the barrier terms from both theories contain the same dependence on l then the resulting dependence on temperature will also be the same. It is therefore important in assessing their validity to be aware of any other implications of each model and to test them against observations. The nucleation models make quantitative predictions for given input parameters which are in agreement with experimental work, however one should always be aware of the range of predictions which could have been obtained with different input parameters. On the other hand, the entropic theory is primarily qualitative, relying on computer simulations to provide evidence of the correct type of behaviour. This is a drawback for its establishment as a theory as there are no predictions for comparison with experimental data which help to substantiate the argument. However, a lack of mathematical tractability is certainly not a good reason for its rejection.

3 Nucleation Theories

3.1 General Outline

Most of the kinetic theories to date have invoked nucleation as the primary barrier to growth. However, even within this area there are different types of nucleation processes which are claimed to be the most important by various authors. This section aims to cover all the nucleation theories in a logical manner and therefore necessarily contains several sections and subsections.

Section 3.2 is an introduction to general nucleation behaviour as first used for low molecular weight materials, that is, non-polymeric substances. This topic is frequently not covered in polymer papers but the extensive work done in this area provides the framework for the theory, and indicates the limitations and problems to be expected in predicting the physical behaviour.

Section 3.3 outlines how this theory has been applied to long chains, defining a 'nucleation rate', i, and a 'filling-in rate', g. A great deal of information on the growth behaviour pertaining in this model may be obtained without specific expressions for i and g [45, 46], and it is shown that there are three predicted regimes of growth, depending on the relative magnitudes of these quantities and the size of the crystal. Section 3.4 contains further considerations of real systems and examines whether nucleation is capable of explaining observations in a self-consistent manner. (Included in this section are concentration effects, molecular weight effects and measurements on twins.)

The explicit forms for i and g are next discussed, which is the point at which the nucleation theories diverge. Section 3.5.1 lays out a simple form of the most widely used theory, pioneered by Lauritzen and Hoffman [4] and modified several times later [47–50]. Although this has often been reviewed elsewhere we include it here both for completeness and because we wish to concentrate on its most basic aspect. At this stage it is described with little comment so that the content of the theory is clear, and in Sect. 3.5.2 we work through the various stages explaining, with either justification or criticism, the assumptions which are involved. The ways in which the results agree or disagree with experiments are discussed in Sect. 3.5.3 and the refinements are described in Sect. 3.6.

We then turn to a more recent approach to the determination of i by Point [51, 52] in Sect. 3.7 explaining how it differs from that of Lauritzen and Hoffman (LH). Section 3.8 covers other proposed nucleation models and we conclude with an overview of nucleation theories and their successes and most notable shortcomings.

3.2 Surface Nucleation Theories

Growth theories of surfaces have received considerable attention over the last sixty years as summarized by Laudise et al. [53] and Jackson [54]. The well-known model of the crystal surface incorporating adatoms, ledges and kinks was first introduced by Kossel [55] and Stranski [56]. Becker and Döring [57] calculated the rates of nucleation of new layers of atoms, and Papapetrou [58] investigated dendritic crystallization.

The processes involved for low molecular weights are diverse and are determined by the material, the orientation of the face, the temperature of the face and the medium from which it is growing. All of the processes will have some effect on the growth rate, however, if any one inhibits growth more than the others then this will be the rate determining factor. These processes can be divided into two categories which characterize the type of dynamics: interface and diffusion controlled growth. In diffusion control the rate determining factor is the transport of mass or heat to or from the moving interface: for example, the rate at which material can be supplied to an interface may limit the growth. For interface control it is the actual processes of attaching and detaching molecules at the surface which determines the growth rate. The 'availability' of the molecules is then treated as unimportant, and a microscopic treatment of the reaction rates is required.

Diffusional processes can be rate limiting in a range of growth processes like crystallization directly from the vapour phase or from a supersaturated solution and solid-state precipitation, but they are most prominent in melt growth. They can be treated via diffusion equations whose solutions give possible instabilities in the interface structure [59, 60] which often lead to the development of dendrites [61–63]. On the other hand, interface control is more likely to be dominant in solution and vapour growth. 'Nucleation' is an important factor in these cases, although there are also many others. In particular, spiral growth at screw dislocations and surface roughening are often the rate determining mechanisms [64, 65].

Interface controlled dynamics, in particular nucleation, have been the primary focus in describing polymer crystal growth. Certainly from solution growth, where the physical difficulties in attaching a long molecule to the surface are much greater than for low molecular weight materials, this seems quite justified. For melt growth it is not at all obvious whether interface effects are able to dominate diffusion effects which are important for low molecular weights. The argument that interface kinetics are indeed rate determining relies on the experimental observation that in the melt the growth rate and the lamellar thickness [66] depend on supercooling in the same manner as in solution growth, and therefore the controlling processes should also be the same. Even if this is a good approximation, we should still expect transport effects to seriously modify the predictions. However, it is only recently that diffusion theories have been applied to polymer melt growth [67–69], and much more work is required in this field. We continue to consider nucleation processes, but emphasize that its bare predictions should only be used for solution growth, and due consideration of diffusion effects are needed before it can also be applied to melt growth (see Sect. 3.6.2 and Disc. Faraday Soc. 68, 1979).

Faces of a crystal can be classified in terms of their equilibrium structure being 'smooth' or 'rough' on an atomic scale. A smooth face is almost planar with few 'steps' or 'vacancies' as shown in Fig. 3.1a. At higher temperatures a smooth face may, though not necessarily, become rough. A typical configuration is shown in Fig. 3.1b. The transition between these two interface profiles is known as the roughening transition occurring at temperature T_R. This phenomenon was first conjectured by Burton, Cabrera and Frank [65]. It can be rigorously defined as occurring when either the step free energy or the height correlation length becomes zero [70, 71]. However, it is most easily considered as the temperature at which the entropic gain from a fluctuation becomes greater than its associated loss in energy. Notice that by its rigorous definition a rough interface does not have to

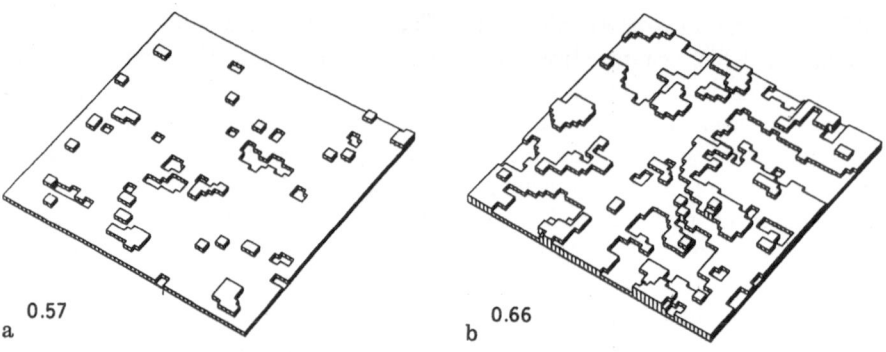

Fig. 3.1 a, b. Interface profiles for simple low molecular weight materials predicted using computer simulation. **a** A 'smooth' surface with few steps or vacancies. **b** A 'rough' surface. Values of the energy of breaking a bond are given in units of kT. (from [163], Copyright 1980 by the AAAS.)

be rough on an atomic level as shown in Fig. 3.1 b, but could consist of largely planar areas separated by long steps. However, computer simulations indicate that this is not normally the case.

The equilibrium structure of a particular face orientation depends upon the strength of the interactions between the molecules and upon the absolute temperature. Various schemes exist [60] to find the equilibrium structure for simplified models. The general approach is to consider the surface as a flat plane and to add onto the plane a number, 'N', of extra atoms. The energy of the plane plus the extra atoms can be calculated and then minimized with respect to N — this gives a measure of the stable surface structure. At low temperatures the minimum occurs for small values of N, but as the temperature increases towards T_R the minimum moves sharply to N equal to half the number of atoms in the plane. For $T > T_R$ there are many fluctuations on the interface and it is very easy for a molecule to attach anywhere. For $T < T_R$ the face is essentially flat and there is a large energetic barrier against the addition of a molecule because of the associated large increase in surface area. Specific predictions for T_R may only be made for substances with particularly simple interactions such as metals, where very reasonable estimates are obtained for T_R. However, little theoretical progress has been made for more complicated structures and, for example, it is not known either experimentally or theoretically whether a roughening transition exists for ionic solids which have long-ranged Coulomb forces.

The equilibrium structure of an interface has a strong influence on the growth kinetics — first we consider growth on a flat surface. At a given supercooling, ΔT, the bulk crystalline phase is stable. However, if an isolated molecule attaches onto a smooth surface, the increase in energy associated with its exposed sides is greater than the decrease in free energy due to the bulk terms, leading to an overall increase in free energy of the system. The molecule is therefore most likely to return to the fluid. Groups of molecules form on the surface due to random fluctuations — the free energy change for any particular grouping will depend both on the number of molecules (proportional to the decrease in bulk free energy) and on

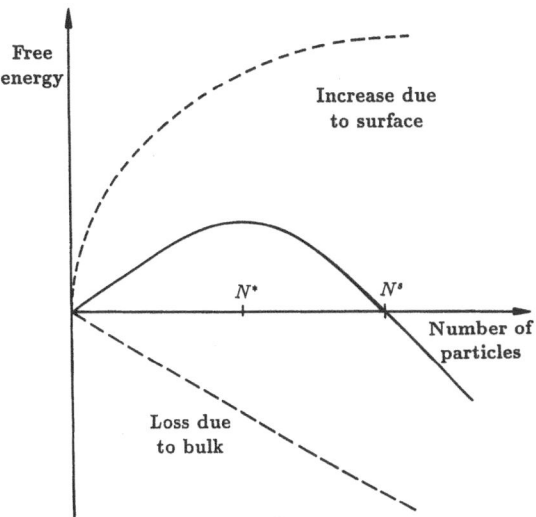

Fig. 3.2. Variation of free energy of a cluster of molecules on a surface as a function of the number of molecules, at a fixed supercooling. N^* is the number of molecules at which the free energy is a maximum. Any cluster larger than N^s is stable. The dashed curves show the contributions from the increase in surface area and the decrease in bulk free energy. Increasing the supercooling shifts all curves towards the origin and decreases the height of the maximum

the shape of the group (related to the increase in surface area). Figure 3.2 shows how the free energy change typically varies with the number of particles, N, arranged as a 'cluster', that is the molecules are arranged such that the increase in surface area for a given N is minimized. Any nucleus of size less than N^* is unstable and is more likely to be removed than to continue growing. For nuclear size greater than N^* an increase in N leads to a decrease in the free energy and fluctuations are most likely to increase the size of the nucleus. However, the nucleus only becomes stable when N reaches N^s and the total change in free energy is negative. The nucleus at N^* is known as the critical nucleus and occurs with probability proportional to $\exp(-\Delta F^*/kT)$, where ΔF^* is the free energy change associated with the formation of an N^*-nucleus. When such a nucleus has formed it is assumed to grow across the surface with a constant velocity v. This model of layer growth is the secondary nucleation model: an island of particles is deposited on a face via a random fluctuation with a certain probability. Having formed, the island grows laterally across the face at a net rate v which is greater than the forward advance of the face.

The most difficult part of the theory lies in obtaining actual values for ΔF^* and v. For a large cluster of N molecules the extra surface tension due to the incremental surface area, σdA, contributes an increase to the total free energy, whilst the bulk free energy per volume summed over the incremental volume, $\Delta F\, dV$, gives a decrease to the total free energy. Hence, ΔF^* can be estimated as the maximum value of $\sigma dA - \Delta F\, dV$ as a function of N. It is found that ΔF^* is proportional to $1/\Delta F$ and hence in general, from Eq. (2.10), to $1/\Delta T$. It is important to realize that the use of a well-defined σ is only valid for a large number of molecules in the nucleus: surface tension used in this way is a macroscopic concept which sums over the most probable, degenerate fluctuations of a surface. For small nuclei all configurations of molecules with their associated increase in energy must be considered separately [72].

The spreading rate v is taken to be the difference between the rate of arrival and departure of molecules at the crystal surface:

$$v = R_A^0 \exp\left(-Q_A/kT\right) - R_D^0 \exp\left(-Q_D/kT\right), \tag{3.1}$$

where Q_A and Q_D are the activation energies for each process, $Q_D - Q_A$ is the latent heat of fusion, L, and R_A^0, R_D^0 are constants assumed to be independent of temperature. At equilibrium, $T = T_E$, there is no net growth, that is $v = 0$, hence:

$$R_A^0/R_D^0 = \exp\left(-L/kT_E\right) \tag{3.2}$$

leading to

$$v = R_A^0[\exp\left(-Q_A/kT\right)]\left[1 - \exp\left(-L\,\Delta T/kT_E T\right)\right]. \tag{3.3}$$

At small ΔT

$$v \simeq R_A^0 \frac{L\,\Delta T}{kT_E T} \exp\left(-Q_A/kT\right). \tag{3.4}$$

The uncertainty in estimating the numerical factors, which depend on the properties of liquid, fluid and crystal in bulk, as well as the modifications an interface may make to these values, means that a wide range of values are all consistent with the theory.

The overall growth rate is given by a suitable combination of v and the nucleation rate $J \propto \exp\left(-\Delta F^*/kT\right)$. The competition between nucleation and spreading leads to the prediction of two growth regimes [73]. If the spreading is relatively fast the growth rate is limited by nucleation, and $G \propto J$. Otherwise more than one nucleus exists on the growth face at any time and the growth rate depends on both terms: $G \propto (v^2 J)^{1/3}$.

The strongest dependence on temperature comes from the nucleation factor, proportional to $\exp\left(-1/\Delta T T\right)$, and predicts negligible growth rate at small supercooling. The finite growth rates seen experimentally are due to screw dislocations [64], impurities and other inhomogeneities which are totally ignored by the theory. For larger supercooling the number of molecules in the critical nucleus, N^*, decreases and its formation becomes more probable. When N^* becomes $O(1)$ there is no real barrier to growth and the growth rate depends only on v — kinetic roughening is said to occur.

In contrast, there is no nucleation barrier for rough surface growth at any supercooling. The growth rate is then simply proportional to v as given by Eq. (3.4), and hence is expected to be linear in ΔT for small undercoolings.

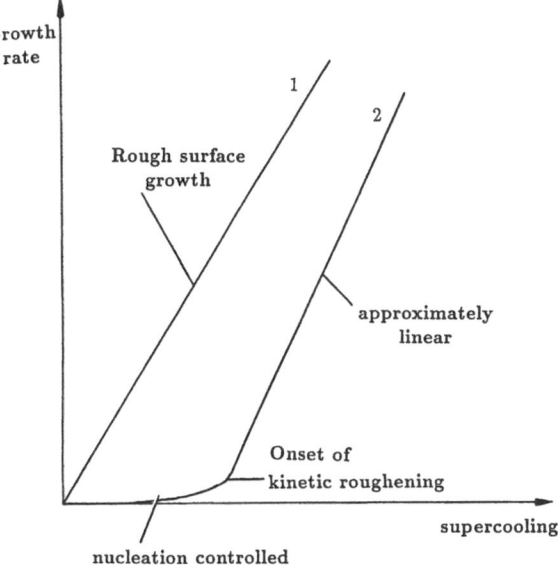

Fig. 3.3. Growth rate versus supercooling for two different face orientations. '1' is above its roughening temperature and is approximately linear. '2' is below its roughening temperature and is nucleation controlled at low supercooling but the growth rapidly increases after kinetic roughening

These predictions have been tested using computer simulations [74] and agreement is generally good. Figure 3.3 shows examples of their results, displaying nucleation control and kinetic roughening for one face orientation, and rough surface growth for a different orientation which is chosen to be above the roughening temperature.

Other important effects, not included in the above theory, are surface diffusion and impurities. Surface diffusion is expected to be important for solution growth because a molecule is unlikely to attach to the surface next to an existing protrusion – this would entail losing all its solution bonds at once which is energetically unfavourable. It is more likely to join at a relatively flat portion of the surface and then diffuse to an energetically more stable position thus giving up its solvent bonds in a piece-meal fashion. Simulations of this effect [74] show that allowing the molecules to diffuse over the surface causes a dramatic increase in the absolute growth rate in the nucleation controlled region, but has little effect for rough surface growth. Also, they allow for impurities which may adhere more or less strongly to the interface than to the growth units. The variation in the strength of bonding with temperature will depend not only on the impurity and material but also on the orientation of the face, and will therefore affect the growth rate of a crystal to a different extent in each direction. Their results showed that impurity effects can overwhelm all other effects predicted by the models, and hence it seems likely that neither pure nucleation nor rough-surface growth will be observed.

3.3 Nucleation Theory for Polymers

In the previous section we described nucleation as the formation of a critical cluster, which decreases in size as the supercooling increases, followed by steady growth in all directions. The lamellar nature of crystalline polymers together with the known facts about chain folding, prohibit such a theory being adapted to high molecular weight materials in a straightforward manner. However, the experimental result that growth rate is proportional to exp $(-1/\Delta T)$ and the observed facetted crystals in solution were strongly suggestive of some sort of nucleation controlled behaviour [47]. It is now shown how the nucleation treatment given for low molecular weights may be extended to long chains in a very general way[1]. We begin with a simple model to illustrate the arguments underlying the derivation of the equations[2]. It is then possible to discuss the applicability of nucleation theory to experimental observations and to derive conditions which must be fulfilled for self-consistency.

The basic model is shown in Fig. 3.4. A nucleus, which is assumed to be a single stem of length l, deposits on a flat surface and therefore creates an excess surface free energy and a reduction in bulk free energy, both of which depend upon l. Subsequent stems are laid down in adjacent positions and spread laterally, forming a strip on the surface. Each addition incurs an extra surface energy due to the fold surface and also a reduction in bulk free energy which is proportional to l. For simplicity only the case where all stems are laid down with the same length will be discussed — fluctuations in the fold length may be included (see Sect. 3.6.1) but they do not change the basic assumptions and only serve to complicate the details. If $l < l_{min}$, the minimum thickness at a given ΔT, there is an overall gain in free energy each time a new stem is deposited, that is, the bulk free energy gain is not large enough to compensate for the energy lost during the formation of the folds. In this case the strip is unstable and will soon fluctuate off the surface. If $l > l_{min}$ each stem deposition after the first will decrease the free energy and the

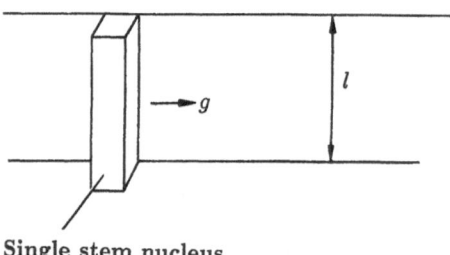

Single stem nucleus

Fig. 3.4. A simple nucleation model for long chains. The first stem is of length l and attaches to the surface. Subsequent stems are deposited adjacent to attached stems, so that the patch spreads in either direction with velocity g

---

[1] There are some recent nucleation theories which do not fit in with this approach and will be discussed in Sect. 3.8.

[2] This model has been modified recently, however, the basic framework remains unchanged and these results are presented in Sect. 3.4.1.

strip will have a net growth in each direction of rate g. This 'spreading rate' is driven by the free energy loss per stem which increases with l, hence g also increases with l. It is the analogue of v, Eq. (3.1), in the low molecular weight theory, but it is a function of two variables, $g(l, \Delta T)$, compared with only one variable for $v(\Delta T)$. The analogue of the critical nucleus is the first stem of thickness $l \geqq l_{min}$ which is deposited on the surface at a net rate per unit substrate length of i — a spontaneous fluctuation must overcome the barrier to its formation, before subsequent additions lead to the stable region. The free energy barrier depends on the length l, becoming greater with increasing l. Recall that l_{min} decreases as ΔT increases, hence the free energy barrier to lay down l_{min} also decreases with increasing ΔT. Notice that this dependence is opposite to that of g, and some competition between the two effects may be anticipated.

First we stress the similarities between this process and conventional nucleation theory: There is a barrier which must be overcome via a spontaneous fluctuation to deposit an initial nucleus. The size of the initial nucleus, that is the length of the first stem for polymers, must be larger than a critical size in order for growth to proceed. Thereafter growth proceeds at a constant rate independent of surface structure, eventually leading to a stable patch on the surface. The critical size of the nucleus decreases with increasing ΔT leading to an increase in the overall growth rate.

The most obvious difference is that the spreading of the stable patch can only occur laterally in the nucleation model for polymers[3], so that in many ways the growth could be considered as one-dimensional [75]. The free energy barrier which must be overcome by the initial nucleus increases with the thickness of the first stem only and cannot increase further with the addition of further units. Hence there is no 'critical' nucleus size in the sense of that depicted in Fig. 3.2. Because the spreading rate, $g(l, \Delta T)$, is a function of the nuclear length it will also contribute to the determination of l. There is some debate as to the validity of calling this process nucleation [76] because of these differences. However, if the important criteria are the existence of an energy barrier to the deposition of a given group of atoms, followed by steady growth of the patch, then it would seem acceptable to use this terminology.

The growth rate of a nucleation-controlled face can be expressed, within certain limits, in terms of i and g without their explicit evaluation. The first limit we shall consider is that of large g so that a nucleus, having formed, spreads rapidly to cover the substrate before another nucleation event occurs, see Fig. 3.5a. If the substrate has a width L, then the rate of nucleation is iL and the growth rate is given by:

$$G_M = biL, \tag{3.5}$$

where b is the thickness of the stem and the subscript M denotes that this limiting growth behaviour is known as mononucleation regime. Notice that the growth

---

[3] The recent theory of sliding diffusion (see Sect. 3.8.2) considers the possibility of growth in all dimensions.

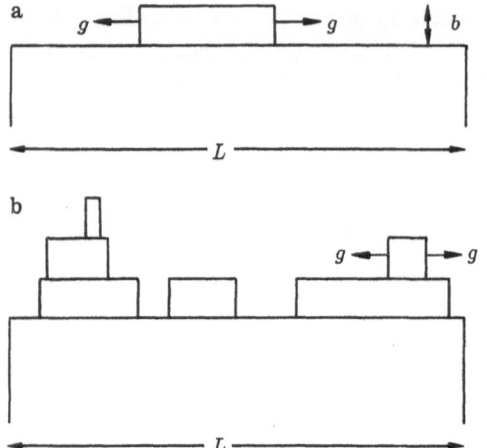

Fig. 3.5a, b. Growth on the surface of lamellae, viewed looking onto the fold plane, showing two different types of growth. In **a** a single strip spreads across the surface before another patch can nucleate. **b** shows several patches in the same layer, which may in turn support further nucleation events

rate is predicted to be proportional to the substrate width, L, and is independent of g. Sometimes this regime is called regime I, however, this terminology is best reserved until after a further modification which will be introduced in Eq. (3.28).

The second regime, known as the polynucleation regime or regime II occurs when i is sufficiently large that there are many patches in the same layer simultaneously, and each of these may in turn support a new nucleation event, so that growth occurs in several layers as shown in Fig. 3.5b. This situation will become more probable as the lamellar thickness decreases, that is at higher supercoolings, because i increases and g decreases (see discussion above). Notice that i must still be regarded as a relatively rare event in order for the model of nucleation followed by spreading to retain any meaning. There are several approaches to this regime: the first we shall give was developed by Sanchez and Di Marzio [49] based on the calculations by Hillig [73] for low molecular weight molecules and is very approximate, but still contains the essential processes; the second is a continuum approach by Frank [45] and is more quantitative. Both approaches are mean-field like in that they consider the behaviour of some averaged i and g and neglect the effects of correlations and fluctuations. The assumption that this is reasonable will be briefly discussed in Sect. 3.4.4. An alternative derivation [77] relies upon dimensional analysis and will not be described. At the end of this section we shall compare Frank's results with other work on the same model.

Consider an isolated initial stem, which is nucleated at time $t = 0$ and grows laterally in both directions for a time t. The length of exposed area which is available on top of the patch for formation of new stems is proportional to $2\,gt$. Hence the rate at which further nuclei will form on this patch is $2\,git$ and so the number of nuclei which form within time dt, $N(t)\,dt$, is:

$$N(t)\,dt \propto 2\,git\,dt\,. \qquad (3.6)$$

The total number of nuclei formed after t_0 seconds is:

$$\int_0^{t_0} N(t)\, dt \propto \int_0^{t_0} 2\, git\, dt = git_0^2 . \tag{3.7}$$

Hence the average time, $\langle t \rangle$, taken to form a new nucleus on a growing patch is:

$$1 \propto gi\langle t \rangle^2 , \tag{3.8}$$

or

$$\langle t \rangle \propto (ig)^{-1/2} . \tag{3.9}$$

This neglects the time taken for nucleation, which will be a good approximation if g is sufficiently small. The rate at which new layers form, $\langle t \rangle^{-1}$, is proportional to $(ig)^{1/2}$, hence the growth rate in the polynucleation regime for stems of thickness b is:

$$G_{II} \propto b(ig)^{1/2} . \tag{3.10}$$

Notice that this result is independent of the length of the substrate but does depend on g.

The continuum theory of Frank [45] considers a substrate of fixed length L[4], with an ensemble average of left and right facing steps given by $l(x)$ and $r(x)$, respectively, with $-L/2 < x < L/2$. In the steady state approximation the time rate of change of l and r vanishes, so that in any length of substrate the rate at which steps are created must equal the rate at which they are annihilated. This gives:

$$i + g\, dl/dx = 2\, glr \tag{3.11}$$

$$i - g\, dr/dx = 2\, glr . \tag{3.12}$$

The terms on the left represent the nucleation and drift of steps respectively whilst that on the right is the annihilation term due to the collision of right and left facing steps. This term neglects correlations between left and right moving steps, and fails to take into account annihilations of steps of more than single height which can occur in the discrete case. However, as pointed out by Frank [45], the physical situation which is being solved prohibits many such effects, and therefore we may take Eqs. (3.11)–(3.12) to be valid for our purposes.

Equations (3.11)–(3.12), together with the boundary conditions:

$$l(x = L/2) = r(x = -L/2) = 0 \tag{3.13}$$

may be solved to give l and r. Equation (3.13) states that no steps enter from outside the limits $x = \pm L/2$. The growth rate can be written as the product of thickness, density, and velocity of the steps:

$$G = b(l + r)\, g \tag{3.14}$$

---

[4] See Sect. 3.6.4 for an extension of the model to increasing $L = L(t)$.

The algebra is not particularly informative, and we shall just give the results:

$$G = bg2[i/2g - c^2]^{1/2} \tan [L(i/2g - c^2)^{1/2}] \tag{3.15}$$

where the constant $c = (l + r)/2$, i.e. Eq. (3.15) is an implicit equation for the growth rate. It simplifies in two limits: for $L^2 i/g \ll 1$

$$G_M = biL[1 - L^2 i/3g] \tag{3.16}$$

$$\simeq biL, \tag{3.17}$$

and for $g/L^2 i \ll 1$

$$G_{II} = b(2ig)^{1/2} [1 - g/4L^2 i] \tag{3.18}$$

$$\simeq b(2ig)^{1/2}. \tag{3.19}$$

The approximate forms of Eqs (3.17) and (3.19) are correct to within 1% for $L^2 i/2 g < 0.01$ and $2 g/L^2 i < 0.01$ respectively. The crossover in behaviour may be taken to lie at $L^2 i/2 g = 1$, giving rise to a maximum deviation of 33%. $L^2 i/4 g$ is often denoted by the dimensionless number z [77]. It is evident that the continuum approach agrees with that of Sanchez and Di Marzio [49], Eq. (3.10), and provides a quantitative footing for the factors involved.

Bennett et al. [78] and Goldenfeld [79] attempted to obtain exact results for this continuum model of nucleation plus growth of steps, without the mean-field approximations inherent in Eqs. (3.11)–(3.12) but still neglecting dissolution processes, and therefore fluctuations in step position (see Sect. 3.4.4). This has only proved possible using periodic boundary conditions which are less realistic for a three dimensional crystal limited by sector boundaries than Frank's absorbing boundary conditions. Toda and Tanzawa [80] have performed a Monte Carlo simulation using both sets of boundary conditions and have compared their results with the theoretical expressions. They find that the agreement between simulation and the exact theory under periodic boundary conditions is extremely good and even for Frank's absorbing boundary conditions the largest error is 4% which occurs for $z = O(10)$. There is also very little difference between the two sets of simulation results with the different boundary conditions, and therefore we conclude that Eq. (3.15) is very satisfactory to describe the growth of this model for all z, and simplifies to Eqs. (3.17) and (3.19) as the limiting cases.

A discrete stochastic approach has been applied by Gates and Westcott [46]. Their results coincide with those of Frank and of Bennett et al. in the continuum limit. However, if the edge structure is serrated on a molecular scale the continuum model fails, as had already been noticed by Frank [45] and Goldenfeld [79]. In this limit Gates and Westcott obtain substantially different growth rates. They find two subregimes of regime II: Regime IIa is equivalent to regime II above whereas regime IIb applies if the surface is 'rough'. (Note the difference between this merely descriptive use of the term 'rough' and the rigorous use in the context of the roughening transition, Sect. 3.2). In this case they find that the growth rate is proportional to the nucleation rate i.

In fact, this behaviour, denoted as regime *III*, had been proposed earlier by Hoffman [13] who recognized the deficiency of the Frank model in the case that the niches get so close that patches have little or no time to spread before colliding with another niche. He argued that in the limit of maximum step density the average spacing between nuclei is constant, so *g* will no longer be important in determining the overall growth rate. A regime *I* like behaviour will result, but with the length of the substrate replaced by the distance between nuclei, which is expected to be of order two to three lattice spacings. This prediction has been tested via Monte Carlo simulation [81], and although the crossover between regimes *II* and *III* occurs at larger supercoolings than predicted, the qualitative picture is correct.

Finally in this section we mention the thickness of the lamella, which is intrinsically connected with the growth rate. Many kinetic theories assume that there is a large number of crystals each having a different thickness and a growth rate which is determined by that thickness. The fastest growing crystals will form the majority and determine the average thickness which is measured − the number of crystals of thickness *l* in a sample is proportional to the growth rate of that thickness. A more refined theory would allow for fluctuations of stem length within a single crystal and calculate the resulting average thickness and growth rate (see Sect. 3.6.1 and 3.7). One would hope that *i* and *g* could be expressed as functions of the average thickness, in which case the above derivations for growth rate will continue to be valid.

3.4 Implications of Nucleation Theory

Although specific calculations for *i* and *g* are not made until Sect. 3.5 onwards, the mere postulate of nucleation controlled growth predicts certain qualitative features of behaviour, which we now investigate further. First the effect of the concentration of the polymer in solution is addressed − apparently the theory above fails to predict the observed concentration dependence. Several modifications of the model allow agreement to be reached. There should also be some effect of the crystal size on the observed growth rates because of the factor L in Eq. (3.17). This size dependence is not seen and we discuss the validity of the explanations to account for this defect. Next we look at twin crystals and any implications that their behaviour contain for the applicability of nucleation theories. Finally we briefly discuss the role of fluctuations in the spreading process which, as mentioned above, are neglected by the present treatment.

3.4.1 Concentration Effects

The growth rate of many crystals is often observed to depend upon temperature in a manner consistent with nucleation theories. If measurements are made on growth from solutions of different concentrations then, at equivalent thicknesses, the dependence of growth rate upon concentration may be determined. Equations (3.16) and (3.17) can be used to predict the concentration dependence of this nucleation approach.

In a first approximation the rate at which molecules approach the crystal surface is assumed to be proportional to the amount of polymer in solution, that is the concentration, C. Therefore, if the barrier to attachment is only determined by the free energy barrier, the rate at which molecules stick to the surface, i, is also proportional to C. Further, if the spreading rate depends only on the free energy barriers to growth of an already nucleated molecule it can be expected to be independent of C. Hence Eqs. (3.17) and (3.19) predict that the growth rate is proportional to C and $C^{1/2}$ in the two regimes. It is not necessary to make any decisions as to the regime of growth to deduce that this behaviour is not observed. Generally the growth rate may be expressed as:

$$G \propto C^{\gamma}, \tag{3.20}$$

where γ is known as the concentration exponent [82, 83]. Experimentally the exponent depends on the supercooling, ΔT, the molecular weight and on C [84, 85]. For very high molecular weights $\gamma \simeq 1/5$ for all ΔT; decreasing the molecular weight increases γ, showing a tendency towards $\gamma = 1/3$, but increasing with crystallization temperature for a given fraction; at low molecular weights $\gamma \simeq 1/2$, and these fractions show the strongest dependence of γ on ΔT. The accuracy of the values obtained may be open to criticism due to the variation of the concentration during growth and fractionation effects which leave the low molecular weight components in the solution [86] — even when sharp fractions are used there is evidence that these can lead to very marked effects [87]. However, it appears safe to say that the observed dependence is not in agreement with the simple model and our main concern will be whether nucleation is capable of describing the wide variation and the general trends which are observed.

The model introduced in Sect. 3.3 considers all nucleation events to occur at the same rate, i, and to occur through the addition of new molecules from the solution. Among the many processes which this neglects are the following:

i) A molecule folding to form a strip will eventually come to an end, creating an 'immobile step', and a new molecule must attach itself in order for the step to continue propagating. Although this is not a 'nucleation' event, its occurrence will certainly depend upon the concentration of polymer.

ii) A molecule which attaches itself to such an immobile step will usually have two ends dangling in solution, both of which may be of significant length. One of these will continue with step propagation along the strip whilst the remaining end, or cilium, will be left dangling, but still attached to the surface of the crystal. It will therefore have an enhanced probability of nucleating the next growth layer at a rate which depends upon the length of the cilium.

iii) When two steps 'collide' as must be the case in regime II, there will be two cilia dangling in solution which may nucleate the next layer, and may even cooperate so that a stable nucleus is formed more easily.

iv) At very low molecular weights the length of chain required to form a stable nucleus may exceed the length of a molecule, and cooperative effects between molecules in solution are required to form a stable patch (but see also Sect. 3.8.2 on the possible need for 'molecular nucleation').

These processes have been assumed to have varying importance in the literature but all lead to a different type of concentration dependence compared with that of Sect. 3.3. Sanchez and Di Marzio [49] considered cases (ii) and (iv) whilst more recently Toda et al. [88] argued that (i) and (iii) are likely to be more important. It is probable that no one process will predominate for all molecular weights, supercoolings and concentrations, and therefore we explain under what conditions the above processes will be important and the dependence on concentration to which they give rise. If several processes contribute then there will be many complicating factors and only the overall trends will be observed.

For case (i), an immobile step may either be annihilated by a mobile step moving in the opposite direction or may be 'regenerated' by the addition of solute molecules. Immobile steps will be predominant for low molecular weights and thick crystals where the distance between nucleation events is of the order of, or greater than, the distance h which can be covered by a folded molecule. Toda et al. [88] incorporated this effect into Frank's equations by allowing the addition of solute molecules at rate i' per immobile step (the notation may be misleading as this addition is not a proper nucleation event) and assuming that i' is proportional to the concentration of polymer. Two types of behaviour are found depending upon the ease with which steps may be regenerated:
a) Immobile steps become mobile: both regimes exist:

$$G_M \propto i \propto C \tag{3.21}$$

$$G_{II} \propto i^{1/2} \propto C^{1/2} \quad \text{for large} \quad i'h, \tag{3.22}$$

and

$$G_{II} \propto (ii')^{1/2} \propto C \quad \text{for small} \quad i'h. \tag{3.23}$$

b) Immobile steps are mostly annihilated: only regime II exists:

$$G_{II} \propto i \propto C \quad \text{for small} \quad ih, \tag{3.24}$$

and

$$G_{II} \propto i^{1/2} \propto C^{1/2} \quad \text{for large} \quad ih. \tag{3.25}$$

For more detailed statements of the results the reader should refer to the original paper. The mononucleation regime only exists when there is only one nucleation event per layer — collisions between steps are unimportant — and therefore can only occur for a). The nucleation rate, i, must be the controlling process and no change in concentration dependence is observed. Still considering a) but now for regime II, if i' is large enough immobile steps will begin to spread again shortly after their creation so the concentration dependence will be unaltered from the original, simple model. Only if i' is small in regime II will the effect of immobile steps be noticeable: more new 'nucleation' events occur within a given layer than would otherwise be the case, and Eq. (3.23) is predicted. A crossover from

Eqs (3.23) to (3.22) is expected with increasing concentration and was indeed observed [89].

If mobile steps are mostly annihilated (case b)) than regeneration does not play a part so i' does not appear explicitly in the growth rate. Two types of behaviour are obtained depending on whether many or just a few immobile steps are formed.

Notice that all the above dependencies assume that i is proportional to C and g independent of C, that is the chain length must be sufficiently long that a nucleation and spreading model is still valid. At very low molecular weights process (iv) will be important and may predominate the effects considered here.

Furthermore, in their work on twin crystals (see Sect. 3.4.4) Toda and Kiho [90, 91] obtained a spreading rate proportional to C, and a nucleation rate independent of it except for very low concentrations. Such a behaviour had not been suspected previously, and it stands in sharp contrast to the assumptions made above. As an explanation Toda and Kiho propose the existence of an adsorption layer which reaches saturation at rather dilute concentrations. Nucleation is then governed by a Langmuir adsorption isotherm. The spreading of a patch, on the other hand, requires more material then is available in the adsorption layer and therefore necessitates volume diffusion. This explains the observed concentration dependence. However, whilst not maintaining that these conclusions are erroneous, the uncertainty in other factors as explained in Sect. 3.4.4 makes general deductions from these experiments alone unjustifiable. Further work on single crystals which is expected to confirm this new picture is still in progress [92].

Sanchez and Di Marzio [49] did not consider the rate of addition of a molecule to a step to be the important factor but rather the enhanced nucleation rate due to the 'primary' cilium in the next layer, that is process (ii). The number of primary cilia in a growth strip is independent of concentration. However, if the cilia are generally not long enough to form a stable nucleus then extra molecules from solution will be required. Hence the cilia nucleation rate will have a concentration dependence which increases with decreasing molecular weight. The trend is therefore the same as for process (i) and it is unlikely that one could distinguish between the contributing factors experimentally.

Toda et al. [88] also discussed case (iii), that is the generation of two cilia due to a collision between steps. This process will operate irrespective of whether a chain is finite or infinite. The average length of cilia formed by collisions is greater than that of the primary cilia of Sanchez and Di Marzio. This, as well as the possibility of cooperative effects between the two chains, lead to the argument [88] that (iii) should be the more important effect for all chain lengths. Certainly for high molecular weight and at large supercoolings where i is large, and hence the distance between nucleation events is relatively small, (iii) will dominate and should be noticeable. If there is a rate of nucleation per unit length due to the cilia, called i'', which is independent of concentration, then again both two regimes are observed. The mononucleation regime occurs for small i so that there is only one nucleation event per layer and i'' is irrelevant. For regime II:

$$G_{II} \propto i^{1/2} \propto C^{1/2} \quad \text{for small} \quad i'', \tag{3.26}$$

and

$$G_{II} \propto (ii'')^{1/3} \propto C^{1/3} \quad \text{for large} \quad i''. \tag{3.27}$$

If i'' is small then the nucleation of cilia is unimportant and simple regime II behaviour is obtained. For larger i'' cilia nucleation predominates and Eq. (3.27) is predicted. This is the explanation put forward to explain the result that γ is close to 1/3 for medium to high molecular weight materials [84].

At very low molecular weights process (iv) will become important [49]. Then at least two molecules are required to form a stable nucleus[5], and they must both join the surface at approximately the same time so that a relatively large increase in concentration is needed to significantly enhance the nucleation rate, that is, $\gamma > 1$. Although Sanchez and Di Marzio made this more quantitative there are too many other complicating effects to make general approximations.

As already stressed, the actual physical situation is unlikely to be any of these limiting cases, and a variety of factors will influence the concentration exponent, including smaller effects not considered here. However, the conclusion of this section is that nucleation is not inconsistent with the experimental trends of concentration, although it would be difficult to make any a priori predictions.

3.4.2 Effects of Crystal Size

It is generally observed that the 'radial' growth rate of single crystals is independent of the size of the crystal. In the mononucleation regime Eq. (3.5) predicts a linear dependence on the length of a growth face. Therefore either this regime is never observed experimentally, or some other explanation must be found. To discuss these size effects it is convenient to define two naturally arising length scales. In the derivations of Sect. 3.3 the length of the substrate, called L, was not given a precise definition. It could be interpreted as the length of a facet. However, it is argued that a real crystal face will contain imperfections which prohibit the spreading of steps and reduce L to an 'effective length' which is considerable less than the width of the crystal face. The average distance between imperfections is known as the persistence length, L_p, and replaces L in Eq. (3.16). That is:

$$G_I = biL_p \tag{3.28}$$

This is known as regime I growth. It is independent of the size of the crystal if $L_p > L$. When growth is argued to take place in regime I, then application of $z < 1$ must lead to a reasonable estimate for L_p. Notice that the introduction of regime I does not affect the concentration arguments of the previous section.

The second quantity is the kinetic length, L_k, and measures the average separation between growth patches in the polynucleation regime. From Eq. (3.14):

$$G_{II} = b\langle N \rangle g \tag{3.29}$$

---

[5] Note, however, that a) the size of a stable nucleus depends on the model used (see Sect. 3.5.1), and b) molecular nucleation theory (Sect. 3.8.1) rules out such a mechanism.

where $\langle N \rangle$ is the number of growth patches per unit length. Use of Eq. (3.19) leads to:

$$\langle N \rangle = (2i/g)^{1/2} . \tag{3.30}$$

Therefore the average distance between nuclei is:

$$L_k = (g/2i)^{1/2} . \tag{3.31}$$

If growth is believed to take place in regime *II*, then Eq. (3.31) must give rise to an acceptable value of L_k. For regime *II* growth, $L_k < L_p$ must hold, and $L_k > 10$ lattice spacings would be a reasonable lower limit for which the present nucleation approach may be expected to work. Below this limit regime *III* growth is predicted.

These two lengths have been discussed in detail by Point et al. [93] and Dosière et al. [94]. They study the size dependence of the growth rate of polyethylene for very small crystals using a decoration technique. The accuracy of their measurements is carefully considered, and they conclude that there is no size dependence of the growth rates for all length scales measured (> 200 nm). Several other claims that there is no size dependence do not seem justified by the number and accuracy of the measurements involved. As shown below, a detailed investigation of these facts would be extremely useful and would enable limits to be placed on the magnitudes of i and g.

First, on a purely physical basis, we determine when departures from a linear growth rate may be expected, that is when the crystal dimensions do not increase proportionally with time, but also depend on the size of the crystal (and maybe even other factors). Then we show how these limits relate to the possible values of i and g.

Consider growth such that for an infinite substrate $L_p < L_k$, that is growth is in regime *I*. Then for large substrates Eq. (3.28) applies and the growth is linear. However, if $L \ll L_p$ the effective facet size is the actual crystal size and growth is in the mononucleation regime. The growth rate, given by Eq. (3.17), increases as the dimensions increase. Therefore L_p is a lower limit for L, below which non-linear growth must be observed.

For regime *II*, $L_p > L_k$ and the growth rate is given by Eq. (3.19). Again consider small L such that $L < L_k$, which implies that on average there is only one nucleus per layer and a mononucleation type growth behaviour will result, given by Eq. (3.17). When the crystal reaches a size $L > L_k$ the polynucleation growth will become dominant and growth will become linear.

The above discussion shows that if a value for L, L_0, can be determined above which the growth is linear, then L_0 is a lower limit for L_p in regime *I*, and for L_k in regime *II*.

For regime *II*, using Eqs. (3.19) and (3.31):

$$g = GL_k/2b < GL_0/2b \tag{3.32}$$

that is, a knowledge of L_0 and the growth rate leads to an upper limit for g — this will be very useful in assessing later theories. Similarly we may deduce that:

$$i > G/bL_0 \tag{3.33}$$

giving a lower limit for i.

We emphasize that we have not developed any new theory, but have deduced conditions which must be fulfilled for self-consistency. It will become apparent later that some theories clearly fail in this respect.

A further consequence of the size dependence of the growth in the mononucleation regime is the existence of a stable (100) facet of constant length L^* in polyethylene crystals [95, 96]. Following Colet et al. [95], consider the triad of faces shown in Fig. 3.6. The lengths of faces α, β, γ are denoted by L_α, L_β, L_γ, respectively, and the growth rates G_α, G_β, G_γ are allowed to depend on the respective lengths. The angles φ and ψ may be measured for a particular morphology and are uniquely defined for crystallographic directions of growth. The components of the rate of growth of the ends of the facet are:

$$v_{yA} = v_{yB} = G_\beta(L_\beta) \tag{3.34}$$

$$v_{xA} \cos \phi + v_{yA} \sin \phi = G_\alpha(L_\alpha) \tag{3.35}$$

$$v_{xB} \cos \psi + v_{yB} \sin \psi = G_\gamma(L_\gamma) \tag{3.36}$$

and the rate of change of the length of the β facet is:

$$v_\beta = v_{xA} - v_{xB} = \frac{G_\alpha(L_\alpha)}{\cos \phi} - \frac{G_\gamma(L_\gamma)}{\cos \psi} + G_\beta(L_\beta)(\tan \psi - \tan \phi). \tag{3.37}$$

In the case of diamond-shaped polyethylene crystals α and γ are (110) facets, β is the (100) facet, and $\psi = \pi - \phi$. This yields:

$$v_{100} = 2 \tan \varphi(G_{110}/\sin \phi - G_{100}). \tag{3.38}$$

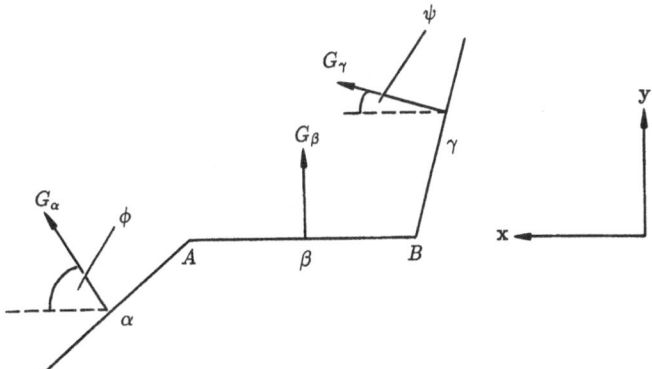

Fig. 3.6. A triad of faces, denoted α, β and γ with growth rates perpendicular to these faces (After Colet et al. [95])

If $G_{100}(\infty)/G_{110}(\infty)$ is less than $1/\sin\phi$ then v_{100} is positive for all lengths and the well-known diamond-shaped crystals are obtained. However, if $G_{100}(\infty)/G_{110}(\infty)$ is greater than $1/\sin\phi$ (which is generally the case in lath shaped crystals) there exists a stable facet length L^*, i.e. $v_{100}(L^*) = 0$ and $dv_{100}/dL|_{L=L^*} < 0$. This length is of the same order of magnitude as the kinetic length L_k. Such (100) sectors of constant width were indeed observed by Point et al. [97] and by Toda [98], and they are probably the strongest evidence so far of mononucleation growth.

3.4.3 Measurements on Twins

The importance of twinned crystals in demonstrating that nucleation is the relevant growth mechanism has been realized since 1949 [64, 99][6]. They were first investigated extensively in polymer crystals by Blundell and Keller [82] and they have recently received increased attention as a means of establishing, or otherwise, the nucleation postulate for lamellar growth [90, 91, 95, 100–102]. The diversity of opinion in the literature shows that it is very difficult to draw definite conclusions from the experimental evidence, and the calculations are often founded upon implicit assumptions which may or may not be justified. We therefore restrict our discussion to an introduction to the problem, the complicating features which make any a priori assumptions difficult, and the remaining information which may be fairly confidently deduced.

In Fig. 3.7 we show some of the more common twins are observed for polyethylene – the particular morphology depends upon the molecular weight, supercooling, concentration and solvent. These diagrams are purely schematic and the real crystals are not well facetted, may be asymmetrical, contain growth spirals and grow as three-dimensional pyramids. Generally the morphologies shown are referred to as 'laths', and are bounded by four (110) planes and either

a) two (100) planes – Fig. 3.7a
b) two (100) planes and two (110) planes – Fig. 3.7b
c) two (110) planes – Fig. 3.7c.

A systematic study of the effect of molecular weight and supercooling upon the morphology was performed by Sadler et al. [102], and the sequence a) to c) is generally obtained with decreasing molecular weight and decreasing supercooling. However, evidence by Dosière et al. [87] shows that fractionation effects can play a large part in changing the growth rate during growth and may affect the various growth faces by different amounts. Therefore quantitative measurements should be treated as trends rather than absolute, and even the final morphologies may be different from those prevalent during growth. A very informative decoration method has been used [94, 95] which allows measurements of growth rate during growth as well as indicating the fold directions in the various sectors, and more systematic measurements along the same lines would be very useful.

The importance of twins lies in the existence of a permanent re-entrant corner at the twin boundary for cases a) and b) – and possibly for c) where it is too small to be detected – and therefore a position at the surface where there is no

---

[6] For an explanation of twins see, for example, [100].

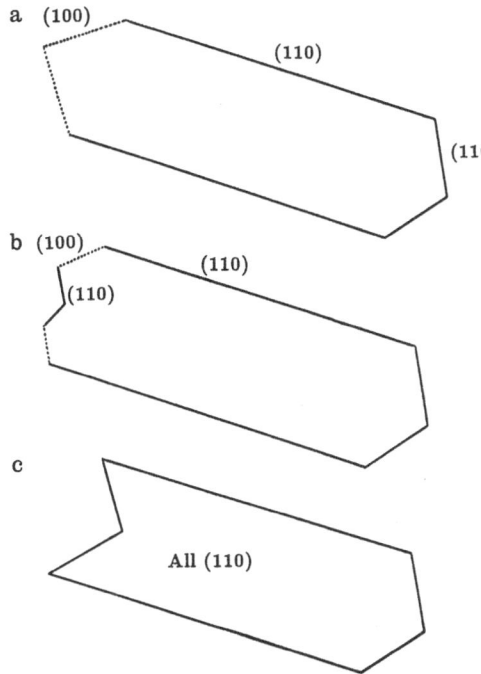

a (100)

(110)

(110)

b (100)

(110)

(110)

c

All (110)

Fig. 3.7. Some of the common twins seen for polyethylene. Continuous and dashed lines depict (110) and (100) growth faces respectively

nucleation barrier to growth. If nucleation is the rate controlling process, there should be significant enhancement of growth along the long lath dimension. If the rate of attachment of new molecules to the re-entrant corner is j, then the total nucleation rate onto a facet of length L bounded by one re-entrant corner is $j + iL$. The extent to which growth is enhanced depends on the relative magnitudes of j and iL, and also the regime in which growth is supposed to take place. The calculations in the literature aim to estimate, from the growth rate of twins compared with the growth rate of 'normal' crystals, the quantity j, how it varies with the crystallization conditions and whether this is compatible with nucleation theory. The evidence is not clear, however, and conflicting conclusions have been reached by different authors. Whereas Toda and Kiho [90, 91] claim that their results are consistent with regime *II* growth of nucleation theory, Sadler et al. [102] conclude that the enhancement of growth is three orders of magnitude smaller than would be expected from nucleation theory. However, as explained below, there is considerable uncertainty in various factors which casts considerable doubt on any deductions from twin experiments alone.

The long chain nature of polymers introduces considerations which may be neglected for low molecular weights. The proximity of two (usually small) faces inclined at an angle less than 180° means that a molecule crystallizing on one face may also be partially adsorbed on the adjacent face. This will strongly affect the spreading rate, g, and may even block further growth of that molecule, as well as preventing solution molecules from reaching the re-entrant corner − hence this process will decrease rather than increase the overall nucleation rate and changes the spreading rate, g, on the re-entrant face. If molecules reach the corner via surface diffusion then the net concentration over the remaining surface will decrease, again reducing the nucleation rate compared with the 'infinite' surface case. Mansfield [103] considered several possible cases of chain diffusion, adsorption and crystallization near a notch. Nucleation rates in the notch proved to be smaller than expected because access of the chain is impeded.

The reentrant planes are not usually straight (110) facets but are inclined at some angle to the crystallographic direction. As decoration methods indicate that folding takes place parallel to the actual growth front, the nucleation rate and spreading rate should be different from those of the pure (110) planes even if other effects are negligible, and will obviously become more different with increasing angle. All of these factors mean that, even apart from the re-entrant corner nucleation rate, j, the growth of the reentrant faces cannot be assumed to depend on the same nucleation and spreading rates as the growth of non-reentrant ones. We believe that these considerations make many of the deductions in the literature unwarranted without further supporting evidence, and now turn to anything that may safely be said.

Following Colet et al. [95] the triad of faces shown in Fig. 3.7 is now taken to represent a re-entrant corner configuration as in Fig. 3.7b. Firstly, α is identified with (100), and β and γ with the (approximately) (110) faces bounded by the re-entrant corner. The rate of change of the length of the (110) facet is then calculated from Eq. (3.37). In most cases there exists a constant and stable facet size L^*_{110}. Secondly, by identifying α with the 'normal' (110) facet, β with (100), and γ with the re-entrant corner (110) the rate of change of the length of the (100) facet can be calculated. It turns out to be protected if G_{110} (corner) greatly exceeds G_{110} (normal), i.e. if there is significant enhancement of the growth rate due to the re-entrant corner.

The experimental observations are all in agreement with this qualitative picture, which may help towards understanding the origin of the morphology. However, other than observing that lath shapes generally imply that a re-entrant corner has enhanced the growth rate, and maybe obtaining qualitative trends, it does not appear that unbiased quantitative predictions can be made.

3.4.4 The Neglect of Fluctuations

All the calculations and simulations of the growth rate discussed in Sect. 3.3 including the so-called 'exact' treatments are based on a 'mean-field' approach in the sense that the spreading of a nucleated patch is described as occurring at an average velocity g. This resultant velocity is the net forward effect of both additions and subtractions at the edge of the patch, which cause fluctuations about the mean position of the step. The magnitude of these fluctuations, that is the deviation of the step from its mean position, depends upon the supercooling: at zero supercooling there is no net growth rate and the step may perform very large fluctuations. However, at large supercoolings the driving force for crystallization is large and only small fluctuations would be expected. If fluctuations are large then some steps will collide before their mean positions coincide and deviations from the growth laws may be expected. However, there will also be steps which collide after their mean positions coincide, and the effects will tend to counterbalance one another. In regime *I*, which occurs for low supercooling and hence large fluctuations, there is only one nucleation event per layer so there are no collisions and this effect is unimportant. In regime *II* the mean distance between steps, L_k, decreases with increasing supercooling, and therefore fluctuations, which are small, may continue to be mostly irrelevant. During any collision of steps chain entanglements and cilia nucleation (Sect. 3.4.2) will play an important role and fluctuation effects add to a very complicated pattern. A simple nucleation theory can only be successful if steps occur infrequently and the fluctuations are small compared with their separation. This criterion is best judged when explicit values for *i* and *g* are developed later.

Sadler [75] investigated the role of fluctuations by simulation of growth on a one dimensional surface. His choice of energy barriers is different from the one generally used in nucleation theory, and therefore his results are mostly inapplicable to the present case. Near equilibrium, i.e. at low supercoolings, however, both approaches converge. It is therefore relevant to note that Sadler finds good agreement between simulation and nucleation model results, but only because two effects which are neglected by the nucleation model cancel each other out; namely the enhancement of step annihilations due to fluctuations and the creation of steps by dissolution of a stem from a flat surface ('cavity creation').

A further result of Sadler's 2D-simulation was a relation between the step density and growth rate on the one hand and the inclination of the surface with respect to the principal axes on the other. From this relation crystal shapes were derived which show considerable curvature. This result of an 'exact' treatment stands in contrast to Frank's mean-field curvature expression which gives essentially flat profiles. We will return to the discussion of curved edges in Sect. 3.6.3.

3.5 The Theory of Lauritzen and Hoffman

In this section we analyse the nucleation theory due to Lauritzen and Hoffman beginning with their derivation of *i* and *g*, which can then be used to predict the growth rate and thickness of a lamella. The emphasis lies on the underlying assumptions and therefore the derivation is for the simplest case of infinite molecular weight and does not allow for fluctuations in the upper and lower surfaces — the 'fold' surfaces. These factors may be incorporated, as explained later in Sect. 3.6.1, and initially only obscure the important processes. At first the theory is simply presented, closely following the review article by Hoffman, Davies and Lauritzen [104], and whilst stating the necessary assumptions no comment is made as to their validity. A more detailed discussion of the various stages is contained in Sect. 3.5.2 covering some theoretical criticisms which have been raised in recent years [51, 52, 75, 76]. Finally, in Sect. 3.5.3 we look at some of the experimental evidence both in support of and against the predictions.

3.5.1 Explicit Forms for i and g

In the LH model a nucleus is formed and subsequently spreads by the addition and removal of complete stems, as shown in Fig. 3.8, where σ_e and σ are the fold and lateral surface free energies, a and b are the width and depth of the chains. The probabilities of addition or removal are reflected in the appropriate rate constants using the following notation:

A_0: The rate constant for adding a new stem to the substrate, that is, the probability per possible stem position per unit time that a new stem will be added.

B_0: The rate constant for removal of an isolated stem.

A_1: The rate constant for addition of a stem next to an existing stem of the same length.

B_1: The rate constant for removal of a stem next to an existing stem of the same length.

The explicit forms for these rate constants will be given later, as even within LH theories the choice is not unique, but any choice depends upon the length of the stem, the supercooling and the surface energy. However, it is assumed that A_1 and B_1 are independent of the number of stems in the nucleated patch, i.e. the rate constants for all stems after the first are the same. The possible transitions associated with these rate constants are shown in Fig. 3.9 a and b. Figure 3.9c shows transitions involving two adjacent stems, with associated rate constants A_2 and B_2, which are neglected. In the following i and g are determined in terms of these quantities.

The spreading rate, g, is the overall rate at which stems are added to a nucleated patch, multiplied by the thickness of the stem:

$$g = a(A_1 - B_1).\tag{3.39}$$

The probability per site of forming a nucleus on an infinite substrate in the absence of other nuclei is taken to be equivalent to the nucleation rate, i. This obviously assumes that neighbouring patches do not collide during their formation, which is fully consistent with the nucleation model (see Sect. 3.4.4).

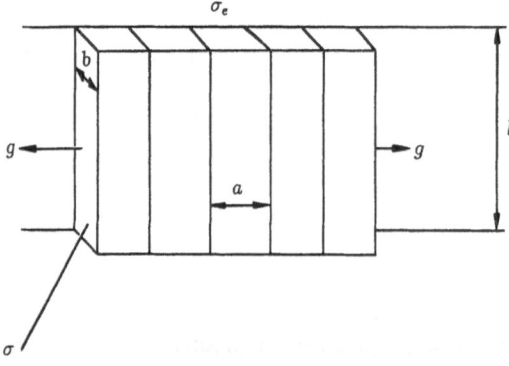

Fig. 3.8. The model of Lauritzen and Hoffman for stems of width a and depth b. The surface tensions on fold and lateral surfaces are σ_e and σ, respectively

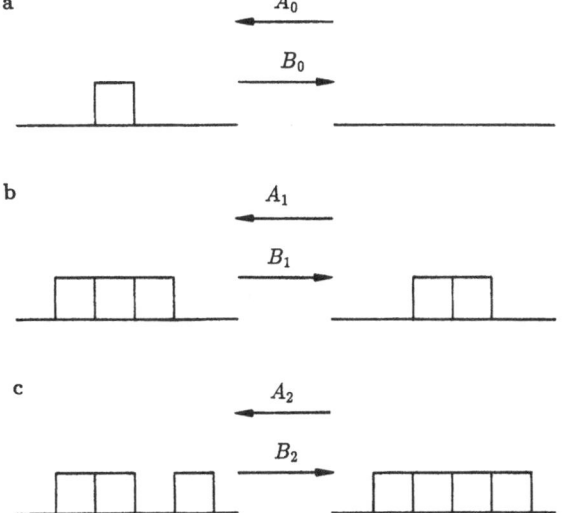

Fig. 3.9a–c. The possible transitions in the LH theory. A's and B's are rate constants for addition and subtraction of a stem and depend upon the number of neighbours of a stem. Transitions of type C are neglected by the model

In an ensemble containing $N_n(t)$ systems with patches of n stems and $N_{n+1}(t)$ systems with patches of $n + 1$ stems at time t the flux (net transition rate) between the n_{th} and $(n + 1)_{th}$ stages is:

$$S_n(t) = A_1 N_n(t) + B_1 N_{n+1}(t) \quad \text{for} \quad n > 1, \tag{3.40}$$

and

$$S_0(t) = A_0 N_0(t) + B_0 N_1(t) \quad \text{for} \quad n = 1. \tag{3.41}$$

In the steady-state approximation $dN_n(t)/dt = 0$, and the flux through all states is the same and can be expressed as [50]:

$$S = N_0 A_0 (A_1 - B_1)/(A_1 - B_1 + B_0). \tag{3.42}$$

Given values for the A's and B's Eq. (3.42) yields the flux onto lamellae of specified thickness. The total flux is obtained by summing Eq. (3.42) over all possible thicknesses and this total flux is then put equal to i.

To make further progress specific forms for the rate constants are required. In the steady state, the principle of detailed balance gives:

$$A/B = \exp\left[-\Delta F_T/kT\right] \tag{3.43}$$

where A and B are the rate constants connecting two states, and ΔF_T is the total free energy difference between those states. For the transition between zero and one stem of length l there is a bulk free energy gain of $abl\,\Delta F$ and a surface free

---

[7] There is a conflict between this assumption and the condition for treating stems as complete units given by Frank and Tosi [105], see Sect. 3.5.2 and Colet at al. [95].

energy loss of $2bl\sigma$, where ΔF is the bulk free energy difference per unit volume. The excess energy owing to the fold is associated with the deposition of subsequent stems and therefore does not contribute to the initial stem free energy barrier[7]. Hence:

$$A_0/B_0 = \exp\left[-(2bl\sigma - abl\,\Delta F)/kT\right]. \tag{3.44}$$

For subsequent stems the loss in free energy is still $abl\,\Delta F$ but there is a gain, independent of l, which is $2ab\sigma_e$ from the formation of a fold. Therefore:

$$A_1/B_1 \exp\left[-(2ab\sigma_e - abl\,\Delta F)/kT\right]. \tag{3.45}$$

Equation (3.43) only determines the ratio A/B and does not specify the absolute values of the rate constants, hence some choice must be made based on the physical processes involved. This problem is referred to as 'apportioning' and has often been discussed [48, 75].

The choices which are generally made are:

$$A_0 = \beta\exp\left[-(2bl\sigma - \psi abl\,\Delta F)/kT\right] \tag{3.46}$$

$$B_0 = \beta\exp\left[-(1-\psi)\,abl\,\Delta F/kT\right] \tag{3.47}$$

$$A_1 = \beta\exp\left[-(2ab\sigma_e - \psi abl\,\Delta F)/kT\right] \tag{3.48}$$

$$B_1 = \beta\exp\left[-(1-\psi)\,abl\,\Delta F/kT\right] \tag{3.49}$$

where β is the number of times per second that a molecule attempts to attach to the surface, and is taken to be the same for both the initial stem and subsequent stems. It depends upon the molecular motions involved in transporting the segments of polymer molecules to the crystallization site and is expected to be very dependent on temperature [13].

The physical reasoning behind these choices is that the free energy gain due to the surfaces ($2bl\sigma$ and $2ab\sigma_e$) acts as a barrier which must be overcome in order to lay down a stem, and is therefore put into the forward rate constants, A_0 and A_1. However, some of the bulk free energy loss is released during the deposition process and acts to reduce this barrier — the actual amount being determined by the factor ψ. The exponential factors in Eqs. (3.46) and (3.47) are known as the activation barriers and are shown for various values of ψ in Fig. 3.10. The detailed balance Eqs. (3.44) and (3.45) then lead to definite values for B_0 and B_1. Notice that this choice of apportioning gives $B_0 = B_1$ so that Eq. (3.42) simplifies to:

$$S = N_0 A_0 (1 - B_1/A_1). \tag{3.50}$$

Substitution of Eqs. (3.46)–(3.49) into Eq. (3.50) leads to an expression for the steady-state flux which depends on l:

$$S(l) = \beta N_0 \{\exp\left[(-2bl\sigma + \psi abl\,\Delta F)/kT\right]\}$$

$$\times \{1 - \exp\left[(-abl\,\Delta F + 2ab\sigma_e)/kT\right]\}. \tag{3.51}$$

$S(l)$ is the nucleation rate for non-interacting nuclei and is further interpreted as the probability distribution for a crystal to have thickness l. Notice that for $2\sigma_e/\Delta F < 1$, $S(l)$ is negative, which corresponds to the statement that a lamella of this thickness is unstable. The total flux, S_T, in an ensemble of crystals is obtained by summing $S(l)$ over all possible values of l:

$$S_T = \sum_{l_{min}}^{\infty} S(l) \simeq \frac{1}{l_u} \int_{2\sigma_e/\Delta F}^{\infty} S(l)\, dl \tag{3.52}$$

where l_u is the monomer repeat unit. This can be evaluated to give:

$$S_T = \frac{\beta N_0 P}{l_u} \{\exp\,(2ab\sigma_e\psi/kT)\} \{\exp\,(-4b\sigma\sigma_e/\Delta F\, kT\} \tag{3.53}$$

where

$$P = \frac{kT}{2b\sigma - \psi ab\,\Delta F} - \frac{kT}{2b\sigma + (1 - \psi)\, ab\, \Delta F}. \tag{3.54}$$

Equations (3.39) and (3.53) can now be used to find explicit forms for the growth rate and thickness. First we briefly summarize the steps which lead to these equations: The nucleating patch grows by the deposition of complete stems and the transitions between successive stems obey detailed balance arguments. The nucleation rate is determined via the total flux onto an ensemble of separated and independent patches and is related to the number of sites on the surface, N_0, transport properties in the fluid, β, the supercooling and the surface free energies. In any experiment there is a range of thicknesses with which a crystal will grow, and the probability of obtaining a given thickness, l', is proportional to $S(l')$.

i) Thickness of the Crystal

By identifying the probability of occurrence of a crystal with the flux corresponding to its thickness as explained above, the average thickness of an ensemble of crystals is given by:

$$\langle l \rangle_{av} = \frac{\displaystyle\int_{2\sigma_e/\Delta F}^{\infty} lS(l)\, dl}{\displaystyle\int_{2\sigma_e/\Delta F}^{\infty} S(l)\, dl} \tag{3.55}$$

$$= \frac{2\sigma_e}{\Delta F} + \delta l \tag{3.56}$$

where

$$\delta l = \frac{kT}{2b\sigma} \frac{2 + (1 - 2\psi)\, a\, \Delta F/2\sigma}{[1 - a\, \Delta F\, \psi/2\sigma]\, [1 + a\, \Delta F(1 - \psi)/2\sigma]}. \tag{3.57}$$

Notice that for small ΔF, δl is approximately constant, and the main temperature dependence of $\langle l \rangle_{av}$ is given by $2\sigma_e/\Delta F$. Using $\Delta F \propto \Delta T$ from Eqs. (2.10) and (3.56) gives:

$$\langle l \rangle_{av} \propto 1/\Delta T + \delta l \qquad (3.58)$$

as observed experimentally. Hence the dependence of the average thickness on supercooling is controlled by that of the minimum thickness permitted for stability. This formula breaks down further away from equilibrium and will be discussed in more detail in Sect. 3.5.2.

The distribution of crystal thicknesses, $\langle (l - \langle l \rangle_{av})^2 \rangle$ can be evaluated as:

$$\langle (l - \langle l \rangle_{av})^2 \rangle = \frac{(kT)^2}{[2b\sigma - ab\,\Delta F\,\psi]^2} + \frac{(kT)^2}{[2b\sigma + (1 - \psi)\,ab\,\Delta F]^2} \qquad (3.59)$$

$$\simeq \frac{1}{2}\,(kT/b\sigma)^2 \quad \text{for small} \quad \Delta T .$$

ii) *Size of a Stable Nucleus*

The deposition of the first stem of thickness $l > l_{min}$ is the critical stage of nucleation. However, further stems need to attach next to it before the patch reaches a stable size (see Fig. 3.10). This size can be determined by setting the change in free energy $\Delta\Phi_v$ of formation of a surface strip possessing v stems to zero:

$$\Delta\Phi_v = -abl\,\Delta F + 2bl\sigma + (v - 1)\,(-abl\,\Delta F + 2ab\sigma_e) . \qquad (3.60)$$

One then obtains with Eq. (3.56):

$$v_s = \frac{4\sigma\sigma_e}{a_0\,\delta l(\Delta F)^2} + \frac{2}{\Delta F}\left(\frac{\sigma}{a_0} - \frac{\sigma_e}{\delta l}\right) . \qquad (3.61)$$

Because of its dependence on σ, σ_e, δl, and ΔF its size can be expected to vary according to the type of nucleation model employed in analyzing a set of experimental data (see Refs. [106 and [107]). The size of a stable nucleus has consequences as far as, e.g., the number of molecules needed to form it and the minimum required substrate length are concerned.

iii) *Growth Rate*

The simplest LH theory neglects any cilia nucleation effects and takes Eqs. (3.17) and (3.19) to be the growth rate in the limits $L^2i/4g \ll 1$ and $L^2i/4g \gg 1$, respectively. However, L should be replaced by the persistence length, L_p, when the crystal size is greater than L_p. The nucleation rate is the total flux per available site, that is $i = S_T/N_0$, and g is given by Eq. (3.39) using Eqs. (3.48) and (3.49) to substitute for A_1 and B_1. A_1 and B_1 are dependent upon l, but only the value which corresponds to the average thickness of the crystal, as defined by Eq. (3.56), is used. Therefore the expression for g reads:

$$g = a\beta Q \exp\left[-2ab\sigma_e(1 - \psi)/kT\right] \qquad (3.62)$$

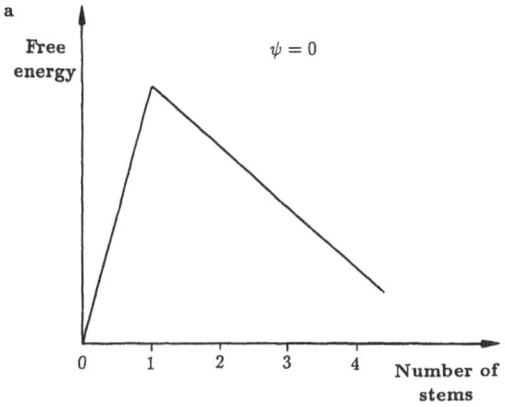

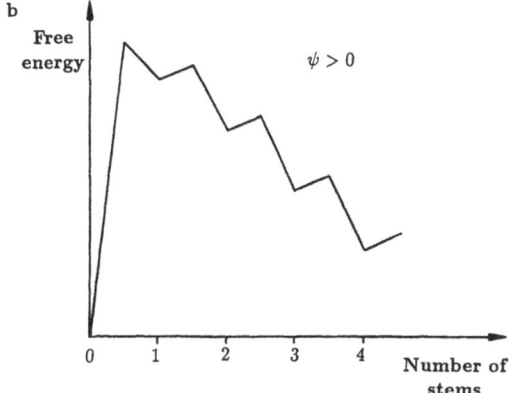

Fig. 3.10a, b. Free energy activation barriers for addition and removal for two values of the apportioning factor, ψ. In **a** there is no barrier to addition after the first stem. Increasing ψ increases both the forward and backward barriers as shown in **b**

where

$$Q = \exp\left[\psi ab\,\Delta F\,\delta l/kT\right]\left[1 - \exp\left(-ab\,\Delta F\,\delta l/kT\right)\right]$$

$$\simeq 1\,, \quad \text{at low undercoolings}\,. \tag{3.63}$$

The temperature dependence of the spreading rate is generally small and can often be neglected against that of the nucleation rate (but see Sect. 3.6.3).

Hence the resulting growth rates are:

$$G_I = \frac{b}{a}\,\beta L_p \exp\left(2ab\sigma_e\psi/kT\right)\exp\left(-4b\sigma\sigma_e/\Delta F\,kT\right) \tag{3.64}$$

$$G_{II} = b\beta \exp\left(ab\sigma_e\psi/kT\right)\exp\left(-ab\sigma_e(1 - \psi)/kT\right)$$

$$\exp\left(-2b\sigma\sigma_e/\Delta F\,kT\right)\,. \tag{3.65}$$

The 'retardation factor' β governs the rate of transport of polymer to the crystal surface and as we are considering the actual crystallization process it is not our immediate concern. However, some reasonable estimate of its magnitude, temperature dependence and possible variance is needed in order to justify the assumption that the crystallization process is rate determining, and also to allow for its effect and deviations when comparing with experimental results [13].

When applied to growth from the melt (despite our reservations in Sect. 3.2), β is normally assumed to take the form

$$\beta = \frac{kT}{h} J \exp\left[-U^*/R(T - T_\infty)\right], \tag{3.66}$$

and from solution

$$\beta = C' \frac{kT}{h} \exp\left[-\Delta H^*/RT + \Delta S^*/R\right] \tag{3.67}$$

where the constants are characteristic of the energetics of the liquid and are independent of temperature. Both of these forms are empirical and further comments on their effect will be made later. T_∞ is the hypothetical temperature where all motion associated with viscous flow ceases and is approximately 60 K below the glass transition temperature.

Although a wide choice for the other parameters occurring in Eqs. (3.64) and (3.65) is possible, their temperature dependence is small in the vicinity of T_m^0. In practice either G/β or $\log(G/\beta)$ is normally plotted as some function of temperature which necessarily entails some choice for these parameters. Each case should be examined individually to ascertain the change a different choice would make, and to only rely on the results within these limits.

3.5.2 Discussion of the Theoretical Basis of the Work of Lauritzen and Hoffman

The theory described in Sect. 3.5.1 has received most justification from its agreement with observations. Ultimately this must be the case for any successful theory. However, the consequences of any assumptions made should be well understood as several independent and adjustable parameters can make an unrealistic approach appear reasonable. In recent years, some of the experimental features not well represented by this theory have been attributed to unrealistic approximations which would invalidate the basis of the theory. Hence we give a careful step-by-step critique, and emphasize the consequences of alternative assumptions. In Sect. 3.5.3 when we investigate the experimental evidence, we should be better able to establish whether various types of behaviour can be accommodated within the theory.

i) The Incorporation of a New Stem as a One-step Process

The attachment of a stem, which is part of a flexible molecular chain, onto a crystal surface is obviously a very complex process which takes place via a large number of intermediate steps. We must therefore question whether we can treat such a process by a single stage with an associated pair of rate constants, and if so whether their forms are correctly given by Eqs. (3.44) and (3.45). The justification usually forwarded [105] derives an expression for the flux of material onto the crystal from a 'fine grained' approach, that is by considering the incorporation of each small segment of chain separately, and compares this with a 'coarse grained' approach in which a whole stem is deposited in one stage. Within conditions normally taken as fulfilled the growth rates are the same.

The two growth processes are illustrated in Fig. 3.11. Lower case letters denote the transition rates and occupation numbers of the 'substages' in the fine grained approach whilst upper case letters refer to those of the stages in the coarse grained treatment with m substages equivalent to one stage. Adjacent substages differ by the addition or substraction of one subunit whilst stages differ by one unit, that is, m subunits = 1 unit. An ensemble of independently growing systems is considered and the probability that any one system is in a certain state is just proportional to the number of systems in that state in the ensemble. The following notation is used which corresponds to the allowed transitions:

v: The number of subunits in a substage, $0 < v < \infty$.

n_v: The number of substages with v subunits in the ensemble.

α_v: The rate constant of addition for a substage with v subunits, that is the probability per unit time that a substage of v subunits will gain a subunit. It may depend upon v but is independent of the shape of the units.

β_v: The rate constant of removal for a substage with v subunits.

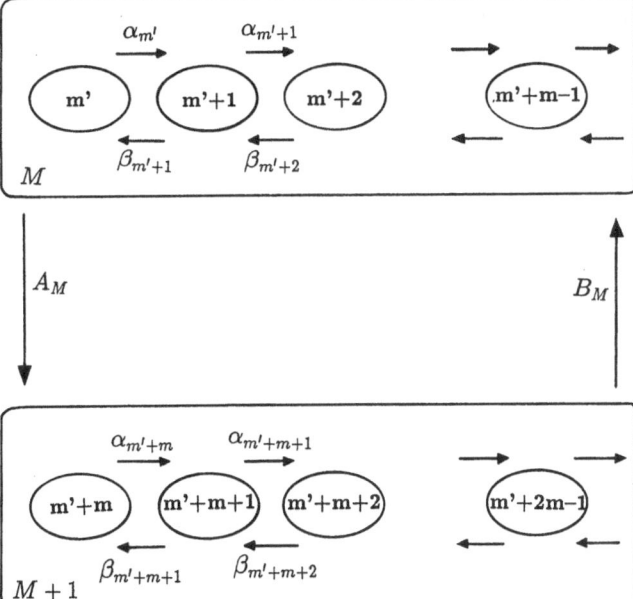

Fig. 3.11. Illustration of fine- and coarse-grained approaches. There are m substages and their associated transition rates (*small letters*) corresponding to one stage

s_v: The rate at which v substages become $v + 1$ substages.
M: The number of units in a stage, $0 < M < \infty$.
N_M: The number of stages with M units in the ensemble.
A_M: The rate constant of addition for a stage with M units.
B_M: The rate constant of removal for a stage with M units.
S_M: The rate at which M stages become $M + 1$ stages.

Notice that the only fine grained transitions which are allowed are those which lead directly to the coarse grained stage. In the specific example we have in mind where the coarse grained stage is a stem of m subunits, the subunits are only allowed transitions along the length of the stem — transitions which involve a new stem, before m units have been deposited, are not allowed. This deficiency was first highlighted by Point [51, 52], and led him to the conclusion that although the LH calculation is technically correct, this implicit assumption means that it is irrelevant for the growth of polymers. We shall be returning to this argument later.

The details of the calculation are algebraically cumbersome, so first we list the steps to be taken so that the physical reasoning is not lost. The net rate at which n_v substages develop into n_{v+1} substages can be written down in terms of the rate constants for the individual subunits which may be v dependent. Assumption of steady-state conditions leads to a general expression for the flux through each substage. Repeating the argument at a coarse grained level would lead to a similar expression for the flux through each stage. These fluxes are almost the same if certain conditions are fulfilled, that is the coarse grained and fine grained approaches are equivalent. In the discussion of how to relate this to polymer growth we consider whether these conditions are indeed likely to be satisfied.

From the definitions of transition rates:

$$s_v = \alpha_v n_v - \beta_{v+1} n_{v+1}. \tag{3.68}$$

In the general case all of these quantities will be time-dependent, and the rate of change of n_v is given by:

$$dn_v/dt = s_v - s_{v-1} = \alpha_v n_v - \beta_{v+1} n_{v+1} - (\alpha_{v-1} n_{v-1} - \beta_v n_v) \tag{3.69}$$

In general Eq. (3.69) cannot be solved to give the time dependence of n_v. However, a characteristic of this equation is that at very long times ($t \to \infty$) the solution becomes time-independent, that is to say a steady-state solution exists. Therefore at long times:

$$s_v - s_{v-1} = 0, \tag{3.70}$$

$$s = s_{v-1} = s_v = \alpha_v n_v - \beta_{v+1} n_{v+1}. \tag{3.71}$$

Multiplication of Eq. (3.71) for each of indices $r = 0, 1, ..., v - 1$ by:

$$\prod_{i=0}^{r} (\beta_i/\alpha_i), \quad \text{including} \quad (\beta_0/\alpha_0) = 1$$

gives the following equivalent set of equations:

$$s(\beta_0/\alpha_0) = \alpha_0 n_0 (\beta_0/\alpha_0) - \beta_1 n_1 (\beta_0/\alpha_0) \tag{3.72}$$

$$s(\beta_0/\alpha_0)(\beta_1/\alpha_1) = \alpha_1 n_1 (\beta_1/\alpha_1)(\beta_0/\alpha_0) - \beta_2 n_2 (\beta_1/\alpha_1)(\beta_0/\alpha_0) \tag{3.73}$$

$$s \prod_{i=0}^{r} (\beta_i/\alpha_i) = \alpha_r n_r \prod_{i=0}^{r} (\beta_i/\alpha_i) - \beta_{r+1} n_{r+1} \prod_{i=0}^{r} (\beta_i/\alpha_i) \tag{3.74}$$

$$s \prod_{i=0}^{v-1} (\beta_i/\alpha_i) = \alpha_{v-1} n_{v-1} \prod_{i=0}^{v-1} (\beta_i/\alpha_i) - \beta_v n_v \prod_{i=0}^{v-1} (\beta_i/\alpha_i). \tag{3.75}$$

The second term on the right hand side of the rth equation is identical, but of opposite sign, to the first term on the right hand side of the $(r+1)^{th}$ equation. Hence adding together all Eqs. (3.72)–(3.75) leaves only the first term of Eq. (3.72) and the second term of Eq. (3.75). That is:

$$s \sum_{r=0}^{v-1} \prod_{i=0}^{r} (\beta_i/\alpha_i) = \alpha_0 n_0 - \beta_v n_v \prod_{i=0}^{v-1} (\beta_i/\alpha_i) \tag{3.76}$$

For infinite growth $(\beta_i/\alpha_i) < 1$, hence $\prod_{i=0}^{v-1} (\beta_i/\alpha_i) \to 0$ as $v \to \infty$. Therefore the only way in which the last term in Eq. (3.76) can remain finite in this limit is if $n_v \to \infty$. This is rejected as being unphysical, and therefore:

$$s = \frac{\alpha_0 n_0}{\sum_{r=0}^{\infty} \prod_{i=0}^{r} (\beta_i/\alpha_i)} \tag{3.77}$$

Having determined s, which is independent of v, by its behaviour as $v \to \infty$, we can then use Eq. (3.76) to determine the occupation numbers for arbitrary v.

$$n_v = s \frac{\sum_{r=v}^{\infty} \prod_{i=0}^{r} (\beta_i/\alpha_i)}{\beta_v \prod_{i=0}^{v-1} (\beta_i/\alpha_i)} \tag{3.78}$$

Equations (3.77) and (3.78) give the solutions to the steady-state problem, given β_i and α_i.

In the coarse grained representation the equivalent statement to Eq. (3.71) in the steady state reads:

$$S = A_M N_M - B_{M+1} N_{M+1} \tag{3.79}$$

This can be compared with the fine grained representation by summing m substage equations from Eq. (3.75). This leads to:

$$s \sum_{r}^{r+m-1} \prod_{i=0}^{r} (\beta_i/\alpha_i) = \alpha_r n_r \prod_{i=0}^{r} (\beta_i/\alpha_i) - \beta_{r+m} n_{r+m} \prod_{i=0}^{r+m-1} (\beta_i/\alpha_i) \quad (3.80)$$

If the two representations are equivalent then Eqs. (3.79) and (3.80) describe how A's and B's must be transformed in terms of α's and β's. (These identities are performed explicitly by Sanchez and Di Marzio, [49]. Frank and Tosi [105] further show that if α's and β's are chosen to satisfy detailed balance conditions, that is equilibrium behaviour, then the occupation numbers of the two representations are only equivalent if the n_r's are in an equilibrium distribution within each stage. This is likely to be true if there is a high fold free energy barrier at the end of each stem deposition, and thus will probably be a good representation for most polymers. In particular, the rate constant for the deposition of the first stem, A_0 must contain the high fold free energy term, i.e.:

$$A_0 = \beta \exp\left[-(2bl\sigma + 2ab\sigma_e - abl\,\Delta F)/kT\right] \quad (3.81)$$

as given by Frank and Tosi. In the LH theory, however, A_0 does not contain a fold free energy term (see Eq. (3.46)). Therefore it does not strictly fulfill the Frank and Tosi criterion for using a coarse grained representation.

The details discussed in this section are the most important approximations involved in the LH theory, and therefore we re-emphasize the steps that have been taken. Given a strictly sequential process, a complete stem may reasonably be described as a complete unit, neglecting all the intermediate deposition steps, within certain conditions likely to be fulfilled for cases of physical interest. However, this very model is open to question as there are many other routes, such as folding and unfolding before a complete stem is deposited, which are not permitted to occur. Although these may only have a small probability of occurring, their inclusion radically changes the formalism and the above results are no longer valid — we present the evidence for this argument in Sect. 3.7.

Although the largest approximation has already been made, we continue to discuss other details of the theory. This is partially justified by the hope that at some future date a form for A's and B's may be found which does not suffer from the above drawbacks. But also, any theory which shows agreement with experiment deserves understanding in its own right.

ii) The Rate Constants

Even within the framework laid out in the previous section, there is still a large choice available for the A's and B's. Detailed balance at equilibrium gives:

$$(\beta_i/\alpha_{i+1}) = \exp\left(-\Delta f/kT\right) \quad (3.82)$$

where Δf is the free energy difference between states n_i and n_{i+1}. Away from, but still close to, equilibrium this equation is still assumed to hold because it also governs fluctuations giving rise to local deviations from equilibrium. It should be a good approximation if there is a large number of 'on' and 'off' events. At large supercoolings Eq. (3.82) will probably cease to be valid, however the lack of rigorous results for nonequilibrium situations makes any other choice rather arbitrary. Taking the Frank and Tosi [105] conditions as fulfilled means that the coarse grained constants can be written down as:

$$A_i/B_i = \exp\left(-\Delta F_T/kT\right) \tag{3.83}$$

where ΔF_T is the total change in free energy upon adding a new stem and is the sum of all the free energies, Δf, associated with the fine grained representation. ΔF_T includes both the bulk free energy decrease as well as the surface free energy increase. If the bulk free energy decrease per unit volume is ΔF, then the bulk contribution for a stem of length l is $abl\,\Delta F$. The difficulties of relating ΔF to the supercooling ΔT have been explained in Sect. 2.3, and here we will retain ΔF as a term which can in principle be evaluated.

The surface energy contribution requires somewhat more consideration. The deposition of a stem has an associated energy due to the increase in surface area. There are many possible arrangements of a stem, all with the same total energy change, and it is the sum over all such configurations, that is the degeneracy, which gives an entropic contribution to the free energy change and hence leads to a macroscopic surface free energy. It is not obvious how to relate the increase in energy due to the lateral sides of a single straight stem to the surface free energy of the equilibrium crystal which includes many partial stems and defects. Also the small size of a stem makes the use of any macroscopic quantities questionable as was discussed in Sect. 3.2 for low molecular weight materials, and it was pointed out that for a small nucleus the contributions from all possible configurations should be considered separately. These comments are even more relevant for the fold surface, where each fold at the end of a stem is considered separately, yet its free energy is still equated with that for the complete surface. Lauritzen et al. [108] demonstrated that within this framework the surface free energy of the fold surface is determined by kinetic factors and may be very different from that of the equilibrium crystal. Therefore there are many inconsistencies in the parameters used for Eqs. (3.44) and (3.45) and at best they should only be regarded as giving an order of magnitude estimate.

Having decided on the form for the free energy difference between two adjacent states, this energy must be apportioned between the forward and backward rate constants. The thickness and growth rate are primarily determined by the shape of $S(l)$ as a function of l, which can be thought of as being the product of two contribution: the first is A_0 which acts as a barrier to large l, and the second is given by the remaining terms in Eq. (3.44) which act as a driving force. However the apportioning is chosen, the driving force will be zero at l_{min} when there is no net growth, and will tend to unity at high undercooling when the off-rates become insignificant compared with the on-rates. It will therefore have the form shown in Fig. 2.4. For a non-infinite average thickness the barrier term must decay with increasing l — hence its exponential decay, also shown in Fig. 2.4. These requirements are satisfied by Eqs. (3.46)–(3.49) if:

$$\psi a\, \Delta F < 2\sigma. \qquad\qquad (3.84)$$

For smaller ψ or ΔF, A_0 is exponentially increasing and an infinite thickness is predicted, and even at smaller undercooling the average thickness is no longer governed by the $1/\Delta T$ behaviour of l_{min}. Many choices other than those given by Eqs. (3.46)–(3.49) could be made, but they should not predict significantly different behaviour in either the thickness or the growth rate until A_0 deviates from its exponential decay. The forms chosen are based on physical arguments. Hoffman et al. [104] argue that values of $\psi < 1$ are related to physical adsorption of the polymer molecule prior to crystallographic attachment. Hoffman and Miller [109] associate low values of ψ with the 'partial stem' character of the activated state involved in substrate completion: the fold (representing the high free energy barrier) is likely to be formed when only about one third of the stem has been put down in the niche. Nevertheless, the value of ψ remains somewhat arbitrary and is often adjusted to give a better fit to experimental data.

Notice that the lateral increase in surface free energy is always included in A_0 so that molecules do not attach to the surface very frequently per number of attempts. In low molecular weight theories the lateral energy is included in B_0 so that the molecules attach frequently but normally come off immediately – this choice would always lead to an infinite thickness for the long chain case. The root of this difference is similar to that of kinetic roughening, mentioned in Sect. 3.2: this occurs when there is effectively no nucleation barrier and the growth rate only depends on the spreading velocity, v. For high molecular weight, if there is no barrier to the formation of the first stem (the nucleus) then the growth again depends only on the spreading velocity, g, but this is also a function of l and favours large l. Therefore it is the dependence of g on two variables, as stressed at the beginning of this chapter, which brings about this change in behaviour.

The way in which the physical results rely so strongly on the choice of apportioning is peculiar to this particular nucleation theory and increases the number of adjustable parameters – this may be regarded as either an asset or a defect.

iii) *The Total Flux*

The nucleation rate onto a crystal is determined by the flux onto an ensemble of substrates. As the nuclei should be widely separated for the nucleation approach to be valid, this does not appear to be unreasonable. However, the subsequent way in which this flux is used to determine the thickness and growth rate seems somewhat inconsistent as explained below. However, a modification of the derivations would satisfy this query, and it is not likely that this will greatly affect the results.

The average thickness is obtained by using $S(l)$ as the distribution function for l in the ensemble. However, the growth of crystals may be either regime I or II, and it is only in the former that the growth rate depends on the supercooling via i alone – in regime II the growth rate is dependent upon both i and g. Therefore a better choice of distribution function for l would be G_I or G_{II} using Eqs. (3.17), (3.19), (3.39) and (3.51). Any difference this would make may well be insignificant compared with the changes incurred through allowing for fluctuations in height between stems within the same crystal.

When $\langle l \rangle_{av}$ has been determined by this method it is then used in the expression for g in the determination of regime II growth. Therefore in Eq. (3.19) i is determined from the total flux, whilst g is determined from the spreading rate at $l = \langle l \rangle_{av}$. It would be more consistent to use both i and g for arbitrary l and then sum over all possible l, or to use both i and g at $l = \langle l \rangle_{av}$. As g is not the controlling factor for low supercoolings (but see also Sect. 3.6.3), again this is probably unimportant for quantitative results.

iv) *Thickness and Growth Rate*

The assumptions involved in determining these physical observables have already been discussed above. We leave comments on the choice of β to someone more qualified. The important points of Eqs. (3.56)−(3.60) and (3.62)−(3.65) are re-iterated here.

At low undercoolings the $1/\Delta T$ dependence of the thickness arises from the dependence of l_{min}, and would occur for any kinetic theory which had a peak in the growth rate as a function of l close to l_{min}. At large undercooling there is a 'δl catastrophe'. Its onset depends on the apportioning of the free energies in the rate constants. Choosing $\psi = 0$ would eliminate this effect, but is argued to be unphysical [104].

The dominant supercooling dependence of the growth rate comes from the final terms in Eqs. (3.64) and (3.65) for regimes I and II, respectively. A plot of log (G) versus $1/\Delta FT$ ($\propto 1/\Delta TT$) should be approximately straight, and the gradient in regime I will be twice that in regime II. This difference by a factor of two comes from $G_I \propto i$, whilst $G_{II} \propto (gi)^{1/2} \propto i^{1/2}$. Hence it depends crucially on the spreading rate g remaining constant over a large range of supercoolings. This is essential to the interpretation of growth rate regimes as given by the LH formalism.

Furthermore, even if we accept the use of A's and B's as rate constants with a ratio determined by detailed balance, there are many problems associated with evaluating the free energy changes (see e.g. Ref. [26]) and it is unlikely that they are simply related to equilibrium crystals.

3.5.3 Comparison of LH Theory with Experiment

The ultimate test of any theory must be comparison with experimental results, and a successful theory will agree within the errors of the theory and the experiment. Despite the large number of parameters in the LH theory there appears to be no systematic study of the range of behaviour which could be encompassed. If this range were large, then agreement with experiments would be neither a surprise nor a justification. On the experimental side there are many difficulties associated with growth studies, and determination of the steady-state behaviour must be extracted from the initial nucleation of crystals, depletion and fractionation effects [93], and reorganisation of chains after growth [66, 110]. As techniques improve the errors involved will hopefully decrease.

The evidence for LH theory usually stems from thickness and growth rate studies [31]. Plots of l versus $1/\Delta T$ and $\log G/\beta$ versus $1/T\, \Delta T$ are often straight lines in accordance with Eqs. (3.58) and (3.64), (3.65). Potential uncertainties entering the analysis at this stage are related to the choice of T_m and T_d (melting or dissolution temperature), and of the prefactor β according to Eqs. (3.66) and (3.67) [9]. There may be a drop in the slope of the growth rate curve by a factor of approximately two at some supercooling ΔT, which is interpreted as a regime I/II transition. Likewise, a rise in slope by a factor of approximately two at high supercooling is interpreted as a regime II/III transition (see Fig. 3.12). A list of such transitions has recently been compiled by Hoffman and Miller [107]. There is a wide range of examples but we nevertheless believe that each individual case should be examined very carefully in the light of other possible causes of such changes in slope.

Note that the 'regime I/II' break usually occurs in melt grown crystals (cf. Sect. 3.6.2) and is often related to changes in morphology (e.g. Ref. [111, 112]). The change in slope is generally very abrupt whereas the theory predicts a more gradual behaviour (Fig. 3.12). Such a gradual change in slope has recently been observed for solution growth [113, 114]. A regime I/II interpretation fits the Organ and Keller data but changes in other factors cannot be ruled out as contributors at this stage. In particular, impurity effects are discussed by Toda as an explanation for breaks in the growth rate because neither the change in slope nor the concentration dependence is found to be in agreement with a regime I/II transition.

Further, there are some experimental results which could be explained on the basis of a regime II/III transition [22, 112, 115–117], although the high super-coolings required necessitate growth from the melt. Even if the LH approach is accepted, then cilia nucleation and entanglements must be non-negligible in this regime and should be incorporated into the equations. We make no further comments on regime III, because of these difficulties and other problems which become more important far from equilibrium, making any verification extremely difficult.

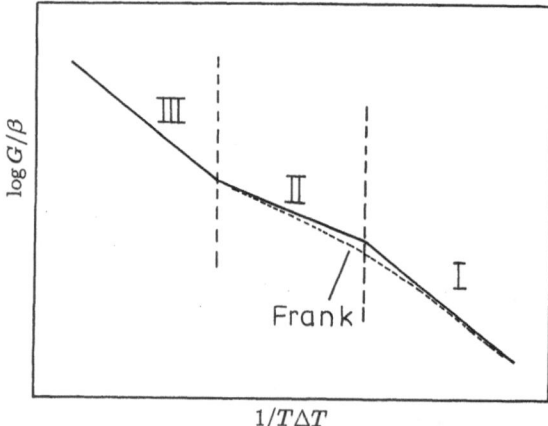

Fig. 3.12. Schematic growth rate plot showing regimes with sharp breaks in slope, as well as the smooth transition predicted by the Frank model

As mentioned above, a regime I/II transition is predicted for any nucleation theory if the temperature dependence is determined primarily by i, i.e. $g = const.$ This precondition of the regime I/II interpretation puts severe restrictions on the substrate length L_p in cases where both regime I/II and II/III transitions are claimed. At the highest supercoolings, in regime III, the kinetic length L_k^{III} is at least the width of one or a few stems. Then, as the growth rate changes from G_{III}, down, often by several orders of magnitude, to G_I at the "regime I/II" transition, so does L_k [115, 118]:

$$L_k^I = L_k^{III} G^{III} / G^I .   \tag{3.85}$$

By definition, this is equal to the substrate, or persistence, length L_p (see Sect. 3.4.2) which by this criterion can easily take values so large that nonlinear growth should be observed for small crystals in regime I.

Another discrepancy involves measurements of the kinetic and persistence length to determine an upper bound for g — Eq. (3.32), [87]. Using decoration techniques Dosière et al. find an upper limit for L_k of 200 nm for polyethylene. This leads to g_{max} ranging from 0.2 nm/s to 0.1 µm/s, which are much lower than those predicted from the LH equations using the values of σ deduced from the growth rate curves. Inconsistencies in the work of Leung et al. [119] have also been pointed out [87, 120] where $g > 30$ m/s would be deduced from the σ's derived, and $L_k > 5$ mm, despite claims that the growth curve is linear for crystals smaller than 1 µm. They conclude that the expressions for i and g are not correctly given by LH theory but, in contrast with Sadler, do not therefore dismiss nucleation as the controlling process (Sect. 3.7).

As a possible resolution of this dilemma Point and Colet [118] suggest that the spreading rate depends on supercooling in much the same way as the nucleation rate does[8]. This would lead to an approximately constant kinetic length, and it follows that a regime I/II transition would not exist. A different explanation for the observed changes in slope would need to be found.

The thickness plot may be used to determine σ_e (or this may be found independently from melting studies [29, 30, 121]), then σ can be determined from the growth rate gradient. As the regime of growth is normally not known in advance it must be established by calculating the z value (Sect. 3.3) in both regimes, and choosing that which makes physical sense. A reasonable value of σ is then taken as 'proof' of the theory. Although these calculations are self-consistent, unless an independent estimate of σ is available they cannot verify the theory.

There is considerable scope for alternative explanations (see Sect. 3.6.3, 3.7, and 4). Most of the deductions from the data ignore the variation in σ_e due to the kinetic nature of the process, despite claims within the theory that this should have a large effect [122]. Equilibrium uncertainties as described in Sect. 2.3 are completely ignored. When the parameters within LH theory are also considered it no longer seems surprising that self-consistency can be obtained.

---

[8] Evidence for this behaviour has indeed been found [98] for (100) faces of polyethylene. It is rather indirect, however, as it draws on a combination of kinetic data from single crystals and twins, so that the results must be treated with great caution, as noted in Sect. 3.4.3.

Finally we would like to draw attention to low molecular weight results and their analysis on the basis of surface nucleation theory. The theory was originally developed for infinitely long chains and cannot easily be applied to extended or once-folded chain crystallization. Therefore any discrepancies in this area would not be surprising and would not discredit the theory at higher molecular weights.

The experimental evidence is derived from melt-crystallization of low molecular weight fractions of poly (ethylene oxide) (PEO) [32] and of solution-crystallization of low molecular weight fractions of polyethylene and a pure n-alkane [119, 123]. The typical growth rate dependence on supercooling shows growth branches with discontinuous transitions. These are believed to relate to changes in the number of folds per molecule. In PEO this expresses itself directly in the lamellar thickness which is always an integral submultiple of the overall chain length. The chain ends prefer to reside in the surface layer rather than in the interior of the crystal or dangle in solution. This is not the case in polyethylene which exhibits a smooth increase in lamellar thickness with temperature.

The LH theory formalism was applied with little comment to the low molecular weight polyethylene data by Leung et al. [119]. They use the Lauritzen z-test (Sect. 3.3) to identify the relevant growth regime without considering that coherent spreading across the growth face cannot be taken for granted for such short chains. Furthermore, in a case where z falls between regime I and II values they simply assume an intermediate growth rate dependence: $G \propto i^{3/4}$, which has no theoretical basis. They obtain widely varying values for the product of surface free energies $\sigma\sigma_e$ (see Eqs. (3.64), (3.65)) which they attribute solely to variations in σ_e, applying a constant σ as given by Hoffman [123]. In the extended and once-folded chain cases σ_e is found to be very much lower than the 'accepted' value, and in fact the growth rates depend linearly on some supercooling $\Delta T = T_{G_0} - T_C$ where T_{G_0} is the temperature at which the growth rates converge to zero. Application of the LH formalism is certainly no longer valid in this case.

Point and Kovacs [32] and later Hoffman [123] modified the theory for extended chain crystals and growth branches of constant thickness. This is simply done by fixing the thickness l in the flux Eqs. (3.50). Note however, that this has no longer any resemblance to the original concept of nucleation as the formation of a critical nucleus due to random fluctuations in nucleus size as described in Sect. 3.2 and 3.3. It can merely be regarded as a formalism which fits the experimental data. It does not give any explanation for the mode of growth (folded or extended) and does not give any insight into the transition between the low and the high molecular weight behaviour. The resulting growth rate dependence on supercooling differs substantially from the long chain, variable thickness case:

$$G \propto 1 - \exp\left[-const\ \Delta T\right]$$

$$\propto \Delta T \quad \text{for small supercoolings}. \tag{3.86}$$

Here ΔT is defined in different ways by Hoffman and Point et al. as discussed in Sect. 2.3. A further difference concerns the free energy barriers (see Eqs. (3.46)–(3.49) and (3.81)). Nevertheless both treatments should lead to at least qualitatively similar results. For a pure n-alkane and a low molecular weight polyethylene fraction Hoffman finds that the lateral surface free energy takes approximately the same value as for chain folded polyethylene. In the PEO case, however, Point and Kovacs obtain such low values for σ that they must be regarded as physically meaningless, reflecting flaws in the theory. In both cases the growth rates seem to be linear with ΔT for small supercoolings. This is formally explained by the modified theory, Eq. (3.86), but it can equally well be regarded as evidence for rough surface growth (see Sect. 2.3), as pointed out by Sadler [124].

3.6 Modifications and Extensions of The LH Theory

The model described in Sect. 3.5.1 is a very crude representation of a true three-dimensional lamella, and over the years modifications have been applied in order to make it more realistic. The major assumptions, however, are still inherent in all of them, that is, the deposition of complete stems is controlled by rate constants which obey Eq. (3.83). No other reaction paths are allowed and the growth rate is then given by nucleation and spreading formulae. We do not give the details of the calculations which are very similar, but more complicated, than those already given. Rather, we try to provide an overview of the work which has been done. Most of this has been mentioned already elsewhere in this review.

3.6.1 Multicomponent Chains

The earliest works extended the treatment to multicomponent chains, so that the rate constants for adding stems to a nucleating strip depended not only on the number of stems already deposited (see Eqs (3.68) and (3.79)), but also on the species of the final stem in the strip and the attaching stem [108]. They derive equations which allow the flux and the average composition of the lamella to be determined if all of the rate constants are known. Lauritzen and Passaglia [48] used this approach to allow stems of any height to be deposited, so that the various 'species' represented the possible lengths the polymer chain may form on folding. The results for the rate of growth and thickness are similar to those of simpler theories, provided that the end surface free energy is replaced by a kinetically determined effective surface free energy, which may be significantly different from that at equilibrium.

Frank and Tosi [105] also included stems of various heights, but only allowed one fluctuation in each growth strip. Successive strips depositing upon one another eventually converge to a stable thickness, however, their treatment is still essentially two-dimensional and as a consequence they predict no growth at low supercoolings. All of these approaches to fluctuating heights in the lamella still predict the δl catastrophe, which is usually discounted on the basis that it occurs for high supercoolings where many other effects will become more important.

There are several further applications of the multicomponent theory: Firstly, cilia have been treated [125, 49], such that there are two possible components attaching to the growth strip — cilia-nucleated molecules or molecules from solution. The consequences of these processes were described in general terms in Sect. 3.4.2, and as indicated there more specific predictions may be made on assuming LH-type rate constants.

Secondly, fractionation effects have been studied [126] where now the components are the various molecular weights in the solution. They conclude that fractionation will be important even at high molecular weights due to the variation in supercooling of the different molecular weights, and hence their different growth rates. Notice that this result could well be affected by the 'definition' of supercooling as described in Sect. 2.3.

Thirdly, the multicomponent model was applied to the case of crystallization of a random A–B copolymer by Helfand and Lauritzen [127]. Their main result is that the composition of A's and B's in the crystal is determined by kinetic, rather than equilibrium considerations: the inclusion of excess B increases with growth rate.

Finally, the amount of adjacent re-entry was studied [128, 129] associating components with the type of folding. The growth problem was elegantly solved by a 'kinetic transfer matrix' technique. For n-paraffins the amount of non-adjacent re-entry was found to be near fifty percent at crystal-liquid equilibrium but falls strongly for lower temperatures and higher dilution. The general validity of these results is limited because of the restriction to small systems, and investigation of longer chains and higher orders of folding would be both desirable and feasible with powerful computers.

3.6.2 Melt Crystallization

The influence of chain diffusion processes on melt crystallization behaviour was first discussed in connection with a debate about the structure of the fold surface: random 'switchboard' or regular adjacent re-entry. Yoon and Flory [14] claimed that adjacent re-entry is an improbable event because molecular motions in the liquid are too slow to allow substrate completion by a 'reeling' process. This objection was addressed by Hoffman, Guttman and Di Marzio [130] by combining the LH nucleation model with the chain reptation concept of de Gennes [131, 132]. They find that the rate of transport in the melt is sufficient to permit a considerable amount of tight chain folding. This result was supported by other papers in Vol. 68 (1979) of the Discussions of the Faraday Society. The consequence of implementing reptation as a part of the crystallization process leads to a chain length dependence of the nucleation rate and the spreading rate: $i, g \propto 1/p$ [133]. Therefore the growth rate is predicted to depend on $1/p$ in both regime I and regime II. Good agreement with kinetic data is claimed if p is based on the z-average molecular weight.

In a recent paper Hoffman and Miller [134] developed the nucleation plus reptation model further. They arrive at a chain length dependence of the growth rate at constant supercooling as $p^{-[y+f(\lambda)]}$, where $f(\lambda)$ is due to the free energy of the first attachment of a chain to the substrate, and $y = 1$ in case of steady state reptation. Based on a range of published growth rate data, y was indeed found to be 1.0 ± 0.2 in the vicinity of the regime I/II transition. In contrast to the original theory, now a different molecular weight dependence in regime I

and *II* is predicted: $G_I \propto p^{-4/3}$ whereas $G_{II} \propto p^{-7/6}$. However, this has not yet been substantiated. In particular, Cheng et al. [112] found that slope changes with molecular weight occur also within one regime.

A further important result of this 'improved' LH theory is a lower substrate completion rate and a smaller substrate length. Instead of the previously assumed value of around 1 μm L_p is now predicted to be only about 50 nm.

This new estimate resulted in further debate between Point and Dosière [106] and Hoffman and Miller [107] which essentially centers around the question of existence of a regime *I* (i.e small L_p and constant g). For further details we refer the reader to the original papers.

The reptation-like molecular weight dependence was also challenged by Point and Dosière [135]. They show a series of published experimental data which clearly does not follow the theoretical curve. After further criticism of the LH and HM theory concerning agreement with experiment and self-consistency Point and Dosière conclude that the Frank model (Sect. 3.3) does probably not hold as a basis for explaining the break in slope of the growth rate in crystallization of polyethylene in the melt[9]. Instead, Point and Dosière suggest other factors as alternative causes: molecular weight segregation, temperature dependence of the interfacial free energies and viscosity, and not least the temperature and molecular weight dependence of the configurational path degeneracy associated with the nucleation process, a constant C_0 in Hoffman and Miller's model [134] (cf. Sect. 3.7).

3.6.3 Application to Curved Edges

The LH nucleation model had been designed to explain the crystal growth of polymers with straight edges. This habit was at the time considered to be the normal 'mode' of polymer crystallization. In recent years, however, curved single crystals (Fig. 3.13) grown from solution or extracted from the melt have been observed in well controlled experimental studies ([113, 136, 137] and further references in Ref. [9]). These observations have been used as an argument against LH theory because the presence of niches on a curved edge obviates the need for nucleation.

In particular, the experimental evidence that the curvature increases with decreasing supercooling is just the opposite of the trend predicted. The LH theory

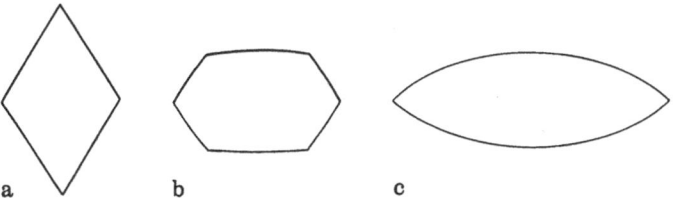

Fig. 3.13a–c. Series of shapes of polyethylene single crystals at increasing temperatures. **a** $T_c = 70\,°C$, from xylene; **b** $T_c = 99\,°C$, from dodecane (courtesy of S. Organ); **c** $T_c = 129\,°C$, from the melt (from [136])

---

[9] This would of course have drastic consequences as far as the validity of interfacial free energies derived from LH analysis of growth rate data is concerned.

is based on a nucleation barrier whose magnitude increases with the thickness of the crystal, i.e. with decreasing supercooling, yet the increasing number of steps on the surface of the crystal would seem to suggest the opposite.

Sadler [76] analysed the outline of the crystal edges and found that the lateral surface tension compatible with such a curvature would be $\sigma \simeq 1 - 3\,kT/\langle l \rangle$. Although he used a very simplified approach, it seems unlikely that its assumptions could lead to $\sigma \simeq 20\,kT/\langle l \rangle$ as predicted by the LH theory.

It has also been noticed [138] that lower supercoolings are achieved through the use of poorer solvents. These act as impurities in the crystallization process obstructing the spreading of steps on the substrate. It is shown that this effect is particularly strong near the edges of the growth face. This 'poisoning' causes a rounded profile as the attachment of a molecule to a niche site is no longer governed by the surface tension. Using growth equations similar to the Frank Eqs. (3.11) and (3.12), but allowing for the immobility of steps, Toda generates solutions and the profiles of crystals under such an assumption.

Another approach to the problem of curved edges is based on a solution of Frank's equations in the case of moving boundaries by Mansfield [139]. Figure 3.14 shows the ellipitical profile which would arise if the sides of a crystal sector move outwards with a constant velocity, h, which is of comparable size to the spreading rate, g. The magnitude of h is supposed to be determined by the growth rate of the adjoining 'dominant' sector.[10]

Given the nucleation theory described so far these conditions could not be fulfilled and straight facets should always be expected. If, however, there exists a mechanism to reduce g preferentially on one of the sectors then a curved face may be created. Recently it has been suggested that this mechanism may be provided via lattice strain [109], which occurs because of the 'bulkiness' of the fold along one growth face and causes the interior of the crystal to expand. The resulting volume strain is translated into a strain surface free energy σ_s. This takes the role of a 'nucleation' barrier as the lateral surface free energy σ is no longer relevant

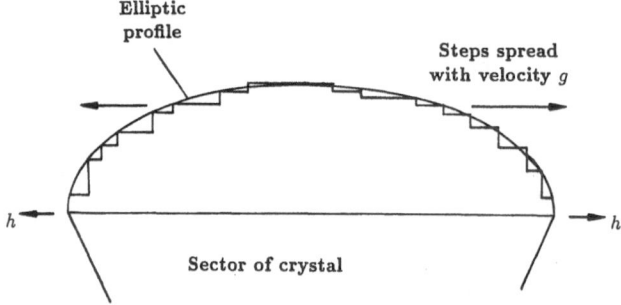

Fig. 3.14. The profile of a lamellar growth face which may be obtained by using moving boundary conditions. The boundaries move with a constant velocity, h

---

[10] The correctness of Mansfields calculations has been questioned [106]. Recently Mansfield supported his case by publishing computer simulation results [140]. The debate remains to be fully settled, however.

on a surface serrated on a molecular level. The nucleation formalism can then be applied in the usual way. But as σ_s is much smaller than σ the substrate completion rate turns out to be significantly reduced compared with the neighbouring unstrained sector and the Mansfield condition for a curved profile is met. Notice, however, that in dropping the lateral surface free energy σ a high density of steps has already been assumed from the onset. Therefore, even if lattice strain plays an important role in causing the curvature, it remains to be explained why and when σ_s takes over from σ as the determining factor.

There are a number of further consequences and predictions of the Hoffman-Miller-Mansfield model which will be discussed briefly below.

i) δl catastrophe
See "Note added in proof".

ii) The dependence of the spreading rate on supercooling
The reduced interfacial free energy ($\sigma_s \ll \sigma$) operating on the curved face leads to a dependence of the spreading rate on supercooling in much the same way as that of the nucleation rate. As a consequence the regime interpretation no longer applies (see Sect. 3.5.2 iv). Growth on (200) is always in regime II[11].

iii) The size of L_p
It is a condition of the Mansfield model that steps are not in principle obstructed from travelling across the entire growth face. This leads to a substrate length on (200) of polyethylene very much larger than that predicted for (110) ($L_{(110)} \simeq 50$ nm).

iv) Melting Points
A different melting point, and hence supercooling, is predicted for the strained sector. This is the basis for a different interpretation of the (200) growth rates: a regime I/II transition occurs on (110) but not on (200). This is despite the fact that the raw data [113] show a similar change in slope when plotted with respect to the equilibrium dissolution temperature (Fig. 3.15). It is questionable whether it is correct to extrapolate the melting point depression equation for finite crystals which is due to lattice strain caused by folds, to infinite crystal size while keeping the strain factor constant.

Furthermore, different fold surface free energies for the different sectors and a maximum crystallization temperature are predicted. For details the reader is referred to the original paper.

In spite of the queries concerning the Hoffman-Miller model, further evidence is gathering for a strong link between curved edges and a combination of moving sector boundaries and slowly spreading steps [141]. When special is taken in the choice of boundary conditions and the proper accounting of step annihilations, the full range of observed shapes can be obtained [142].

---

[11] It is of interest to notice the similarity between this result and the view held by Point that (albeit for different reasons) g is supercooling dependent and regime I does not exist.

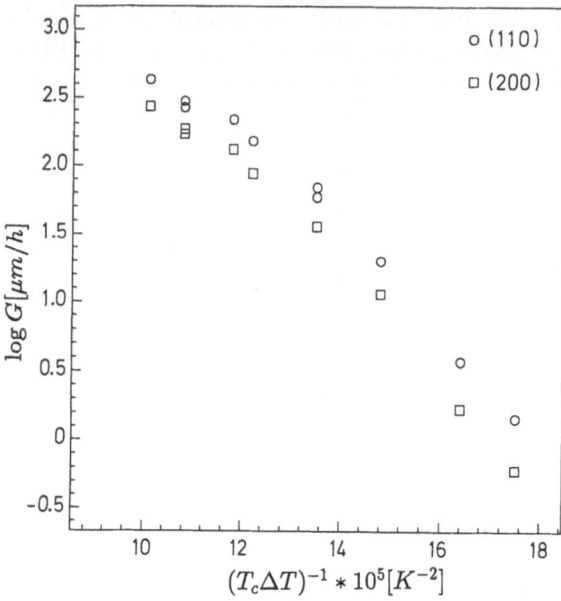

Fig. 3.15. Growth rates of (110) and (200) sectors of polyethylene single crystals grown from solution in tetradecanol (courtesy of S. Organ)

3.7 Point's Multi-path Approach

The nucleation theory explained in Sect. 3.3–3.6 views the deposition of stems as the attachment of whole units rather than the 'zippering down' of individual segments to form a complete stem. Although this shortcoming is recognized it is justified by the calculation of Frank and Tosi [105], described in Sect. 3.5.2, which concludes that the rate of deposition of m units is equivalent to the deposition of one unit if the final free energy barrier is large enough and the units are in equilibrium proportion to one another. These conditions are assumed to be satisfied for long chains because of the high fold energy incurred at the end of each completed stem deposition. Since this initial work several authors [51, 52, 143] have suggested that whilst the conclusions of Frank and Tosi are correct for the particular situation they consider, the situation itself does not accurately model the processes occurring during the attachment of stems. In particular the segments which join the stem are allowed to fluctuate forwards and backwards, but only when they have reached the lamellar thickness are they allowed to fold. A more realistic picture would allow the chain to fold at any stage during the formation of a stem (stem 1), and possibly depositing several segments of another stem (stem 2), before fluctuating backwards to continue along stem 1, which must at least reach a minimum size for thermodynamic stability. In Figure 3.16a we show a series of fluctuations which would be allowed by Frank's model, and in Fig. 3.16b and c a series of events which would not be allowed, but might be expected to occur. Even if these events are thought to be very unlikely, their presence is sufficient to prevent the 'δl catastrophe' which is inherent in all the LH theories. If the

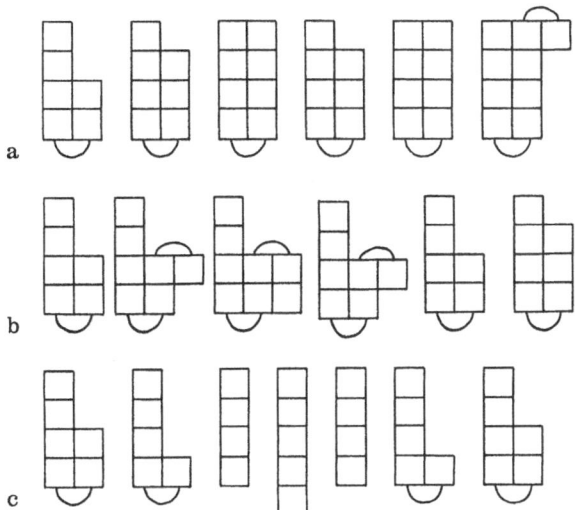

Fig. 3.16a–c. a A series of fluctuations which are allowed in Frank's model, that is, attachments and detachments are both allowed but the chain is not permitted to fold until it has reached the lamellar thickness. b and c show a series of events which may also be expected to occur

probability of folding at any stage is reasoned to be likely, so that the chain is envisioned as folding and unfolding as it finds its thermodynamically and kinetically preferred stem length, then the simple form of the LH free energy barrier will not be applicable at any supercooling.

In Sect. 3.7.1 we show how a very simple picture [143] can be used to demonstrate the prevention of the 'δl catastrophe'. Section 3.7.2 describes the theory due to Point [51, 52] in which he allows for folding during the deposition of the initial stem of a molecule onto the crystal substrate, but thereafter considers subsequent stems to join as one unit with a stem length equal to that of the initial stem. The problem then becomes that of determining the rate of nucleation of the first stem, i, which is no longer related in a simple manner to the surface free energies. The filling in velocity, g, is given by the same expression as in the Lauritzen and Hoffman theory − Eq. (3.39).

3.7.1 Limitation of the Lamellar Thickness

This very simple and rather unphysical model [143] is intended to illustrate the basic limiting process. An initial stem is laid down one segment at a time, such that the probability of continuing is a_c, and at each level the chain may also bend, that is, fold, with probability a_b. It does not take any detaching or unfolding events into account. Hence:

$$a_c + a_b = 1, \tag{3.87}$$

The process is strictly sequential, so the probability of the stem achieving a length l is a_c^{l-1} (the first step must have been a deposition), and the probability of reaching length l followed by a fold is:

$$p_l = a_c^{l-1} a_b \tag{3.88}$$

with

$$\sum_{l=1}^{\infty} p_l = 1 . \tag{3.89}$$

The expected stem length is:

$$\langle l \rangle = \sum_{l=1}^{\infty} l p_l = [1 - a_c]^{-1}$$

$$= 1/a_b$$

$$= 1 + a_c/a_b \tag{3.90}$$

which is always finite. It is probably misleading to calculate the detailed behaviour of the average thickness from this model in view of its neglect of backward fluctuations. However, at high supercoolings where detachment events become unlikely it would seem reasonable to assume that both a_c and a_b have the same supercooling dependence (the bulk free energy gain for both is the same), and hence $\langle l \rangle$ would be independent of supercooling. This behaviour has indeed been observed in experiments [144].

3.7.2 Point's Nucleation Barrier

Figure 3.17 illustrates the processes allowed in Point's model, and defines the rate constants to be used. The approach is very similar to that used in Sect. 3.5.2, and we shall use any results derived there which are applicable without repeating the calculation. The first stem can be of any length, l, and the number of such stems in an ensemble is N_l. The net current between N_l and N_{l+1} is S_l which depends on the forward and backward rate constants for a segment, A and B. Subsequent stems are of the same length and the current between the kth and $(k + 1)^{th}$ stem of length l is J_{lk} and depends on the rate constants for a complete stem, A_{lk} and B_{lk}, and on the number of such stems, M_{lk} and M_{lk+1}. The time dependent equations are:

$$dN_l/dt = S_{l-1} - S_l - J_{l1} \quad \text{for} \quad l \geq 1 , \tag{3.91}$$

$$dM_{lk}/dt = J_{lk-1} - J_{lk} \quad \text{for} \quad l, k \geq 1 , \tag{3.92}$$

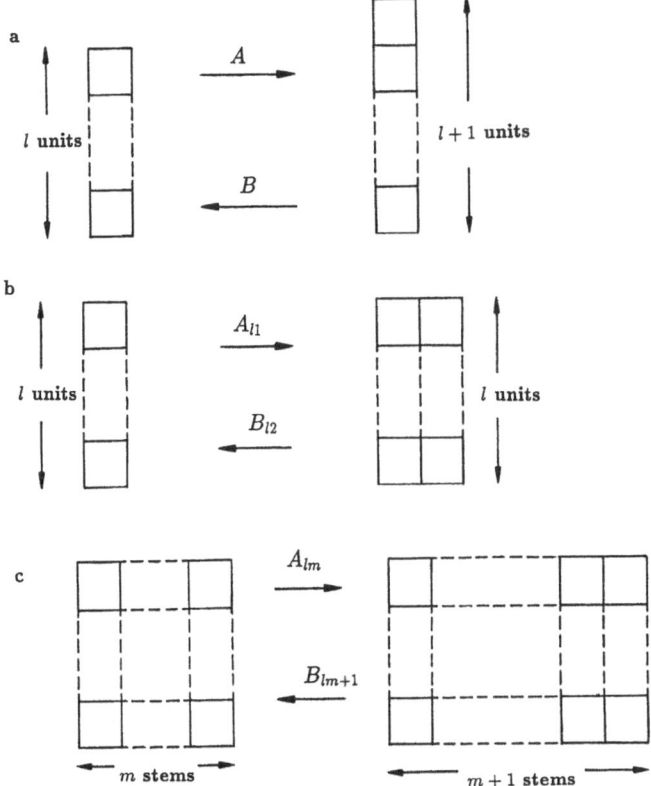

Fig. 3.17. The processes and associated rate constants in Point's model. During the deposition of the intial stem units can add or subtract with rate constants A and B. After the first stem only complete stems may add

and we can write expressions for the J's and S's in terms of the occupation numbers as:

$$S_0 = AN_0 - BN_1$$

$$\vdots$$

$$S_l = AN_l - BN_{l+1} \quad \text{for} \quad 0 \leqq l \leqq \infty$$

$$J_{1,1} = A_{11}N_1 - B_{12}M_{12}$$

$$\vdots$$

$$J_{l1} = A_{l1}N_l - B_{l2}M_{l2} \quad \text{for} \quad 0 \leqq l \leqq \infty$$

$$J_{lk} = A_{lk}M_{lk} - B_{lk+1}M_{lk+1} \quad \text{for} \quad 1 \leqq l \leqq \infty, \quad 1 \leqq k \leqq \infty.$$

$$\text{(3.93)}$$

As in Sect. 3.5.2 we take the steady state solution as the same as the $t \rightarrow \infty$ solution. Equations (3.91), (3.92) and (3.93) are identical in form to Eqs. (3.68) and (3.69), and so immediately we have the result:

$$S_{l-1} = S_l + J_{l1} \tag{3.94}$$

$$J_{lk} = J_{lk+1} = J_l \tag{3.95}$$

$$J_l = N_l A_{l1} / \sum_{j=0}^{\infty} \prod_{k=0}^{j} (B_{lk}/A_{lk}) = C_l N_l. \tag{3.96}$$

Substituting back into Eq. (3.91) gives:

$$A N_{l-1} - (A + B + C_l) N_l + B N_{l+1} = 0. \tag{3.97}$$

This is Point's [51] equation (1), which he derived by simply postulating a net forward rate for folding, C_l. We followed Di Marzio and Guttman's [143] derivation because it illustrates the way in which C_l is connected both with the microscopic forward and backward rate constants.

A general solution for Eq. (3.97) is not possible, however it may be possible to deduce the qualitative behaviour of the solutions for the growth rate and distribution of thickness. This requires a reasonable estimate of the way in which C_l varies with i from Eq. (3.96). The simplest LH type approach would assume that:

$$B_{lk}/A_{lk} = \exp\left[(-\Delta F \, abl + 2\sigma_e ab)/kT\right] \quad \text{for} \quad l > 1 \tag{3.98}$$

that is, the stem can be treated as a complete unit and is independent of the number of stems, k, already laid down. This is just the type of assumption which this approach tries to avoid. However, it would be a reasonable first step to assume independence of k, indeed it would be difficult to postulate any other type of dependence. Also from thermodynamics Eq. (2.3) we get:

$$B_l/A_l > 1 \quad \text{for} \quad l < l_{min} \tag{3.99}$$

Although this is true in some sort of averaged sense, in that the net forward rate is less than the net backward rate for $l < l_{min}$, the length of the individual stems may fluctuate about l_{min} because of surface entropy effects. Using Eq. (3.99) in Eq. (3.96) shows that:

$$C_l = 0 \quad \text{for} \quad l < l_{min} \tag{3.100}$$

because $(B_l/A_l)^r \rightarrow \infty$ as $r \rightarrow \infty$, for $B_l/A_l > 1$.

For $l > l_{min}$, $B_l/A_l < 1$ and C_l can be written:

$$C_l = A_{l1}/[1 + (B_l/A_l) + (B_l/A_l)^2 + (A_l/B_l)^3 + \ldots]$$

$$= A_l(1 - B_l/A_l)$$

$$= A_l - B_l \tag{3.101}$$

which would be expected to be an increasing function of stem length, l, and supercooling, ΔT.

We now turn to obtaining estimates for the expected crystal thickness, that is the solution of Eq. (3.97), for various values of C_l where $C_l = 0$ for $l < l_{min}$ and increases with l otherwise. The case $C_l = constant$ can be used to show that the finite probability of folding is sufficient to obtain a finite thickness at all supercoolings thus avoiding the δl catastrophe, which was demonstrated in Sect. 3.7.1. This case is unphysical and was only considered because of its mathematical simplicity. It leads to the prediction that the thickness, though finite, increases with ΔT.

In an ensemble of crystals, each of thickness l, the mean value of l is given by:

$$\langle l \rangle = \sum_{l=0}^{\infty} lJ_l / \sum_{l=0}^{\infty} J_l$$

$$= \sum_{l=l_{min}}^{\infty} lN_lC_l / \sum_{l=l_{min}}^{\infty} N_lC_l \tag{3.102}$$

as $C_l = 0$ for $l < l_{min}$. This uses the fact that the probability that a segment resides in a stem of length l is proportional to J_l. As C_l increases with both l and ΔT we can choose an l^* not much greater than l such that for $l \geq l^*$ and sufficiently high supercoolings the following inequality holds:

$$C_l \gg (A + B) \tag{3.103}$$

and hence from Eq. (3.97)

$$N_{l-1} \gg N_l \gg N_{l+1} \tag{3.104}$$

$$N_l/N_{l-1} = A/C_l . \tag{3.105}$$

Point chooses rate constants

$$C_l = 2 \sinh \{(ab \, \Delta F \, l/2 - \sigma_e ab)/kT\} \tag{3.106}$$

$$A = \exp \{(\Delta F \, abl_u/2 - \sigma bl_u)/kT\} \tag{3.107}$$

$$B = \exp \{(-\Delta F \, abl_u/2 + \sigma bl_u)/kT\} \tag{3.108}$$

where l_u is the length of a single segment. He shows that the contribution to $\langle l \rangle$ from all $l \geq l^*$ is approximately l^*, that is, the contribution to $\langle l \rangle$ from large values of l is negligible and only the first few terms in (3.102) need be considered. The choice of C_l in Eq. (3.106) treats the deposition of stems as complete units and therefore its accuracy may be questionable (as in Sect. 3.5.2). However, any choice of C_l which satisfies Eq. (3.103) would lead to a similar result and leads to the important conclusion that allowing the initial stem to fold at each stage in its deposition not only prevents the δl catastrophe, but prevents l from increasing with ΔT. Whether l decreases with increasing ΔT or remains approximately constant requires a specific choice for the rate constants and a numerical solution is necessary. Using rate constants given by Eqs. (3.106)–(3.108) Point fitted such a numerical result to experimental data for polyethylene, polystyrene and Nylon 6.6, and finds that the best fit values for the parameters l_u, σ_e and σ are close to their expected values: for example for polyethylene $l_u = 13.3 \times 10^{-10}$ m, $\sigma_e/k = 70.2 \times 10^{20}$ m^{-2} K, $\sigma/k = 5.7 \times 10^{20}$ m^{-2} K.

The total growth rate is given by:

$$S = \sum_{l=0}^{\infty} J_l = \sum_{l=l_{min}}^{\infty} N_l C_l. \tag{3.109}$$

The calculation of the dependence of S upon ΔT does not seem to have been performed, however, as explained in Sect. 3.3 any theory which has a growth rate proportional to $\exp\{-(l - l_{min})/kT\}$ will display the correct temperature dependence. As Eq. (3.109) should be dominated by values of l close to l_{min} the correct functional form for the growth rate follows directly.

We conclude this section by emphasizing that although specific choices for the rate constants were made in the original paper which could be questioned on the same basis as those of LH theories, this is not necessary to deduce the general physical behaviour of this model. Models which alow for a wide range of crystallization paths are capable of predicting a wide range of physical behaviour.

3.8 Other Nucleation Theories

In this final section on nucleation theories we briefly discuss two alternative approaches which require different frameworks for their descriptions. A characteristic of the theories detailed so far is that the basic nucleus consists of a single stem crossing the growth face of the lamella — the discrepancies arise in the origin of the barrier against the formation of this nucleus. In the following, the assumption of a multi-stem nucleus is used, and hence although the nucleation and subsequent growth rates are still theoretically valid quantities, they are not the natural parameters for the problem.

First we consider the theory initiated by Wunderlich and Mehta [145] which they termed 'molecular nucleation'. The we turn to recent work by Hikosaka [146, 147] who introduced the idea of 'sliding diffusion'.

3.8.1 Molecular Nucleation

It is well-established that fractionation occurs during crystallization for broad band molecular weight homopolymers [86, 148]; that is, segregation occurs such that the lower molecular weight components remain in the melt (or solution) and only the higher molecular weight components become crystalline. The rejected components depend upon the growth conditions, which indicates that fractionation is directly linked to the crystallization process. Hence a study of the conditions which affect the rejection should provide important information on the factors governing polymer crystallization. In a series of papers [145, 149–151] detailed experiments on polyethylene and PEO with results on the segregation and the growth rates have been reported.

Using a broad band molecular weight distribution of polyethylene [145, 149] a 'critical molecular length' is found − this is the limiting length above which molecules of the original sample are accepted into the crystalline phase and below which they are rejected and depends upon both the crystallization temperature and the distribution of molecular weight. A typical experimental curve is sketched in Fig. 3.18 − a similar curve shifted in temperature is also obtained for solution growth. This curve may then be compared with a theoretical curve for those components which may be included under equilibrium conditions using a Flory-Huggins expression − this curve is also shown in Fig. 3.18. At low supercoolings there is a wide discrepancy between this curve and that determined experimentally. This result stands in contrast to Sadler's [86] analysis of fractionation on crystallization. He found that the distribution of chains between crystal and solution could be fitted best by an equilibrium curve.

Cheng and Wunderlich [150] studied the crystallization and melting behaviour of sharp fractions and combinations of fractions of PEO. These results also led them to the conclusion that the melting of low molecular weight components can be well represented by equilibrium theory, but that molecular segregation on crystallization cannot. In a subsequent study [151] they measured the molecular weight dependence of the growth rate as a function of supercooling − typical results are shown in Fig. 3.19. Thermodynamic effects of chain lengths are largely eliminated by calculation of ΔT from T_m for a given fraction, and as all fractions should have a similar structure and morphology there should be no major

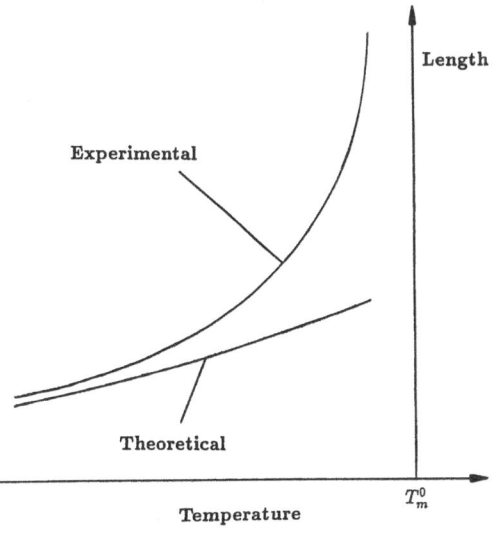

Fig. 3.18. Temperature dependence of the critical length determined experimentally and theoretically from equilibrium theory

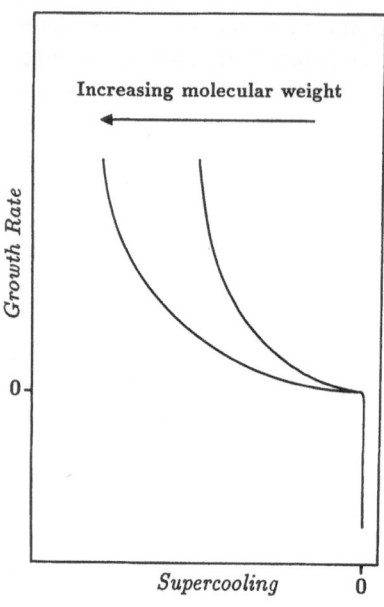

Fig. 3.19. Growth rate as a function of supercooling for sharp fractions of different molecular weights

differences in primary and secondary nucleation. The cause of the increasing barrier must therefore be looked for in molecular weight.

These observations led to the conclusion that there must be a dynamic process able to segregate the different molecular weights. This process was postulated to be molecular nucleation; each molecule must form a stable nucleus consisting of several adjacent stems, which is of sufficient size to overcome fold and chain-end effects, before crystallization. Fractionation is then governed by the ability of a molecule to form a stable nucleus, rather than the requirement that its extended length has the required lamellar thickness. On the basis of such a nucleation stage increasingly longer, flexible chains need increasing numbers of cooperative elementary steps to go from the melt to the crystal — hence the growth rate decreases with increasing molecular weight for a given supercooling. Computer simulations [152] have been performed to show that the nucleation rate would indeed vary as expected with molecular weight.

The results on fractionation are a very interesting problem which cannot be explained using a simple LH theory, which would predict all lengths allowed by equilibrium considerations to be incorporated into the crystal (although at different rates because of their different supercoolings). Although molecular nucleation would explain the results, other possible solutions should also be considered. Any kinetic process which depends upon molecular weight (which is to be expected) may lead to a similar observation. For example, multi-step processes, diffusion on the crystal surface and entanglements will all be very dependent on the length of the molecules. Also, molecular nucleation poses the problem of how extended chain crystals are formed — this and other possibilities must be thoroughly explored before accepting or rejecting this hypothesis.

3.8.2 Sliding Diffusion

As experimental techniques have improved there have been increasingly more observations which must be accommodated by theory. Such is the case for recent growth studies performed on polyethylene under high pressure [153]. A diamond anvil cell allows direct observation by optical microscopy as well as the recording of X-ray diffraction patterns during crystallization under constant pressure. Both folded and extended chain crystals are obtained − having orthorhombic as well as hexagonal structures (see also Refs. [154–157]). This work is still in progress and new developments are constantly being reported, therefore the explanations are very much in their infancy and may yet undergo several modifications. However, the foundations of a new nucleation theory have been laid [146, 147, 157, 158], so we explain the basic concepts and the way in which they differ from those of the previous sections.

The most important assumption of the theory is that the nucleus may grow two-dimensionally − this is reminiscent of the nucleation models for low molecular weight. Hence, when several stems have deposited on the growth surface they may not only spread laterally but also lengthen along the stem direction, increasing the thickness of the nucleus. Thermal equilibrium favours this thickening process so as to minimize the ratio of the excess nuclear surface area to its volume, however, during growth equilibrium cannot strictly apply and kinetic factors are also important. It is considered that macroscopic crystals grow through nuclei for which the growth rate is larger than for many other types of nuclei. As the growth rate is mainly determined by the largest nucleation rate, the nucleus with the largest nucleation rate will be observed.

The elementary process of growth is treated as the attaching or detaching of one repeating unit on the surface. There are two possible ways in which a unit may add to a nucleus, which are shown in Fig. 3.20 (from Ref. [146]). A unit may diffuse from the liquid to the side of the nucleus with a small activation energy compared with kT. However, it is very difficult for a new unit from the liquid to add directly onto the fold surface, and the thickening of the nucleus is due to the

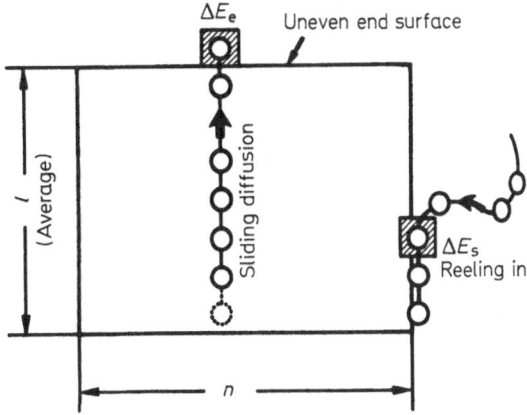

Fig. 3.20. A two-dimensional nucleus which may grow either by the addition of a new unit onto the lateral surface or by sliding diffusion within the nucleus (from [146] by permission of the publishers, Butterworth-Heinemann Ltd. ©)

diffusion of units already in the nucleus to the surface. The free energy of activation for this 'sliding diffusion' depends very strongly on the thickness of the nucleus as well as on its lattice structure (hexagonal or orthorhombic), its defect density and its chain conformation.

The relative magnitude of these two activation free energies determines the size and shape of the critical nucleus, and hence of the resulting crystal. If sliding diffusion is easy then extended chain crystals may form; if it is hard then the thickness will be determined kinetically and will be close to l_{min}. The work so far has concentrated on obtaining a measure for this nucleus for different input parameters and on plotting the most likely path for its formation. The 'δl catastrophe' does not occur because there is always a barrier against the formation of thick crystals which increases with l.

As already stated, we shall not explain the details but refer the reader to the literature for further developments. However, we would stress that there are very large intrinsic differences between one- and two-dimensional nucleation, and these are likely to be important for highly mobile phases such as the hexagonal phase in polyethylene.

3.9 Comments on Nucleation Theories

It is difficult, and probably misleading, to make generalized comments on nucleation theories as applied to polymer growth. There are so many differences between the various approaches that they should each be assessed on their own and appropriate conclusions drawn. They should obviously be compared with experiments, however this alone is not enough to establish validity, as it has been shown that very different assumptions may lead to the same functional form for thickness and growth rate. One must also check the self-consistency of a theory via, for example, persistence and kinetic length calculations, concentration dependence and morphology. Some theories, as they stand, clearly fail in this respect whilst others still need to be checked. However, the basic model of nucleation, that is, a model in terms of a general i and g, appears capable of encompassing experimental observations. The exact calculation of these factors and the inhibiting factors which are supposed to lead to curved crystals are still open questions.

4 Rough Surface Growth

4.1 General Outline

The premise that nucleation was always the rate controlling factor in kinetic theories was first disputed by Sadler in 1983 [44]. The disagreement arises from a comparison of the morphologies which would be obtained using the free energies from the Lauritzen-Hoffman theory with those observed experimentally. The compatibility of the LH theory and reality has been discussed in Sect. 3.5.3. The purpose of this section is to explain Sadler's rough surface approach as a separate topic, as acceptance of his nucleation criticism does not necessarily validate his alternative proposals. This Introduction explains his ideas for crystal growth and the way in which alternative barriers may restrict the thickness. Section 4.2 describes the models used to test the theory, which are usually only amenable to study via computer simulation, and the results obtained from them. In Sect. 4.3 we assess the applicability of such an approach.

It is now widely accepted that many low molecular weight materials exhibit roughening (see Sect. 3.2): below a certain temperature, T_R, which varies with the material and the orientation of the face, the surface of a crystal is essentially smooth on a macroscopic scale and hence appears facetted. Above T_R the 'face' is observed to be macroscopically rounded because the free energy barrier to large fluctuations disappears. This transition is usually, though not always, accompanied by an increase in roughness on a microscopic level. The effect of such a transition on the growth rate has been discussed in Sect. 3.2 and only the relevant and important points are summarized here. Below T_R a face, in the absence of impurities or dislocations, grows via a nucleation mechanism and remains facetted until the supersaturation is large enough to cause kinetic roughening, that is, the critical nucleus consists of just one atom. Above T_R there are always steps present on the surface, so there is no barrier to growth and the growth rate increases linearly with the supercooling. The faces which are observed in a crystal are the slowest growing ones — therefore rough faces are only seen when all directions are rough, else faces of the close-packed directions are seen. These predictions are borne out by observations on simple materials such as metals, and by computer simulation of nearest neighbour lattice models [159, 160]. However the existence, or otherwise, of T_R, in more complicated materials with complex and long-range interactions, such as ionic solids, has still to be decided.

When nucleation theories for polymer growth were first developed most of the crystals grown from solution were facetted at all supercoolings covered. However, since then, many crystals have been observed [136, 113] to exhibit rounded faces and even 'leaf-shaped' morphologies (Fig. 3.13). These shapes tend to occur at higher temperatures than the facetted crystals, although it has recently been pointed out that higher temperatures are obtained by using different solvents [114], and therefore it is not an automatic conclusion that the temperature is the controlling factor. However, gradients of temperature and concentration do not appear to be a general explanation, since curvature is not always associated with high growth rates, nor is kinetic roughening a likely factor as these effects are seen with a decrease in supercooling. Therefore experience with low molecular weight materials led Sadler [44] to propose that there is a thermodynamic driving force for roughness arising from a surface entropy contribution, and the existence of roughness on a macroscopic scale.

In Fig. 4.1 we show three different types of roughness which may occur on the surface of a lamella. Figures 4.1 a and b show stems which are laid down as complete units, as taken to be the case in LH theory (fixed length) and its modification by Lauritzen and Passaglia [48] (variable length), respectively. In a)

a

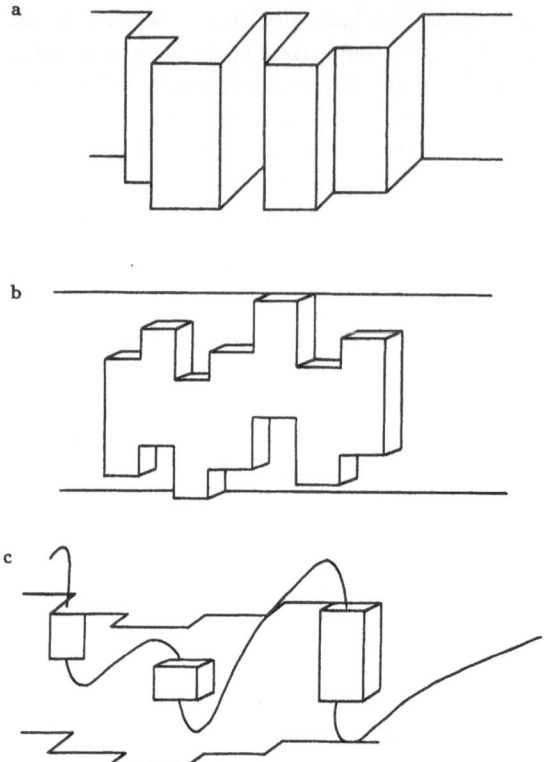

b

c

Fig. 4.1a–c. Various types of roughness which may occur on the growth face of a lamella. (The details of the polymer are not shown). **a** All growth units are complete stems which are of the same length. **b** Stems may be of differing lengths. **c** Growth units are only parts of complete stems

all the stems are of the same length whilst in b) they may be different. Figure 4.1c has taken the individual units to be a part of a stem. These small segments may add anywhere on the surface, only restricted by the connectivity of the polymer chain. If the polymer units are complete stems as in Fig. 4.1 a and b then the free energy involved in creating a step is high compared with kT and the surface can never reach equilibrium roughness. If the units are only part of a stem as in Fig. 4.1c their energy of interaction − the enthalpy per unit − may be comparable to kT and roughening could occur.

In order to test this hypothesis Sadler estimated the enthalpy per CH_2 group in polyethylene, which is expected to be 2.1 kJ mol^{-1}. A simple cubic (100) face is known to be rough for $kT/\varepsilon \simeq 0.6$ where ε is the pairwise nearest neighbour attractive energy. Treating the polyethylene units as independent (that is, neglecting connectivity) and assuming isotropic interactions leads to an estimate of six CH_2 groups in a unit for $kT/\varepsilon \simeq 0.6$. Although such an approach is very simplistic, its intention' was to demonstrate that a rough surface may be consistent with identifying growth units with small chain segments. The argument that this is indeed the case then relies on morphological evidence.

If there is a roughening transition, then instead of occurring at a single temperature T_R (obtained for the theoretical infinite surface consisting of independent units) it must be spread over a range of temperatures, ΔT, because of both the finite size of the surface and also the chain connectivity, which has not been included in the theory. We retain T_R for that temperature above which there is no free energy barrier to step creation, so T_R will be the higher limit of this temperature range ΔT. The mechanism by which roughening could explain habits as observed in polyethylene is as follows: At low temperatures (for polyethylene $T \simeq 70\,°C$) as in Fig. 3.13a the (110) faces are significantly below their roughening temperature, T'_R, and hence appear as facets. The (100) faces have a smaller interstem spacing so the attractive interaction between the units is less, and hence these planes have roughened below T'_R. The rough growth faces, having a higher growth rate, grow out leaving a facetted crystal. At higher temperatures, but still $T < T'_R$, although (110) faces are not rough there will be a significant number of surface steps which increases their growth rate compared with (100) and permits both faces to appear. The (100) faces are curved because they are above their roughening temperature giving crystals as shown in Fig. 3.13b. At temperatures greater than T'_R the (110) faces disappear giving rise to the leaf-shaped morphologies of Fig. 3.13c. Around T_R the complex kinetic motion involved in laying down a macromolecule will determine the growth rates, and the face which is seen depends on subtle details of the molecular interactions.

If roughness is sufficient to describe the crystal morphology, then the effects of such a proposition on the growth rate should also be examined. In low molecular weight materials (see Sect. 3.2) the absence of any nucleation barrier leads to a growth rate which is proportional to the supercooling. However all the crystals shown in Fig. 3.13 have log *(growth rate)* proportional to $1/T\,\Delta T$, which implies some barrier to growth. This barrier is suggested to be entropic in origin rather than energetic [44]. In the remainder of this section we give a qualitative account of how such a barrier may be expected to arise. The models and simple analytical arguments presented in Sect. 4.2 show that this barrier leads to the functionally correct growth laws.

A growth face in contact with its fluid will constantly have units attaching and detaching from it. The crystal grows by the net accumulation of material which, for a low molecular weight material, can take place via many pathways anywhere on the surface. For macromolecules we need to take account of the fact that the adding units are connected to one another and that therefore there will be restrictions on the places at which a new unit can add: a stem which has folded over will be unable to lengthen by the addition of new units onto the fold surface, and if the length of that stem is less than that required for thermodynamic stability (i.e l_{min}) then the fold must 'undo' in order to increase the stem length. In this way the formation of a small stem has effectively blocked further growth of this part of the crystal. The growth may also be blocked if a chain forms stems at two well separated surface sites such that neither stem can lengthen without the removal of the other. Figure 4.2a shows examples of unfavourable configurations which may impede the deposition of further material. This is illustrated further in Fig. 4.2b where a two-dimensional slice of crystal has been drawn showing some of the many configurations which may occur. Most of these cannot be incorporated into the bulk of the crystal; only a minority of 'squared off' profiles are favourable – these are the viable configurations for growth. The fluctuations which occur

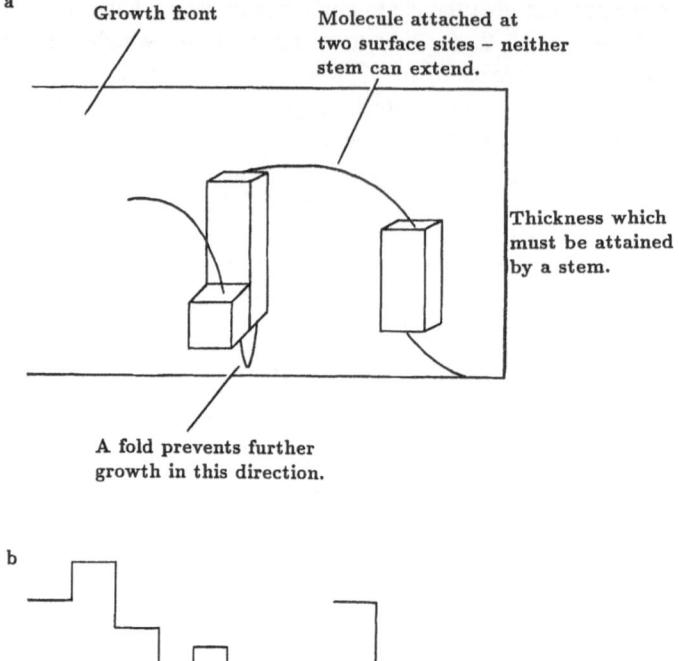

a

Growth front

Molecule attached at
two surface sites – neither
stem can extend.

Thickness which
must be attained
by a stem.

A fold prevents further
growth in this direction.

b

Fig. 4.2a, b. Configurations of molecules which are unfavourable to further growth. **a** The growth face of a lamella is shown, on which a molecule has deposited but is prevented from reaching the length required for stability by other attachments elsewhere. **b** Two examples of possible cross-sections perpendicular to the growth front. The outermost depositions must be removed before further growth of the stable crystal

into and out of the unviable states create a barrier to growth which is of an entropic nature because it arises from a consideration of all the pathways available to the system. The probability of reaching a viable configuration for growth will decrease with increasing l, as the number of unfavourable states will increase, and the next section describes various models and approximate solutions which indicate that the barrier increases exponentially with l, leading to the correct growth form.

4.2 Models for Growth of Rough Surfaces

The models can be split broadly into three categories. The first to be described is a computer simulation of a simple three-dimensional crystal. The second treats growth of a two-dimensional slice as depicted in Fig. 4.2b, and writes down all the possible attachments and detachments as a series of rate equations. The solution of these equations must be performed numerically. Finally a simplified two-dimensional model is treated analytically [161]. Although the simplicity of these models may appear to be a disadvantage, the fact that they all lead to a similar result is one of the strengths of the theory. The basic physical process in all cases is the same, and the other physical parameters, such as type of interaction, conformation, etc., may be important for quantitative results, but are not fundamental in determining the functional form of the growth behaviour.

4.2.1 Monte Carlo Simulation

The computer simulation [162] consists of a Monte Carlo program which treats the crystallizing units as simple 'blocks' which interact with their nearest neighbours in a simple cubic lattice. Standard techniques are used to determine rates at which units attach to and detach from the surface [163]. The probability of an 'event' (either 'on' or 'off') is determined by the free energy change associated with that event, and the site at which it occurs is chosen by a random number. Detailed balance gives only the ratio of the rate constants for on to off events and not their absolute values — as was discussed for the Lauritzen and Hoffman theories in Sect. 3.5.1 — and some choice for the apportioning must be made. In line with earlier growth simulations for low molecular weight materials the 'on' rates were taken to be independent of lattice site and set equal to unity. Hence the units attach to the surface very frequently, but if they incur a large energy by doing so they will soon dissolve again. The opposing scenario would be one in which there is a high energy barrier to attachment, so that attachment events would be infrequent, but having attached, the units are likely to remain. At equilibrium both of these choices, and all other possibilities should lead to the same result. However the various choices of rate constants in a dynamic situation are expected to lead to entirely different results [75]. This effect has not been investigated for the present model, but the importance of a proper energy barrier crossing in a similar dynamic Monte Carlo situation was demonstrated by Kang and Weinberg [164].

The geometry of a growth face is shown in Fig. 4.3. Cyclic boundary conditions apply along y, i.e. the long lateral dimension of the growth face, and interruptions along a stem, in the z-direction, are not allowed. This assumes that two separate molecules will not separately grow at one stem position. The resultant growth is in the x-direction.

In order to model the restrictions imposed by chain connectivity additional rules are required. Two different sets of rules are used, both of which lead to similar results. The first set derives from the inability of a chain to extend once it has been folded over. The Monte Carlo simulation does not explicitly include folds, but any stem which is completely surrounded by other stems is assumed to have folded and additional units are unable to add in the z-direction. In Fig. 4.3 we denote by * all those positions where a new unit may add, all other surface sites are blocked.

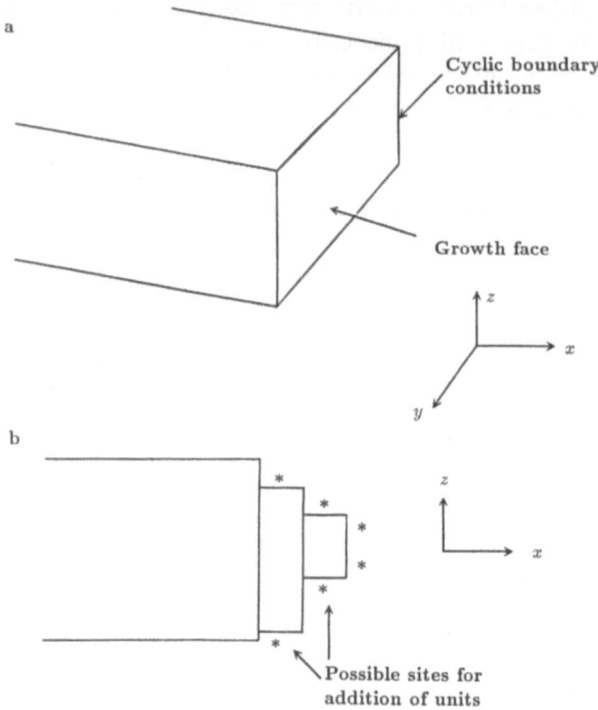

Fig. 4.3.a, b. The geometry of the crystal used in the 3D Monte Carlo simulation. **b** Illustration of one set of rules which mimic the connectivity of the chains. Any stem which is completely surrounded by other stems is assumed to have folded and therefore cannot lengthen; * denotes sites where new units may not be added

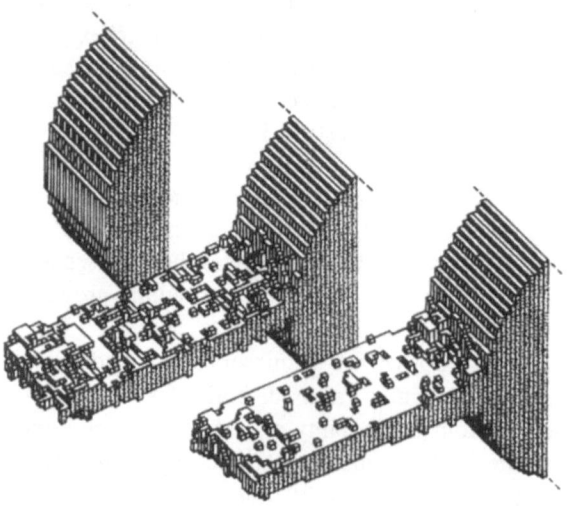

Fig. 4.4. Typical growth configurations from the simulations with $kT_m^0/\varepsilon = 0.7$ *(left)*, and $kT_m^0/\varepsilon = 0.55$ *(right)*. (from [162] by permission of the publishers, Butterworth-Heinemann Ltd. ©)

The second set of rules mimic the inability of a stem to extend because it is attached to the surface elsewhere. A restriction was created in a random way such that a new stem will have a maximum and minimum z value to which it can grow.

Typical growth configurations from the simulations are shown in Fig. 4.4, for $kT_m^0/\varepsilon = 0.7$ and $kT_m^0/\varepsilon = 0.55$, respectively ($\varepsilon$ is the interaction energy between adjacent units, and T_m^0 is the equilibrium melting temperature). Notice the increased roughness of the former which has the lower binding energy compared with the temperature.

The average thickness may easily be measured as the average stem length in the 'bulk' part of the lamella, and the growth rate is the mean distance moved by the growth front during a number of Monte Carlo time-steps, n_t, divided by n_t.

In Fig. 4.5 the thickness and growth rate results are shown qualitatively. The dashed line in Fig. 4.5a is a curve of l proportional to $1/\Delta T$ for comparison

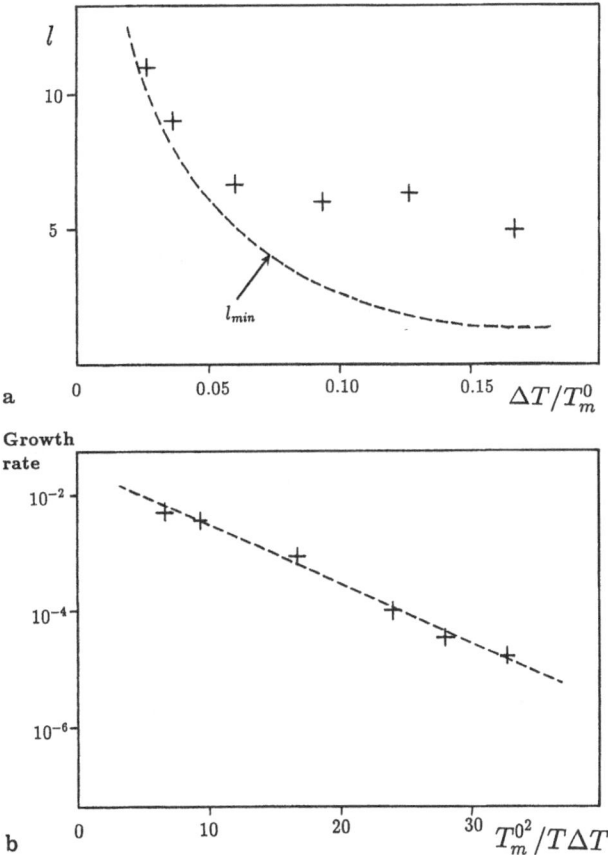

Fig. 4.5 a, b. Typical thickness and growth rate results from the simulations. The *dashed line* in **a** shows *l* proportional to $1/\Delta T$

purposes. It is clear from these plots that the simulation reproduces the main experimental trends, and could conceivably be fitted to a particular polymer by a suitable choice of binding energy.

4.2.2 Rate Equation Approach

For rough surface growth, as opposed to nucleation, there is only a weak correlation between stems in a direction parallel to the growth front (along y in Fig. 4.3a — a niche is no longer a 'special' site which will propagate along the growth front. Therefore a very simple model for growth could be one which only examines the growth of a two-dimensional 'slice' cut out of the lamella perpendicular to the growth front (Fig. 4.6), independent of the interactions with the neighbouring slices [165].

The effect of chain continuity was included in a way similar to the three-dimensional Monte Carlo approach. Additions and removals can occur only at the outermost stem position and the stems must initiate at the bottom of the previous stem and grow upwards. Figure 4.6 shows an arbitrary configuration of stems which may occur and the possible processes through which it may transform. These rules have the great advantage that there is a strictly sequential set of processes which can be expressed as a set of rate equations. These equations can be solved numerically to give accurate values for the lamellar thickness and growth rate and hence avoid the statistical uncertainty inherent in the Monte Carlo simulations.

The rate equations determine the rate of change of the probability of a particular configuration, α, within an ensemble of growing crystals. They must include the rate constants for adding or subtracting units, which are assumed to obey microscopic reversibility. The net flux between configurations α and α' which occur with probability $P(\alpha)$ and $P(\alpha')$ respectively, and differ by one unit is:

$$k^+ P(\alpha) - k^- (\alpha') P(\alpha') . \tag{4.1}$$

The rate constant for addition, k^+, is taken to be independent of α and also independent of the site of addition, either continuing the outermost stem or creating

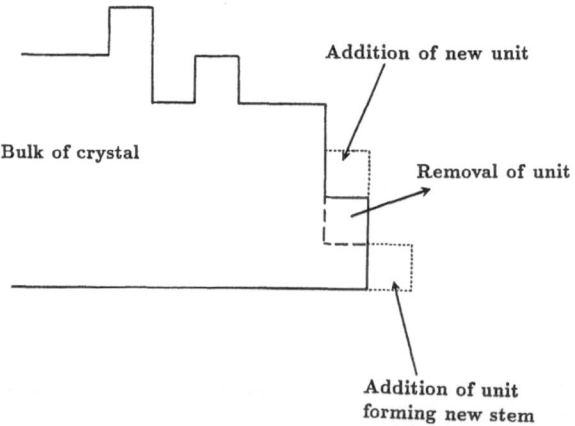

Bulk of crystal

Addition of new unit

Removal of unit

Addition of unit forming new stem

Fig. 4.6. A possible configuration for the 2D rate equation model and the transitions allowed

a new stem. All such factors must therefore be included in $k^-(\alpha')$. There is only one possible unit which may be removed from a configuration, therefore k^- is uniquely defined by α'. The net rate of change of probability $P(\alpha)$ is the difference between the flux out of α and the flux into α.

The following notation is used in the analysis:

i, j: lengths of stems.

n: position of a stem behind the front of the crystal; $n = 1$ is the outermost stem.

$C_n(i)$:   The fraction of stems of length i at position n.

$P_n(i, j)$:   Probability of having stems of length i at n and j at $n + 1$.

$f_n(i, j)$:   Conditional probability that the $(n + 1)^{th}$ stem is of length j given that the nth stem is of length i. (This turns out to be independent of n in the steady state which is our concern at present.)

From these definitions it follows that:

$$P_n(i, j) = f(i, j)/C_n(i). \qquad (4.2)$$

The rate equations are:
for $i > 1$:

$$dP_1(i, j)/dt = k^+ P_1(i - 1, j) - k^+ P_1(i, j) + k^-(i + 1, j)\, P_1(i + 1, j)$$

$$- k^-(i, j)\, P_1(i, j) - k^+ P_1(i, j) + k^-(1, i)\, P_1(1, i)\, f(i, j) \qquad (4.3)$$

and for $i = 1$:

$$dP_1(i, j)/dt = k^+ C_1(i) - k^+ P_1(1, j) + k^-(2, j)\, P_1(2, j)$$

$$- k^-(1, j)\, P_1(1, j) - k^+ P_1(1, j) + k^-(1, 1)\, P_1(1, 1)\, f(1, j). \qquad (4.4)$$

These equations can be solved iteratively until $dP_n(i, j)/dt = 0$, i.e. steady state has been achieved. The growth rate is the difference in attachment and detachment rates for stems at $n = 1$.

$$G = k^+ \sum_{i=1}^{N} C_1(i) - \sum_{i=1}^{N} k(1, i)\, iC_1(i). \qquad (4.5)$$

The average stem thickness is the average stem length at some position n, sufficiently far away from the growth front that the stem length is constant with n:

$$\langle l \rangle = \sum_{i=1}^{N} iC_n(i) / \sum_{i=1}^{N} C_n(i) \quad \text{for} \quad n \simeq 20. \qquad (4.6)$$

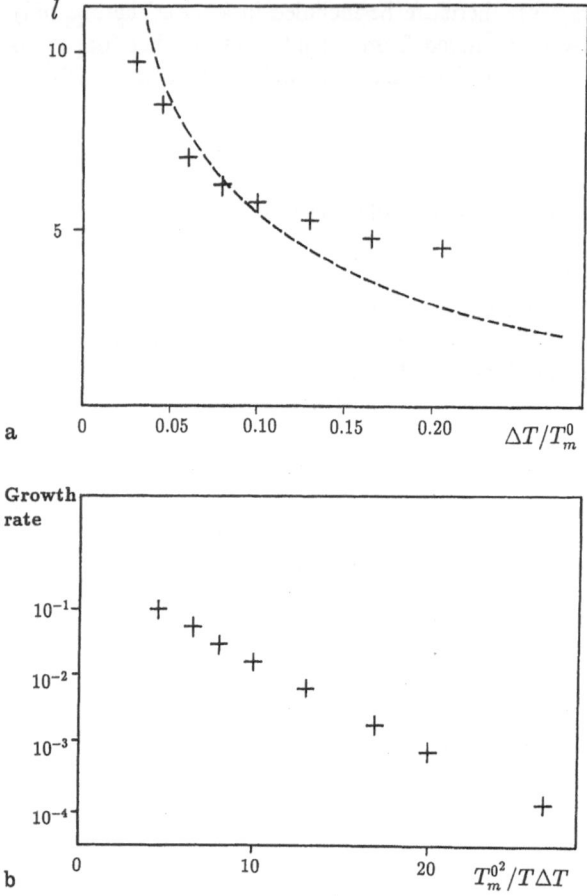

Fig. 4.7a, b. Results for thickness and growth rate from the rate equations with $\varepsilon/kT^0_m = 1.8$

The results for $\varepsilon/kT^0_M = 1.8$ are shown in Fig. 4.7. Again, their importance lies in the 'correct' qualitative features for polymer growth. However, it has recently been demonstrated that even with this simple model a fit of experimental growth rate and thickness data (of the polymer PEEK) can be achieved [166].

A further paper [167] explains the lamellar thickness selection in the row model. The minimum thickness l_{min} is derived from the similation and found to be consistent with equilibrium results. The thickness deviation $\delta l = l - l_{min}$ is approximately constant with l. It is established that the model fulfills the criteria of a kinetic theory: Firstly, a driving force term (proportional to δl) and a barrier term (proportional to l) are indentified. Secondly, the competition between the two terms leads to a maximum in growth rate (see Fig. 2.4) which is located at the average thickness l obtained by simulation. Further, the role of fluctuations becomes apparent when the dependence on the interaction energy ε is investigated. Whereas 'downwards' (i.e. decreasing l) fluctuations are approximately independent of ε, upwards fluctuations are less likely for large bond energies because of the high cost of projections above the substrate. This leads to a decrease in δl and therefore the growth rate with ε via the driving force. This stands in contrast to the behaviour of the nucleation model where the side surface free energy σ is closely related to the barrier term.

Recent results [168, 169] on long-chain paraffins have shown that in certain circumstances one can observe minima in growth rate on increasing ΔT. The thickness of the crystals does not vary smoothly with ΔT, but increases in a stepwise fashion — the measured thickness correspond to twice folded, once folded and extended chains. This is because it is energetically very unfavourable for a 'dangling' end to exist outside the crystal, and the paraffin chain ends will therefore prefer to reside in the surface of the crystal. The crystal thickness should then always be some integer fraction of the total chain length. The growth minima coincide with the transition between two different thicknesses. The possible relevance of these experiments will be referred to again in Sect. 5.

Sadler and Gilmer [170] mimicked the 'preferred' thickness in two ways: in the first they introduced an extra binding energy for units which added at stem lengths of 4, 8, and 12 units, compared with the bulk isotropic binding energy everywhere else. The second method used anisotropic coupling so that the binding between the stems was greater than the binding between units in the same stem. The results are shown in Fig. 4.8. The 'locking in' to preferred values is not unexpected, however the associated minima in growth rates were not predicted beforehand, though in retrospect it is very reasonable. The distribution of probabilities at the growth front shows that as ΔT increases for fixed $\langle l \rangle$ the probability of achieving the outermost stem of the required length decreases, whilst the probability of obtaining the next preferred thinner thickness increases. In effect, the shorter unstable length blocks the growth of the stable crystal close to a change in thickness — this was called 'poisoning' by Ungar and Keller [169].

These results are made somewhat more quantitative by Sadler and Gilmer and may be interpreted in terms of a barrier to growth. It is argued that for normal isotropic energies the barrier to growth increases uniformly with l and can be approximated by an 'equilibrium' (that is, zero growth rate) expression. However, the kinetics of the preferred length problem act as to drive the system away from equilibrium and the resulting barrier term exhibits minima as a function of thickness. This interpretation may help to give a physical appreciation of the process, however, it does not yield any new information.

A very recent application of the two-dimensional model has been to the crystallization of a random copolymer [171]. The units trying to attach to the growth face are either crystallizable A's or non-crystallizable B's with a Poisson probability based on the comonomer concentration in the melt. This means that the 'on' rate becomes thickness dependent with the effect of a depletion of crystallizable material with increasing thickness. This leads to a maximum lamellar thickness and further to a melting point depression much larger than that obtained by the Flory [172] equilibrium treatment.

Using Monte Carlo methods the case of partial comonomer inclusion has also been studied [173]. It was found that the amount of B inclusion is determined by

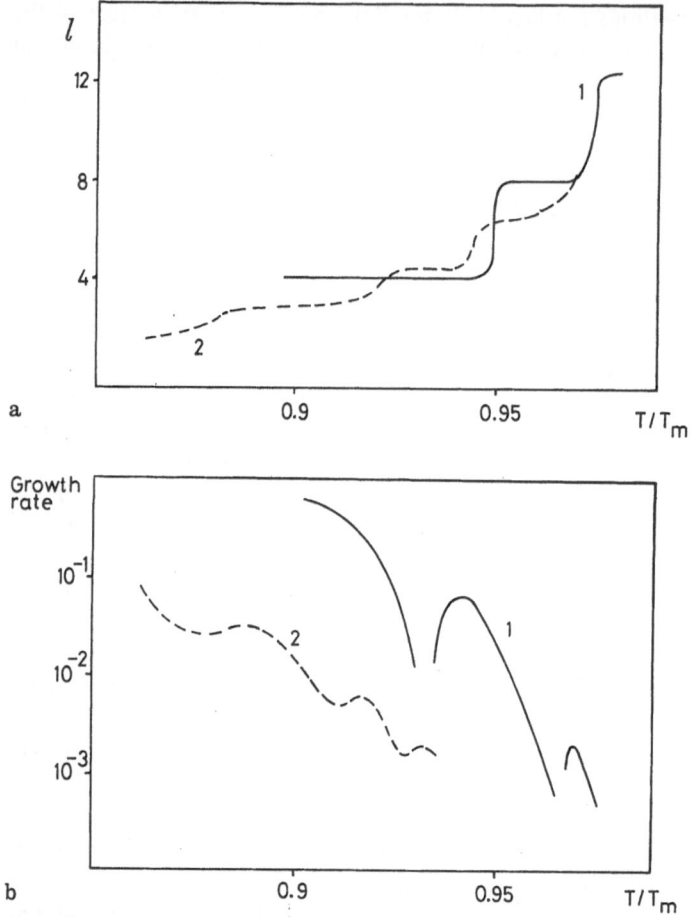

Fig. 4.8a, b. Results from rate equations showing locking-in at preferred thicknesses and the associated minima in the growth rate. Curves labelled '1' have an energy gain at thicknesses 4, 8 and 12. Those labelled '2' have anisotropic energy interactions

kinetics, increasing with supercooling. This is in agreement with the result of Helfand and Lauritzen [127] (see Sect. 3.6.1).

We leave the rate equation approach as having given results which are certainly consistent with observations.

4.2.3 Theoretical Approximations

Both of the numerical approaches explained above have been successful in producing realistic behaviour for lamellar thickness and growth rate as a function of supercooling. The nature of rough surface growth prevents an analytical solution as many of the growth processes are taking place simultaneously, and any approach which is not stochastic, as the Monte Carlo in Sect. 4.2.1, necessarily involves approximations, as the rate equations detailed in Sect. 4.2.2. At the expense of introducing more simplifications it has been possible to obtain mathematical tractability in certain limiting cases [161]. There are several reasons for pursuing this course:

a) The fact that the thickness and growth rates which it predicts are again of the required form shows that it is a very basic property of the model which leads to this behaviour, and that many other physical properties are not vital to its prediction.
b) Numerical approaches do not easily give insight into the physical causes of the resulting growth. An analytical solution allows us to deduce the factors which are governing the behaviour.
c) Although the numerical graphs might be fitted to any particular polymer there is no simple expression which shows how the properties will vary as the physical parameters are changed. For example, how the growth rates on various faces differ, or the growth rates between different polymers. Any theory appears more convincing, and hence may become accepted, if some predictive aspect can be incorporated.

These are the motivations for introducing the analytical model — it is not claimed that the results will be quantitatively correct.

Following Sadler [161] (the details were not included in his work) consider a row model in which all the stems except the outermost are of the same length, h, as shown in Fig. 4.9. A new stem can only be initiated when the outermost stem has reached a height h. A set of flux equations may be written as:

$$S = k^+ C(h) - k_2 C(1)$$

$$S = k^+ C(h-1) - k_1 C(h)$$

$$\vdots \tag{4.7}$$

$$S = k^+ C(1) - k_1 C(2)$$

where $C(i)$ is the probability that the outermost stem is of height i; the 'on' rate constants are all the same, k^+; the rate constant for reduction is k_1, and that for

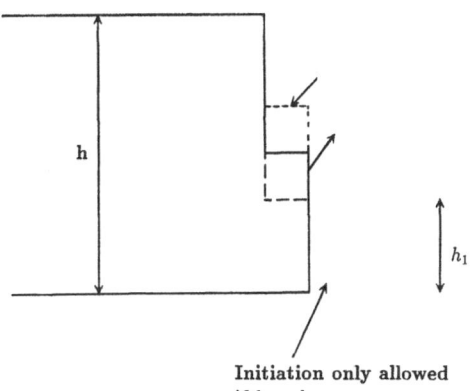

h

h_1

Initiation only allowed
if $h_1 = h$

Fig. 4.9 Illustration of a 2D model used for an analytical argument. The bulk crystal is of uniform thickness and additions and removals are allowed only at one site

stem removal is k_2. Equations (4.7) can be solved following the method in Sect. 3.5.2, leading to:

$$C(j) = \frac{S}{k^+ - k_1} + \left(\frac{k^+}{k_1}\right)^{j-1}\left[C(1) - \frac{S}{k^+ - k_1}\right] \quad \text{for} \quad 1 \leq j \leq h.$$

(4.8)

Using Eq. (4.7) with Eq. (4.8) for $j = h$ leads to:

$$C(1) = S\frac{(k^+/k_1)^h - 1}{k^+[1 - (k_1/k^+)][(k^+/k_1)^h - (k_2/k_1)]}.$$

(4.9)

The minimum possible thickness for the lamella, h_{min}, can be determined by the condition $S = 0$, therefore:

$$k_2/k_1 = (k^+/k_1)^{h_{min}}.$$

(4.10)

Equation (4.9) can then be written as:

$$S = C(1)\frac{[1 - k_1/k^+][1 - (k_1/k^+)^{h - h_{min}}]}{[1 - (k_1/k^+)]}.$$

(4.11)

For h close to h_{min} the main variation of S arises from $1 - (k_1/k^+)^{h - h_{min}}$. Using $k_1/k^+ = \exp(-\Delta f/kT)$, where Δf is the bulk free energy difference for a unit we finally arrive at:

$$S/k^+ \simeq \Delta f(h - h_{min})/kT \quad \text{for} \quad h \simeq h_{min}$$

(4.12)

that is, the flux is linear in $h - h_{min}$. It is important at this stage to be clear about what we have calculated — it is the growth rate for a row of stems all of length h which can only fold when the previous stem has been completed. If $h > h_{min}$ then the growth rate increases with h — this is just equivalent to the filling in rate in nucleation theories, when the opportunity to fold at each stage is neglected.

In the full row of stems model the state with $l(n = 2) = h$ is a necessary profile of the crystal which permits further growth. There are many other profiles which effectively 'block' the growth because they must be removed before further growth can occur. Some of the possible profiles are shown in Fig. 4.2. The probability of achieving a 'squared off' end which is viable for growth depends upon h. For large h there are many possible alternative configurations and the probability is very low. For small h there are not as many configurations and a square end will be achieved more easily. The total growth rate of a crystal of thickness h is the product of the probability that the crystal has a square end and the growth rate of a crystal of that thickness, S, as worked out above.

The probability of achieving a viable configuration can be estimated by associating a Boltzmann weight to all the configurations according to their energies

– this assumes an equilibrium distribution. The probability of a square configuration is then:

$$P(h) = \prod_{z=1}^{l-1} \frac{1}{(1 + (k_1/k^+)^{h-z}}.$$
(4.13)

$P(h)$ is just the barrier term referred to in Sect. 2, and has the form shown in Fig. 2.4. Also, $S(h)$ takes the place of the driving force with the resultant growth rate the product of the two factors. Hence, a theoretical understanding of the factors leading to the computer results may be obtained.

4.3 A Critique of Rough Surface Growth

The basic assumption of the above approaches is that the growth front may be rough, and they set out to show that an entropic barrier can give acceptable growth behaviour. If the initial assumption is perceived as invalid than the remaining work becomes mostly irrelevant in its present form. Nevertheless, the configurational path degeneracy of the attachment process and its related entropy may still play an important role in the crystallization kinetics, and computer models of the type developed by Sadler and Gilmer may well be needed to study these effects.

Unfortunately, even for low molecular weight material it is difficult to obtain clear experimental evidence for a roughening transition [71]. This is mainly due to the fact that during growth the interface generally assumes a metastable shape and relaxation times are long and increase with crystal size. Therefore we certainly cannot expect a definitive answer for macromolecules. We shall therefore just make several comments which hopefully will be of use when reading the literature.

a) Any phase transition temperature is only defined for an infinite system: a roughening temperature can strictly only exist on an infinite surface. On a real but large surface the 'roughening temperature' will appear to be smeared over a range of temperatures. For a lamella of finite thickness, which is certainly not wide, the ability of even a low molecular weight to exhibit a transition is not certain.

b) Any roughening theories have been applied to Kossel type crystals which have only nearest neighbour interactions, and which are free to add or substract units anywhere on the surface. To extend this treatment to connected chains without modification or explanation appears unjustifiable. The number-of-monomers calculation in a polyethylene unit required for roughening must be treated with the greatest caution.

c) Although curved crystals are certainly observed and cannot be explained within the original nucleation theory or by gradient effects, this does not automatically lead to thermal roughening. Impurities in the form of low molecular weight components, misaligned attachments or solvents molecules, may all be expected to contribute and could well lead to the curvature observed.

Despite the above comments on the application of roughening to thin polymer crystals it is certainly possible that thermal effects could cause a significant number of surface steps, and that this is all that is required to lead to a breakdown of the nucleation argument.

The particular models used to demonstrate the theory obviously have many drawbacks as true representations of polymer crystals. These could include the lack of a fold energy, no distinction between new molecules and those already attached, neglect of chain ends, a somewhat arbitrary choice of pinning rules etc. However, they all serve their purpose in that they show that an energetic free energy barrier is not necessary to obtain the experimental curves. A truly representative growth picture can probably only be achieved via molecular dynamics.

5 Concluding Remarks

It has been the aim of this review to present the theories of polymer crystal growth and to compare their different assumptions. We have stressed their strengths and failings and tried to show how these are related to the underlying models. However, it is left to the reader to judge for him/herself the 'correctness' or otherwise in each case. To conclude we first make some cautionary remarks, then summarize the models and finally suggest avenues for future research.

All theories try to explain reasons for experimental results, hence the results themselves must be as accurate as possible. Even the raw data, such as growth rate and morphology, are open to misinterpretation, as they may vary during the course of an experiment and allowances must be made for these variations. When plotting the data, such as growth rate or thickness as a function of temperature, more physical parameters are often involved and it is tempting to give those which give the desired result, for example a straight line, instead of investigating other possibilities. It was highlighted in Sect. 2 that even the melting temperature, which is fundamental to all theories, may have different interpretations and the effect this could have on the growth rate curves is not clear.

The earlier models were based on secondary nucleation [45] which was developed from the well-established low molecular weight theories. Hence, there was assumed to be a nucleation barrier which must be overcome, followed by uniform growth. Lauritzen and Hoffman [4] proposed an energetic barrier due to the deposition of a single stem and predicted the observed growth rate and thickness dependence on temperature (excluding the 'δl catastrophe'). However, descrepancies in its prediction for characteristic lengths lead Point [51, 52] to suggest that the first stem could not be regarded as the deposition of a complete unit, but rather as the attachment of subunits with the opportunity to fold at each stage. Wunderlich [145] and Hikosaka [146, 147] both viewed the critical nucleus as two-dimensional (rather than a single stem), although this is the extent of the similarity.

There have been questions raised as to the suitability of using nucleation to explain, for example, the morphology or concentration behaviour. However, Toda

[138] and Mansfield [139] have shown that these can be accommodated within the theory by invoking cilia nucleation, the finite size of long chains, and by inhibiting the growth along the growth surface. The way in which these factors should be made more quantitative is far from clear.

A drastic departure from nucleation theory was made by Sadler [44] who proposed that the crystal surface was thermodynamically rough and a barrier term arises from the possible paths a polymer may take before crystallizing in a favourable configuration. His simulation and models have shown that this would give results consistent with experiments. The two-dimensional row model is not far removed from Point's initial nucleation barrier, and is practically identical to a model investigated by Dupire [35]. Further comparison between the two theories would be beneficial.

We must look ahead and try to find ways of obtaining further insight into the crystallization process. The work of Hikosaka has already been mentioned and hopefully will lead to an understanding of the role that chain mobility plays during growth. The crystallization of paraffins [169] shows distinct growth branches, in which the thickness is constant, with minima in between. These branches correspond to extended, once and twice folded chains. It is suggested that, for example, once folded chains act as defects ('poisoning') in the growth of extended chain crystals and cause a sharp drop in the growth rate. There appear to be similarities with work on polyethyleneoxide [174] where growth minima are also observed and these are accompanied by a rounding of the morphology. This is fully consistent with the changes expected when growth along a surface is inhibited by external factors [138]. The increasing amount of experimental evidence in these systems, together with the constancy of the lamellar thickness within the growth branches, appears a promising area for further investigation.

Finally we mention a completely different line of work by Patel and Farmer [175], who study the energetics of attaching a stem to (110) and (100) faces of polyethylene, and investigate the barriers to translations and rotations. They find that substantially more energy is gained by a chain nucleating on a (100) face, but the mobility is greater on (110) which leads to a higher lateral spreading rate. They argue that these competing factors and their temperature dependence can lead to the morphologies which are observed. The effect of such subtle differences between the different growth faces had not been considered in any of the crystallization theories until the (100) lattice strain was incorporated into nucleation theory by Hoffman and Miller [109].

Clearly there are still many opportunities and challenges to be undertaken. It is hoped that this review may provide a useful background and provide stimulus for new ideas in the future.

Acknowledgements: The authors are greatly indebted to DSM for the support to which this review owes its inception. One of us (G. G-W) acknowledges support by the SERC.

We wish to thank Prof. A. Keller, Prof. J. J. Point and Dr. A. Toda for encouragement, helpful suggestions and comments.

6 References

1. Keller A (1957) Phil Mag 2: 1171
2. Till PH (1957) J Polym Sci 24: 30
3. Fischer EW (1957) Z Naturf 12a: 753
4. Lauritzen JI Jr, Hoffman JD (1960) J Res Nat Bur Stds 64A: 73
5. Woodward AE (1988) Atlas of polymer morphology. Hanser, Munich
6. Keller A (1968) Rep Prog Phys 31: 623
7. Bassett DC (1981) Principles of polymer morphology. Cambridge Univ Press, London
8. Dosière M (1989) in: Cheremissinoff NP (ed) Handbook of polymer science and technology, vol 2. Marcel Dekker, New York, p 367
9. Phillips PJ (1990) Rep Progr Phys 53: 549
10. Point JJ, Colet MC (1990) Ann Chim Fr 15: 221
11. Frank FC (1958) Disc Faraday Soc 25: 205
12. Frank FC (1979) Disc Faraday Soc 68: 7
13. Hoffman JD (1983) Polymer 24: 3
14. Yoon DY, Flory PJ (1977) Polymer 18: 509
15. Mansfield ML (1983) Macromolecules 16: 914
16. Sadler DM, Harris R (1982) J Polym Sci Polym Phys Ed 20: 561
17. Spells SJ, Sadler DM (1984) Polymer 25: 739
18. Sadler DM (1984) in: Hall I (ed) Structure of crystalline polymers. Elsevier, Barking, p 125
19. Spells SJ, Keller A, Sadler DM (1984) Polymer 25: 749
20. Keller A (1979) Disc Faraday Soc 68: 145
21. Wittman JC, Lotz B (1985) J Polym Sci Polym Phys Ed 23: 205
22. Barham PJ, Keller A, Otun EL, Holmes PA (1984) J Mat Sci 19: 2781
23. Buckley CP, Kovacs AJ (1976) Coll Polym Sci 254: 695
24. Gibbs JW (1928) The collected works of JW Gibbs, vol 1. Longmans, Green and Co, New York
25. Mutaftshiev B (1982) in: Mutaftshiev B (ed) Interfacial aspects of phase transformations. Nato Advanced Study Institute, 29 Aug–9 Sep 1981. Reidel, Dordrecht, p 63
26. Suzuki T, Kovacs A (1970) Polym J 1: 82
27. Sanchez IC, Di Marzio EA (1971) Macromolecules 4: 677
28. Wunderlich B (1980) Macromolecular physics, vol 3. Academic Press, London
29. Leung WM, Manley RStJ, Panaras AR (1985) Macromolecules 18: 746
30. Leung WM, Manley RStJ, Panaras AR (1985) Macromolecules 18: 753
31. Hoffman JD, Frolen LJ, Gaylon SR, Lauritzen JI Jr (1975) J Res Nat Bur Std 79A: 671
32. Point JJ, Kovacs AJ (1980) Macromolecules 13: 399
33. Flory PJ, Vrij A (1963) J Amer Chem Soc 85: 3548
34. Buckley CP, Kovacs AJ (1975) Progr Coll Polym Sci 58: 44
35. Dupire M (1984) Etude critique des models sequentiels de la cristallisation des hauts polymeres lineaires. Thesis, Université de l'Etat à Mons, Mons
36. Peterlin A (1960) J Appl Phys 31: 1934
37. Peterlin A, Fischer EW, Reinhold C (1960) Z Physik 159: 272
38. Peterlin A, Fischer EW, Reinhold C (1962) J Chem Phys 37: 1403
39. Peterlin A, Reinhold C (1965) J Polymer Sci A3: 2801
40. Huggins ML (1961) J Polymer Sci 50: 65
41. Huggins ML (1966) Makromol Chem 92: 260
42. Zachmann HG (1967) Kolloid Z Z Polym 231: 504
43. Allegra G (1980) Ferroelectrics 30: 195
44. Sadler DM (1983) Polymer 24: 1401
45. Frank FC (1974) J Cryst Growth 22: 233
46. Gates DJ, Westcott M (1988) Proc Royal Soc Lond A416: 443, 463
47. Hoffman JD, Lauritzen JI Jr (1961) J Res Nat Bur Std 65A: 297
48. Lauritzen JI Jr, Passaglia E (1967) J Res Nat Bur Std 71A: 261

49. Sanchez IC, Di Marzio EA (1971) J Chem Phys 55: 893
50. Lauritzen JI Jr, Hoffman JD (1973) J Appl Phys 44: 4340
51. Point JJ (1979) Macromolecules 12: 770
52. Point JJ (1979) Disc Faraday Soc 68: 167
53. Laudise RA, Carruthers JR, Jackson KA (1971) Annu Rev Mat Sci 1: 253
54. Jackson KA (1974) J Cryst Growth 24/25: 130
55. Kossel W (1927) Nachr Ges Wiss Göttingen: 135
56. Stranski IN (1928) Z Phys Chem 136: 259
57. Becker R, Döring W (1935) Ann Phys Lpz 24: 719
58. Papapetrou A (1936) Z Kristallogr 92: 89
59. Cahn JW (1967) in: Peiser HS (ed) Crystal growth. Suppl J Phys Chem Solids, p 681
60. Woodruff DP (1973) The solid-liquid interface. Cambridge Univ Press, London
61. Langer JS (1980) Rev Mod Phys 52: 1
62. Langer JS (1987) in: Souletie J, Vannimenus J, Stora R (eds) Chance and matter: Ecole d'été de physique théorique, Les Houches (46th, 1986). Elsevier, Amsterdam, p 629
63. Saito Y, Goldbeck-Wood G, Müller-Krumbhaar H (1988) Phys Rev A 38: 2148
64. Frank FC (1949) Disc Faraday Soc 5: 186
65. Burton WK, Cabrera N, Frank FC (1951) Phil Trans Roy Soc 243: 299
66. Barham PJ, Chivers RA, Keller A, Martinez-Salazar J, Organ SJ (1985) J Mat Sci 20: 1625
67. Keith HD, Padden FJ Jr (1986), Polymer 27: 1463
68. Keith HD, Padden FJ Jr (1987), J Polym Sci Polym Phys Ed 25: 229, 2265, 2371
69. Goldenfeld N (1987) J Cryst Growth 84: 601
70. Weeks JD (1980) in: Riste T (ed) Ordering in strongly fluctuating condensed matter systems. Plenum, New York, p 293
71. van Beijeren H, Nolden I (1987) in: Schommers W, von Blanckenhagen P (eds) Topics in current physics, vol 43: Structure and dynamics of surfaces II. Springer: Berlin, p 259
72. Weeks JD, Gilmer GH (1979) Adv Chem Phys 40: 157
73. Hillig WB (1966) Acta Met 14: 1868
74. Gilmer GH, Jackson KA (1977) in: Kaldis E, Scheel HJ (eds) Crystal growth and materials. North Holland, Amsterdam, p 79
75. Sadler DM (1987) J Chem Phys 87: 1771
76. Sadler DM (1987) Polymer 28: 1440
77. Lauritzen JI Jr (1973) J Appl Phys 44: 4353
78. Bennett CH, Buttiker M, Landauer R, Thomas H (1981) J Stat Phys 24: 419
79. Goldenfeld N (1984) J Phys A 17: 2807
80. Toda A, Tanzawa Y (1986) J Cryst Growth 76: 462
81. Guttman CM, Di Marzio EA (1983) J Appl Phys 54: 5541
82. Blundell DJ, Keller A (1968) J Macr Sci Phys B 2: 337
83. Keller A, Pedemonte E (1973) J Cryst Growth 18: 111
84. Cooper M, Manley RStJ (1975), Macromolecules 8: 219
85. Organ SJ, Keller A (1987) J Polym Sci C 25: 67
86. Sadler DM (1971) J Polymer Sci A 2: 779
87. Dosière M, Colet MC, Point JJ (1986) in: Sedlacek B (ed) Morphology of polymers. de Gruyter, Berlin, p 171
88. Toda A, Kiho H, Miyaji H, Asai K (1985) J Phys Soc Jpn 54: 1411
89. Toda A, Miyaji H, Kiho H (1986) Polymer 27: 1505
90. Toda A, Kiho H (1987) J Phys Soc Jpn 56: 1631
91. Toda A, Kiho H (1989) J Polym Sci Polym Phys Ed 27: 53
92. Toda A (1990) private communication
93. Point JJ, Colet MC, Dosière M (1986) J Polym Sci Polym Phys Ed 24: 357
94. Dosière M, Colet MC, Point JJ (1986) J Polym Sci Polym Phys Ed 24: 345
95. Colet MC, Point JJ, Dosiere M (1986) J Polym Sci Polym Phys Ed 24: 1183
96. Passaglia E, Di Marzio EA (1986) Polymer 27: 510
97. Point JJ, Colet MC, Dosière M (1986) in: Sedlacek B (ed) Morphology of polymers. de Gruyter, Berlin, p 153

98. Toda A (1989) J Polym Sci Polym Phys Ed 27: 1721
99. Stranski I (1949) Disc Faraday Soc 5: 69
100. Khoury F, Passaglia E (1976) in: Hannay NB (ed) Treatise on solid-state chemistry, vol 3. Plenum, New York, p 335
101. Sadler DM (1984) Polym Commun 25: 196
102. Sadler DM, Barber M, Lark G, Hill MJ (1986) Polymer 27: 25
103. Mansfield ML (1989) Polymer 30: 1623
104. Hoffman JD, Davies GT, Lauritzen JI (1976) in: Hannay NB (ed) Treatise on solid-state chemistry, vol 3. Plenum, New York, p 497
105. Frank FC, Tosi M (1961) Proc Royal Soc A 263: 323
106. Point JJ (1989) Macromolecules 22: 3501
107. Hoffman JD, Miller RL (1989) Macromolecules 22: 3502
108. Lauritzen JI Jr, Di Marzio EA, Passaglia E (1966) J Chem Phys 45: 4444
109. Hoffman JD, Miller RL (1989) Macromolecules 22: 3038
110. Martinez-Salazar J, Barham PJ, Keller A (1985) J Mat Sci 20: 1616
111. Vasanthakumari R, Pennings AJ (1983) Polymer 24: 175
112. Cheng SZD, Chen J, Janimak J (1990) Polymer 31: 1018
113. Organ SJ, Keller A (1986) J Polym Sci B 24: 2319
114. Toda A (1987) Polymer 28: 1645
115. Lovinger AJ, Davis DD, Padden FJ Jr (1985) Polymer 26: 1595
116. Phillips PJ, Vatansever N (1987) Macromolecules 20: 2138
117. Phillips PJ, Lambert WS (1990) Macromolecules 23: 2075
118. Point JJ, Colet MC (1990) Ann Chim Fr 15: 221
119. Leung WM, Manley RStJ, Panaras AR (1985) Macromolecules 18: 760
120. Point JJ (1986) Macromolecules 19: 929
121. Holland VF, Lindenmeyer PH (1962) J Polym Sci 57: 589
122. Hoffman JD, Lauritzen JI, Passaglia E, Ross GS, Frolen LJ, Weeks JJ (1969) Kolloid Z Z Polym 231: 564
123. Hoffman JD (1985) Macromolecules 18: 772
124. Sadler DM (1985) J Polym Sci Polym Phys Ed 23: 1533
125. Di Marzio EA (1967) J Chem Phys 47: 3451
126. Sanchez IC, Di Marzio EA (1972) J Res Nat Bur St 76A: 213
127. Helfand E, Lauritzen JI Jr (1973) Macromolecules 6: 631
128. Di Marzio EA, Passaglia E (1987) J Chem Phys 87: 4901
129. Passaglia E, Di Marzio EA (1987) J Chem Phys 87: 4908
130. Hoffman JD, Guttman CM, Di Marzio EA (1979) Disc Faraday Soc 68: 177
131. de Gennes PG (1971) J Chem Phys 55: 572
132. de Gennes PG (1976) Macromolecules 9: 591
133. Hoffman JD (1982) Polymer 23: 656
134. Hoffman JD, Miller RL (1988) Macromolecules 21: 3028
135. Point JJ, Dosiere M (1989) Polymer 30: 2292
136. Labaig JJ (1978) Variation de la morphologie et de la structure des cristaux de polyéthylène et de leur vitesse de croissance en fonction de la température et de la longeur des chaînes. Thesis, Université Louis Pasteur, Strasbourg
137. Khoury F, Bolz LH (1980) Proc Ann Meet Electron Microsc Soc Am 38: 342
138. Toda A (1986) J Phys Soc Jpn 55: 3419
139. Mansfield ML (1988) Polymer 29: 1755
140. Mansfield ML (1990) Polym Commun 31: 283
141. Point JJ, Villers D (1991) J Cryst Growth, in press and Point JJ (1991) Polymer, in press
142. Toda A (1991) Polymer 32: 771
143. Di Marzio EA, Guttman CM (1982) J Appl Phys 53: 6581
144. Jones DH, Latham AJ, Keller A, Girolama M (1973) J Polym Sci Polym Phys Ed 11: 1759
145. Wunderlich B, Mehta A (1974) J Polym Sci Polym Phys Ed 12: 255
146. Hikosaka M (1987) Polymer 28: 1257
147. Hikosaka M (1990) Polymer 31: 458

148. Dlugosz J, Fraser GV, Grubb D, Keller A, Odell JA, Goggin PL (1976) Polymer 17: 471
149. Mehta A, Wunderlich B (1975) Colloid Polym Sci 253: 193
150. Cheng SZD, Wunderlich B (1986) J Polym Sci Polym phys Ed 24: 557
151. Cheng SZD, Wunderlich B (1986) J Polym Sci Polym Phys Ed 24: 595
152. Cheng SZD, Noid DW, Wunderlich B (1989) J Polym Sci Polym Phys Ed 27: 1149
153. Hikosaka M, Kawabata H, Keller A, Rastogi S (1991) J Macromol Sci B (accepted)
154. Bassett DC, Block S, Piermarini GJ (1974) J Appl Phys 45: 4146
155. Yamamoto T, Miyaji H, Asai K (1977) Jpn J Appl Phys 16: 1891
156. Hikosaka M, Tamaki S (1981) J Phys Soc Jpn 50: 638
157. Hikosaka M, Seto T (1982) Jpn J Appl Phys 21: L332
158. Hikosaka M, Seto T (1984) Jpn J Appl Phys 23: 956
159. Leamy HJ, Gilmer GH, Jackson KA (1975) in: Blakely JM (ed) Surface physics of materials. Academic Press, London, p 121
160. van der Eerden JP, Bennema P, Cherepanova TA (1978) Prog Crystal Growth Charact 1: 219
161. Sadler DM (1987) Nature 326: 174
162. Sadler DM, Gilmer GH (1984) Polymer 25: 1446
163. Gilmer GH (1980) Science 208: 355
164. Kang HC, Weinberg WH (1989) J Chem Phys 90: 2824
165. Sadler DM, Gilmer GH (1986) Phys Rev Lett 56: 2708
166. Goldbeck-Wood G, Sadler DM (1989) Molec Sim 4: 15
167. Sadler DM, Gilmer GH (1988) Phys Rev B 38: 5684
168. Ungar G, Keller A (1986) Polymer 27: 1835
169. Ungar G, Keller A (1987) Polymer 28: 1899
170. Sadler DM, Gilmer GH (1987) Polym Commun 28: 243
171. Goldbeck-Wood G, Sadler DM (1990) Polym Comm 31: 143
172. Flory PJ (1955) Trans Faraday Soc 51: 848
173. Goldbeck-Wood G (1990) Polymer 31: 586
174. Kovacs AJ, Gonthier A (1972) Kolloid Z Z Polym 250: 530
175. Patel AK, Farmer BL (1980) Polymer 21: 153
176. Point JJ, Villers D (1991) Polymer, in press

Editor: H.-H. Kausch
Received February 22, 1991

Note Added in Proof

The following note is due to Prof. J. J. Point, and we are very grateful to him for communicating the following arguments [176] concerning the δl catastrophe in the Hoffman-Miller nucleation-model for curved crystals [109].

In section 3.6.3 we mentioned that in growth on a curved face the strain surface free energy σ_s takes the role the lateral surface free energy σ played in the flat surface case, namely that of a barrier to the formation of the first stem. This analogy cannot be made since, in contrast to σ, the strain surface free energy is associated with the deposition of any stem. Therefore and because of its physical origin (the volume strain) it is closely linked with the free energy of fusion. This is indeed recognized by Hoffman and Miller by introducing the effective free energy difference:

$$\Delta F_s = \Delta F - \frac{a + b}{ab} \sigma_s \qquad (N1)$$

and the related supercooling $\Delta T_s = \Delta F_s T_m^0 / \Delta h_f$. The ratio of the rate constants for the first stem then reduces to:

$$A_0 / B_0 = \exp\left(abl\, \Delta F_s / kT\right). \qquad (N2)$$

From comparison with Eqn (3.44) we can see immediately that there is no nucleation barrier in place of σ. Alternatively, we can calculate the free energy difference made by the deposition of v stems (cf. Eqn (3.60)):

$$\Delta \Phi_v = -abl\, \Delta F_s + (v - 1)\left(-abl\, \Delta F_s + 2ab\sigma_e\right). \qquad (N3)$$

This reveals that for the deposition of the first stem ($v = 1$) the free energy always decreases and the more so the longer it is, i.e. there is a δl catastrophe at any supercooling ΔT_s! Hoffman and Miller, however, give the following expression for the critical supercooling:

$$(\Delta T_s)_c = \frac{2\sqrt{2\sigma_s T_m^0}}{a\, \Delta h_f} \frac{1 - \psi}{\psi}. \qquad (N4)$$

This was constructed by introducing an artificial barrier into the on-rate constant A_0:

$$A_0 = \beta \exp\left[-\left((1 - \psi)\, 2\sqrt{2}bl\sigma_s - \psi abl\, \Delta F_s\right)/kT\right]. \qquad (N5)$$

The second factor in the exponent follows directly from the detailed balance equation (N2) by introducing apportioning in the usual way. The first factor, however, has no relation to the actual free energy difference. It is an artificial barrier which is simply added to both on and off rate constants. The critical supercooling in then derived from setting the exponent of A_0 to zero. We notice, however, that in this way any ΔT_c could be constructed. There are limits and consequences, however. Firstly, in contrast to the flat surface case ΔT_c is always zero for $\psi = 1$ (see Eqn (N4)). Secondly, the definition (N5) of the rate constants leads to an overcrowding of the substrate. This can be seen by going back to the derivation of the steady state flux (Eqns (3.40)–(3.42)). As shown by Lauritzen and Hoffman [50], the essential that the number of nuclei comprising v stems, N_v, must be bounded with v requires that $S = N_1(A - B)$. With Eqn (3.50) it follows that $N_1 = N_0 A_0 / A$. In LH theory we find that N_1 is always less than N_0 as long as $l > l_{min}$ and $\Delta T < \Delta T_c(LH)$. The Hoffman-Miller definition of the rate constants, however, leads to:

$$N_1 = N_0 \exp\left(2ab\sigma_e / kT\right), \qquad (N6)$$

so that N_1 is always much greater than N_0, i.e. there are predicted to be more single stems than substrate sites. This is just another expression of the fact that there is a δl catastrophe at any supercooling in the strain-model of Hoffman and Miller.

Properties and Failure of Polymers with Tailored Distances Between Crosslinks

Michael Fischer
Materials Research, Ciba-Geigy AG, CH-1701 Fribourg, Switzerland

Crosslinked polymers are rather peculiar materials in that they never melt and they exhibit entropic elasticity at elevated temperatures. The present review on the influence of crosslink density is structured around model polymers of *uniform composition* but with widely varying numbers of crosslinks. The degree of crosslinking in the polymers was verified by use of the theory of rubber elasticity.

The existence of crosslinks has little effect on the behaviour of the glassy polymers at low levels of strain. Fox's equation was found appropriate for those minor effects. In contrast to their low significance at small strains, crosslinks determine the dimensions of the deformation zone ahead of the tip of a crack propagating through the polymer. Proportionality between the crack opening displacement (Dugdale-Barenblatt-model) and the chain contour length of the strands of the molecular network was established. In this way, the molecular architecture of the polymer determines the toughness of the bulk material.

Polymer fracture studies reported in the recent literature are shown to be in agreement with the conclusions derived from the present model polymers.

List of Symbols and Abbreviations

Capital Letters

B	breadth
C	number of crosslinks in a representative piece of polymer
E	Young's modulus
G_I	energy release rate
G_{IC}	critical energy release rate
H	enthalpy
K	bulk modulus
K_I	stress intensity factor
K_{IC}	critical stress intensity factor
$\bar{M}_C$	effective molecular mass between crosslinks
$\bar{M}_R$	averaged molecular mass between crosslinks as calculated from stoichiometry
N	number of molecular chains between crosslinks (network strands) in a representative piece of polymer
N_L	Avogadro number, $N_L = 6.023 \times 10^{23}\,mol^{-1}$
R	gas constant, $R = 8.31\,J\,K^{-1}\,mol^{-1}$
S	shear modulus
T	absolute temperature
T_g	glass transition temperature
T_{goo}	glass transition temperature of a thermoplastic
$T_{\lambda max}$	temperature of damping maximum
V_C	volume of a molecular strand
W	width
X	property

Lowercase Letters

a	crack length
c	specific heat
f	functionality of network junction
k	Boltzmann constant, $k = 1.381 \times 10^{-23}\,J\,K^{-1}$
l_C	chain contour length
m_i	mass of molecular fragment i
p	constant
s	length of deformation zone
t	time
u	activation energy
v	activation volume
w	half crack opening displacement

Greek Letters

α	coefficient of thermal expansion
Γ	constant
γ	shear rate
δ	crack opening displacement
δ_C	critical crack opening displacement
ε	strain
ζ	chain contour length per interval
η	viscosity
$\varkappa$	constant
Λ	draw ratio
λ	logarithmic decrement
μ	Poisson's ratio
ν_i	molar fraction of component i
ϱ	density
σ	stress
σ_C	cohesive stress
σ_n	necking stress
σ_y	yield stress
τ	shear stress
ϕ	empirical constant
Ω	denotes a molecular chain segment

Abbreviations

BA	Bisphenol A
DDS	Diaminodiphenylsulfone (DADPS)
DGEBA	Diglycidyl ether of bisphenol A
DSC	Differential scanning calorimetry
MDA	4,4′-Methylenedianiline
NMR	Nuclear magnetic resonance
PC	Poly(carbonate)
PES	Poly(ether sulfone)
PGCBA	Polyglycidyl compound of bisphenol A
PMMA	Poly(methyl methacrylate)
PPO	Poly(phenylene oxide)
PS	Poly(styrene)
PVC	Poly(vinyl chloride)
TGCBA	Tetraglycidyl compound of bisphenol A

1 Introduction

A broad variety of structural polymers is nowadays available that are suitable for applications as different as carbon fiber reinforced materials, encapsulation of electronic devices or adhesive bonding. Each of these polymers belongs to one of two classes: thermosets or thermoplastics.

On a molecular scale, the difference between the two classes of materials is rather small. Thermoplastics consist of individual long chain molecules not connected with each other. Addition of a few crosslinks results in an infinite network structure that is the characteristic of thermosets.

The difference between thermosets and thermoplastics is more obvious when processing is considered:

- Thermoplastics are processed by reshaping the finished product in the form of a viscous melt at high temperatures.
- Thermosets are synthesized and shaped simultaneously generally from low viscosity liquids at elevated temperatures.

Both techniques have their advantages and their limitations with respect to process time, process temperatures, and process costs. However, the crucial question is: How much does crosslinking contribute to the desired properties of the material? The performance of the final product is, of course, the major issue. A lot of information on crosslinked polymers is available in the literature. There have been several attempts in the past [1–7], and also more recently [8–10], to sort out this accumulation of scientific data. Yet, it is neither simple nor particularly rewarding to undertake such a venture due to the multitude of variables which make direct comparisons difficult, and to the incidence of apparent contradictions.

Therefore, a different approach was followed in the present paper in order to improve the understanding of the relationship between the structure and the behavior of crosslinked polymers. A series of directly comparable model polymers were prepared with crosslink densities varying from high (thermoset) to zero (thermoplastic). Five polymers with well defined crosslink densities [11] were tested at various levels of deformation. This approach produced a small but assessable and fairly consistant body of results. Basic relationships derived from these results were related to corresponding results from the literature.

2 Materials

2.1 Starting Molecules

The polymers of the present study were all based on the molecule bisphenol A (2,2-di(4-hydroxy-phenyl)propan). The molecular chains between crosslinks were built up as uniformly as possible in order to eliminate effects imposed by variations of the chemical composition.

I Bisphenol A (abbreviated BA)

molar mass: 0.228 kg/mol
high purity product of Bayer AG, Leverkusen (polymer quality).

II Diglycidyl ether of bisphenol A (abbreviated DGEBA)

molar mass: 0.340 kg/mol
epoxide equivalent: 0.172 kg/mol

III Polyglycidyl compound of bisphenol A (abbreviated PGCBA)
        epoxide equivalent: 0.137 kg/mol.

The PGCBA resin is composed of various components, but the important one is the Tetra glycidyl compound of bisphenol A (abbreviated TGCBA)

molar mass: 0.452 kg/mol
epoxide equivalent: 0.113 kg/mol

Based on results of NMR and of gel permeation chromatography studies, a simplified composition for the PGCBA that yielded the epoxide equivalent of 0.137 kg/mol was assumed. The molar fraction of tetra functional molecules was taken as $v_4 = 0.45$ and of trifunctional molecules was taken as $v_3 = 0.55$.

2.2 Networks

The polyaddition reaction in stoichiometric mixtures of glycidyl ethers and bisphenol A resulted in macromolecular chains. Although these chains lack any fixed order in space, their composition is remarkably regular since each epoxy group reacts with one phenol and each phenol group reacts with just one epoxy group (Fig. 2.1). Side reactions are much less favoured. The reaction is essentially complete.

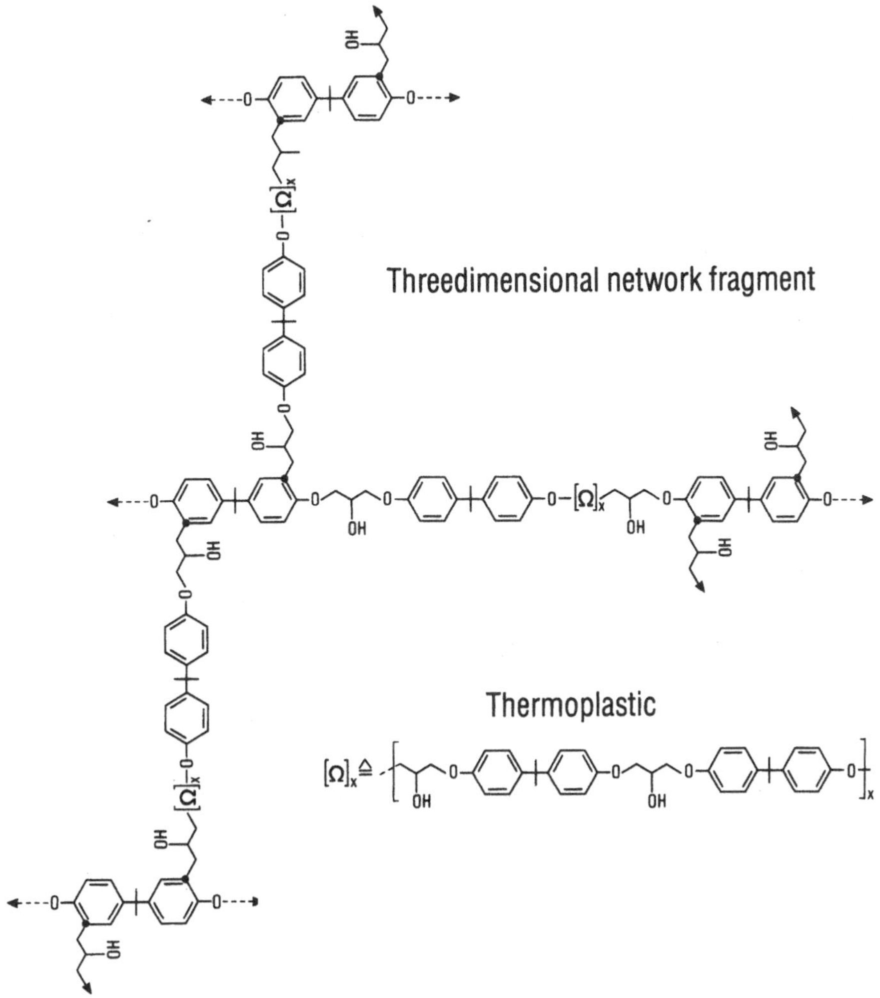

Threedimensional network fragment

Thermoplastic

Polymer	$\overline{M}_R/kg \cdot mol^{-1}$	X
A	0.398	0
B	0.825	0.94
C	1.63	2.71
D	3.24	6.25
E	?	?
Phenoxy* PKHJ	Thermoplastic	≈ 60

Fig. 2.1. Scheme of the molecular network of the studied polymers. The sequence of chemical groups is strictly as shown, although the molecular chains are randomly oriented in space

Crosslinks were introduced in the polymers by adding molecules with more than two reactive groups to the mixture e.g. PGCBA. After the reaction, three or more chains are connected to those molecules. Therefore, the concentration of PGCBA molecules in the resin mixture determines the density of the crosslinks in the cured polymer. The polymers consist of one giant molecule (theoretically infinite) since all molecular chains are linked with each other in the completely cured polymer. Details of the preparation of the polymers are given in the appendix.

All of the studied polymers are described by the network scheme in Fig. 2.1. Consequently, the composition of the polymers is uniform but the molecular chains between crosslinks differ in length. The molecular mass between crosslinks is therefore a dominant parameter for the characterization of the networks.

In order to calculate the average molecular mass $\bar{M}_R$ of the network strands, the mass $\Sigma\, m_i$ of the giant macromolecule is devided by the number N of strands in the network.

$$\bar{M}_R = \frac{\Sigma\, m_i}{N} \qquad\qquad (2.1)$$

The number N of molecular strands is related to the number C of crosslinks:

$$N = \tfrac{3}{2}\, C \qquad\qquad (2.2)$$

assuming each strand connects two crosslinks and each crosslink joins three strands.

This reasoning does not account for segments that are not properly attached to the network nor for additional crosslinks and any deviation from the outlined mechanism of the phenol curing reaction. Side reactions are not considered.

The effective molecular mass $\bar{M}_C$ of the network strands was determined experimentally from the moduli of the polymers at temperatures above the glass transition (Sect. 3) [11]. $\bar{M}_C$ was derived from the theory of rubber elasticity. $\bar{M}_C$ and the calculated molecular mass $\bar{M}_R$ (Eq. 2.1) of the polymers A to D are compared in Table 3.1.

2.3 Thermoplastics

Table 2.1 provides some information on the two polymers that are supposed to be free of chemical crosslinks. Phenoxy PKHJ is a product of Union Carbide Corporation. Plates were produced by injection moulding after drying of the granulate. On the other hand, polymer E was prepared by casting as were the other experimental polymers but without adding PGCBA to the mixture.

2.4 Cooling Conditions

The usual cooling rate was about 100 K/h, but some samples were cooled at two extreme rates: 10^4 K/h and 0.25 K/h. Do crosslinks impede the effects of annealing? And, does the cooling rate affect the properties of the polymers to a similar degree as crosslinks do?

Table 2.1. Comparison of the thermoplastic Phenoxy with the reaction product Polymer E

	Phenoxy PKHJ	Polymer E
molecular mass	~ 30 kg/mol	∞
melting	T $>$ 130 °C	no
density at 23 °C ϱ/Mg m^{-3}	1.1807	1.1837
glass transition temperatures:		
T_{gDSC}	87 °C	93 °C
$T_{\lambda max}$	85 °C	101 °C
tensile tests:		
Young's modulus E/GPa	2.6	2.6
yield strength σ_y/MPa	60	65
elongation at yield ε_y/%	4.4	5.2
post yield:		
true stress in the necking zone σ_n/MPa	90	72
draw ratio Λ	2.1	1.7

Inorganic glasses are annealed in order to release residual stresses. As a result, the density increases by up to 1% [12]. In polymers the change in density is fairly small (<0.2%). Nevertheless, annealing improves creep resistance [13, 14] and increases yield strength impressively [15, 16], but reduces the toughness of the polymers [17]. Apparently, annealing causes some subtle changes in the local order of the molecular segments. These changes are easily detected by mechanical tests or thermal analysis, but are generally too small to be observed by X-ray analysis.

3 Experimental Verification of the Polymer Structure

3.1 Rubber Elasticity

Although the basic concept of macromolecular networks and entropic elasticity [18] were expressed more then 50 years ago, work on the physics of rubber elasticity [8, 19, 20, 21] is still active. Moreover, the molecular theories of rubber elasticity are advancing to give increasingly realistic models for polymer networks [7, 22].

Small deformations of the polymers will not cause undue stretching of the randomly coiled chains between crosslinks. Therefore, the established theory of rubber elasticity [8, 23, 24, 25] is applicable if the strands are freely fluctuating. At temperatures well above their glass transition, the molecular strands are usually quite mobile. Under these premises the Young's modulus of the rubberlike polymer in thermal equilibrium is given by:

$$E = \frac{3\varrho}{\bar{M}_c}\left(1 - \frac{2}{f}\right)RT \tag{3.1}$$

ϱ   density of the polymer,
$\bar{M}_C$ averaged molecular mass of strands between crosslinks,
f   functionality of junction (f = 3),
R   gas constant, $R = 8.31 \, J \, K^{-1} \, mol^{-1}$,
T   absolute temperature.

Two consequences are of particular interest for the present work:

— Moduli of rubberlike materials $(T > T_g)$ rise as the temperature is increased. In this respect rubbers behave in contrast to most materials which usually exhibit decreased rigidity at elevated temperatures.
— Every quantity in Eq. (3.1) is known or measurable except $\bar{M}_C$. Therefore, if experiments furnish the modulus of a rubberlike network, $\bar{M}_C$ of the polymer can be derived by means of the above equation.

The correctness of the supposed structures for the four polymers (A, B, C, D, as given in Fig. 2.1) can be checked by such experiments. The comparison between the $\bar{M}_C$, derived from physical experiments and the $\bar{M}_R$, derived from the proposed chemical structures provides a critical test on the validity of the assumptions involved.

At temperatures well above the glass transition of the polymers, the molecular segments are highly flexible and slip past each other almost without restriction. They behave like the molecules of a liquid except for the fact that their ends are linked with each other. Just the existence of crosslinks distinguishes rubberlike materials from ordinary liquids. The bulk moduli K of liquids and of rubberlike materials are of similar magnitude, e.g. K = 1 to 2 GPa [26].

The shear modulus S of an isotropic material is related [27] to the Young's modulus E and the bulk modulus K:

$$S = \frac{E}{3 - E/3\,K} \tag{3.2}$$

As the Young's modulus E of a rubberlike material is low, $E \ll 3\,K$, the shear modulus S of such a material is well approximated by

$$S = \frac{E}{3} = \frac{\varrho}{\bar{M}_C}\left(1 - \frac{2}{f}\right)RT \tag{3.3}$$

The bracket $(1 - 2/f)$ was introduced into the theory of rubber elasticity by Graessley [23], following an idea of Duiser and Staverman [28]. Graessley discussed the statistical mechanics of random coil networks, which he had divided into an ensemble of micronetworks.

A mechanistic argument yields this factor $(1 - 2/f)$ in a simple way [24]. The N strands of the rubberlike network are not independent, since they are linked with their ends in the junctions. The sum of the forces operating at each junction has to cancel in order to guarantee the mechanical equilibrium of the network. The number C of junctions with functionality f is:

$$C = \frac{2N}{f} \tag{3.4}$$

Therefore, the N forces in the N strands have to satisfy C equations. Not all of the N strands are elastically active, but just N − C:

$$N - \frac{2N}{f} = N\left(1 - \frac{2}{f}\right) \tag{3.5}$$

The argument assumes that the forces act only at the crosslinked ends of the strands. No interactions between the strands exist in this simplistic picture. The interaction of molecular segments, well above the glass transition temperature is usually rather small.

3.2 Moduli in the Rubbery State (T > T$_g$)

Young's moduli were determined in tensile tests using samples of 4 mm thickness. Slow cyclic loading (frequency 0.01 Hz) with small strain amplitudes (ê < 3%) was used for the tests in order to maintain the thermal equilibrium as much as possible. The temperature range was limited to 260 °C as thermal decomposition became noticeable above this temperature [11].

The results of tests on the polymers A, B, C, D are plotted versus the absolute temperature in Fig. 3.1 in order to facilitate comparison with Eq. 3.1. Tests on polymer E were spoilt by plastic deformation. Straight lines were drawn through the points in Fig. 3.1 and through the origin (T = 0 K). Such lines correspond with the Eq. (3.1). At temperatures below the glass transition where the polymers

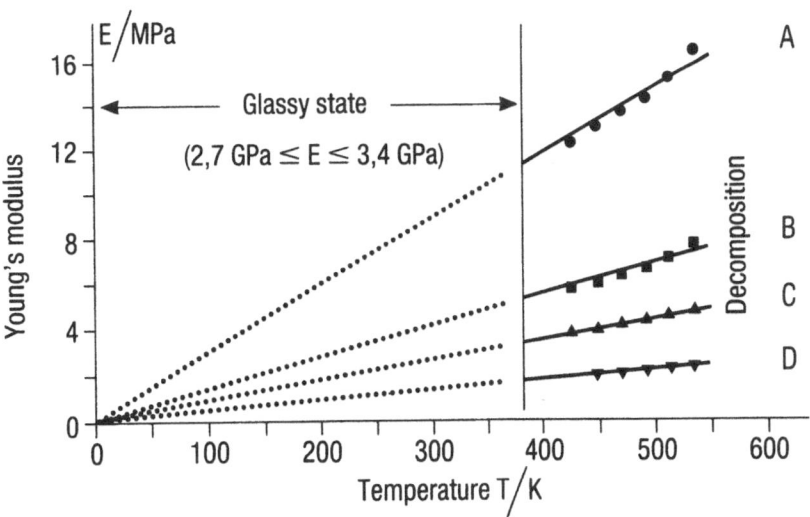

Fig. 3.1. Young's moduli E of the polymers A, B, C, D in the rubbery state against absolute temperature T (test frequency 0.01 Hz). Entropic elasticity is indicated by the proportionality of E to T [11]

Table 3.1. Young's modulus E at T = 500 K and average molecular mass between crosslinks

Polymer	Experimental data from tensile tests:		Estimation from chemical structure in Fig. 2.1	
	E_C/MPa	effective: $\bar{M}_C$/kg mol^{-1}	E_R/MPa	calculated: $\bar{M}_R$/kg mol^{-1}
A	14.6	0.327	12.1	0.398
B	6.9	0.693	5.8	0.825
C	4.4	1.09	3.0	1.63
D	2.2	2.17	1.5	3.24

were in the glassy state, the theory of rubber elasticity is, of course, not applicable. Therefore the lines are dotted in this range.

The effective molecular mass $\bar{M}_C$ of the network strands was determined from the slope E/T of the straight line using Eq. (3.1). $\bar{M}_C$ values are presented in Table 3.1 together with the values $\bar{M}_R$ that were calculated from the concentration of chemical crosslinks. The agreement between $\bar{M}_C$ and $\bar{M}_R$ is encouraging,

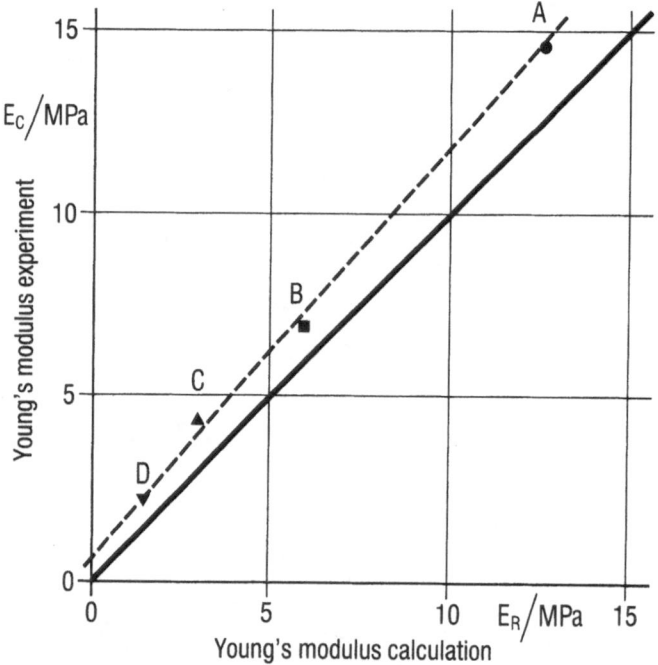

Fig. 3.2. Young's moduli E_C from experiments at T = 500 K are plotted against the calculated moduli E_R (Eq. 3.1; T = 500 K). The agreement is good, although apparently more crosslinks are formed than expected from the calculated crosslink density. *Solid line:* $E_C = E_R$

considering the simplifications involved in the use of the theory of rubber elasticity on one hand and in neglecting side-reactions during the preparation of the polymers on the other hand.

The consistency is readily seen in Fig. 3.2. There the experimental data from tests at 500 K are plotted versus the results calculated from Eq. (3.1) with $\bar{M}_R$ from Table 3.1, also at T = 500 K. The moduli from the experiments are systematically higher than the calculated moduli. Apparently, a few more crosslinks exist in addition to the chemical crosslinks that were introduced into the polymers by the PGCBA resin. These additional crosslinks may originate from entanglements between molecular chains or from chemical bonds which may, for instance, have resulted from neglected side-reactions. To distinguish between chemical links and entanglements is not possible at present. For the sake of simplicity the additional crosslinks will be assigned collectively to entanglements.

In the Eqs. (3.1) and (3.3), a second term can be introduced to account for the average molecular mass $\bar{M}_E$ between entanglements of chains [5, 29]:

$$ E = \varrho RT \left(\frac{1}{\bar{M}_R} + \frac{1}{\bar{M}_E} \right) = \varrho RT \frac{1}{\bar{M}_C}. \tag{3.6} $$

The dashed line in Fig. 3.2 corresponds to a linear regression calculation yielding $\bar{M}_E = 8$ kg/mol for the average molecular mass between entanglements if no chemical crosslinks are present ($M_R \to \infty$). This result agrees reasonably with values for various thermoplastics as determined from elasticity measurements on melts [30, 31, 32]. Examples are given in Table 3.2.

Dynamic shear moduli are conveniently determined with automated equipment, for instance, with the torsion pendulum. However, moduli derived from dynamic tests are often higher than the results from static tests for lack of relaxation. Examples are shown in Table 3.3. Young's moduli of the polymers A, B, C, D, derived from tensile tests (frequency 0.01 Hz) are compared with shear moduli S determined with the torsion pendulum (frequency >1 Hz). For rubberlike materials is 3S/E = 1, according to Eq.

Table 3.2. Entanglements of molecular chains

Polymer	Molecular mass $\bar{M}_E$ between entanglements $\bar{M}_E$/kg mol^{-1}	Ref.
Poly(carbonate)	2.5	[31, 32]
Phenoxy	5	[37]
Poly(dimethyl siloxane)	8.1	[36]
Poly(methyl methacrylate)	9.2	[31, 32, 33]
Poly(styrene)	19–32	[31, 32, 34]
Poly(vinyl acetate)	30	[35]

Table 3.3. Comparison of results from torsion pendulum and from tensile tests (quasistatic) at T = 500 K

Polymer	Dynamic shear modulus (frequency >1 Hz) S/MPa	Quasi-static Young's modulus (frequency 0.01 Hz) E/MPa	Ratio* 3S/E
A	6.9	14.6	1.4
B	3.6	6.9	1.6
C	2.6	4.4	1.8
D	1.5	2.2	2.0

* 3S/E = 1 for rubberlike materials according to Eq. 3.3

(3.3). Apparently, the dynamic elasticity is not purely entropic but is enhanced for various reasons [5]:

— energy elastic contributions,
— relaxation times of the polymers longer than period of oscillation,
— sliding entanglements of molecular chains,
— deviation from thermal equilibrium.

The mass of the network strands is, therefore, more appropriately deduced from static tests than from dynamic tests.

4 Glass Transition

4.1 Glass Transition Temperature

Upon cooling, molten and rubberlike polymers pass the glass transition and solidify as glassy materials. The temperature T_g of the glass transition depends on the chemical nature of the polymer as well as on the number of crosslinks between the molecular chains. Two different test methods were used for the determination of the glass transition range:

a) Shear modulus was measured by the torsion pendulum [38]. T_g was identified with $T_{\lambda max}$, that is the temperature of the highest logarithmic decrement λ (heating rate 1 K/min, frequency ≈ 1 Hz).

b) Differential scanning calorimetry (DSC) was used to determine the specific heat of the sample. T_g was identified with the temperature of the inflection point in the DSC diagram (heating rate 10 K/min.).

The results of the tests are plotted in Fig. 4.1 according to Fox and Flory [39] against $\bar{M}_C^{-1}$, the inverse molecular mass between crosslinks, which was determined from the rubbery modulus. The two sets of results agree basically, although the DSC results are consistently 8 K lower than the temperatures $T_{\lambda max}$, derived from

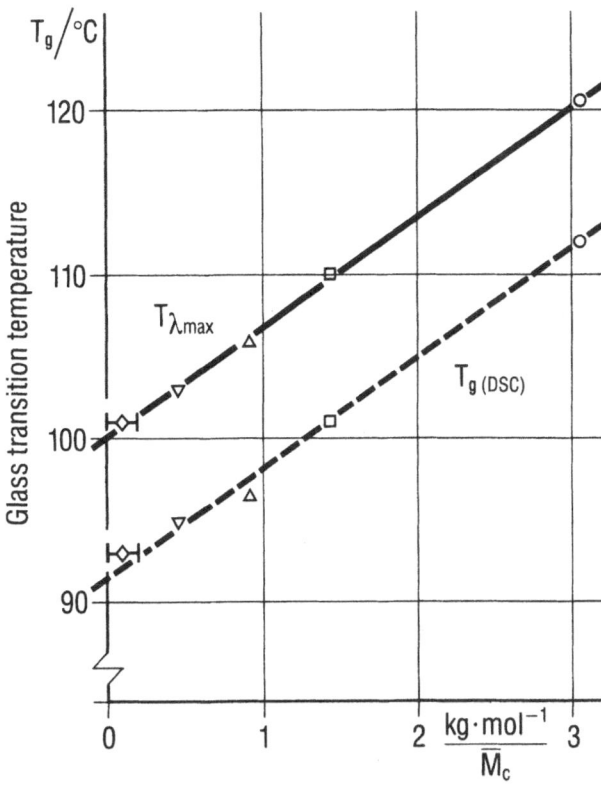

Fig. 4.1. Glass transition temperatures of the polymers are plotted against $1/\bar{M}_C$, that is the inverse molecular mass between crosslinks.

$T_{\lambda\,max}$:   temperature of the highest logarithmic decrement λ
$T_{g\,(DSC)}$:   temperature of the inflection point in the DSC diagram
The *diamond* represents polymer E. Its $\bar{M}_C$ is high, but not determinable by the described technique

the mechanical tests. In both cases the crosslinks are responsible for the shift ΔT_g of the glass transition to higher temperatures. The slope of the two parallel lines in Fig. 4.1 yields the relation:

$$\Delta T_g = \left(7\,\frac{\text{kg K}}{\text{mol}} \right) \frac{1}{\bar{M}_C} \tag{4.1}$$

The relation (4.1) corresponds to the frequently cited empirical equation of Fox [40]:

$$T_g - T_{g\infty} = \frac{\phi}{\bar{M}_C} \tag{4.2}$$

$T_{g\infty}$ glass transition temperature of the uncrosslinked polymer
$\bar{M}_C$ average molecular mass between crosslinks
ϕ empirical factor

Nielsen [1, 41] averaged data from the literature that, admittedly, did not agree very well, and proposed

$$\phi = 39 \frac{kg}{mol} K$$

A few examples for ϕ are listed in Table 4.1.

Obviously, ϕ is not a "universal" constant. ϕ depends on the number of chains that are linked at a crosslink (junction functionality f) and probably on the chemical composition as well.

Estimation of $\bar{M}_C$ is another subject open to debate in connection with Eq. (4.2). To this end, many scientists [e.g. 1, 9, 37, 42, 43, 90, 105, 106, 110] use the equation

$$\bar{M}_C = \frac{\varrho RT}{S} = \frac{3\varrho RT}{E} \tag{4.3}$$

S   shear modulus
E   Young's modulus
ϱ density

Comparison with Eq. (3.1) reveals that the influence of the junction functionality f is neglected in this simplified treatment. Such a simplification is doubtful for $f = 3$.

On the other hand, the equilibrium moduli are significantly lower than dynamic moduli, that are frequently used (Table 3.3). The two simplifications seem to compensate each other to some extend. Nevertheless, this uncertainty in the estimation of the $\bar{M}_C$ contributes to the uncertainty in ϕ.

Table 4.1. Examples for the empirical factor ϕ, that quantifies the effect of crosslinks on T_g

Comments:	$\phi/K\ kg \cdot mol^{-1}$	Ref.
Nielsen's suggestion	39	[1, 41, 42]
	22	[43]
calculated from $788\ K \cdot M_1$ with		
$M_1 = 0.02\ kg \cdot mol^{-1}$ (Sect. 7)	16	[44, 45, 46]
present work	7	
junction functionality		[47]
$f = 3$	13	
$f = 4$	30	
$f = 6$	55	

4.2 Effects of Cooling Rate

DSC tests revealed a difference of about 2 K between the glass transition temperatures as determined for two sets of samples: one set was quenched from 150 °C into liquid nitrogen, the other set was cooled over six days from 120 °C to 84 °C (rate 0.25 K/h). The slowly cooled polymers exhibited consistently higher glass transition temperatures than the reference samples when reheated. In addition, tiny endothermal peaks were observed in the DSC-diagrams. The area of the peaks corresponded to 1.8 J/g in all polymers of this study including Phenoxy. However, none of the quenched samples showed the endothermic effect. Probably, the aging effects could be amplified by extended annealing at an appropriate temperature [48] e.g. during ten days [49]. The polymer was amorphous before and after treatment. Perhaps, the conformations of the chain molecules had changed and some adjacent chain segments became more favourably arranged during the annealing process. As the segments moved closer together, the van der Waals' forces between those segments increased, and the density of the polymer also increased (Sect. 5.1). Irrespective of their crosslink densities all polymers showed similar annealing effects.

5 Properties of the Glassy Polymers

This section is based on tests that can be performed repeatedly with the same sample. A sample is not altered permanently as the tests apply no strain or only small strains. Three examples are dealt with subsequently:
— density,
— elasticity,
— thermal expansion.

5.1 Density

The densities of the polymers at 23 °C were determined using liquids in density-gradient-columns [50]. The tiny samples (about 10 mg) were floated in the columns over a period of one week. The absorption of liquid by the polymer was corrected for by an extrapolation of the readings back in time to the start of the experiment [51].

In Fig. 5.1, the densities of the annealed and of the quenched polymers are plotted against $\bar{M}_C^{-1}$, the inverse molecular mass of the network strands. All the annealed samples were denser by about 0.15% than the quenched ones. Bero and Plazek [52] observed an effect of similar magnitude between quenched samples and samples cooled at 0.2 K/h.

Apparently, annealing was not impeded by crosslinks (Fig. 5.1). The density effects observed agree with the results of the glass transition temperature measurements (Sect. 4.2). There, the T_g of the annealed (and therefore denser) sample was consistently higher by about 2 K than the T_g of the quenched polymer.

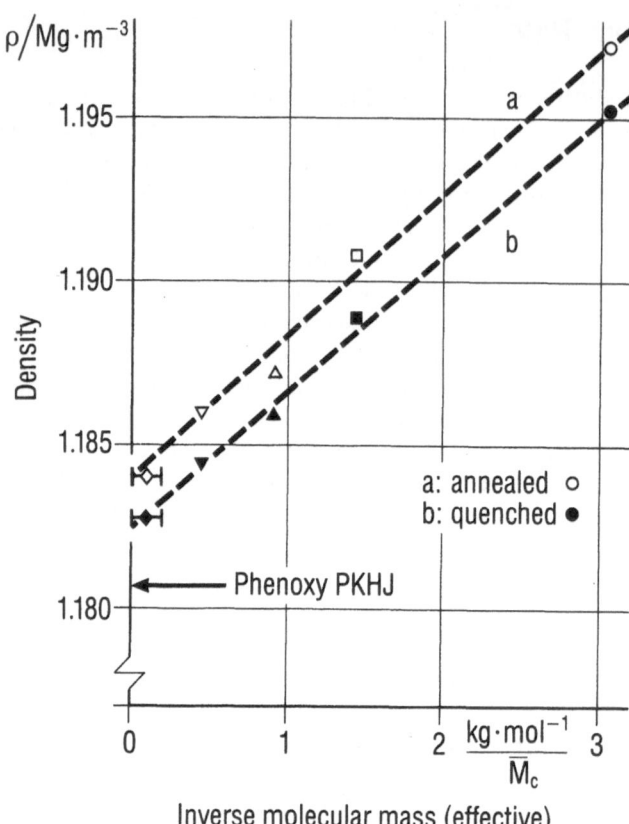

Fig. 5.1. Densities of annealed and of quenched polymers are plotted against $1/\bar{M}_c$, that is the inverse molecular mass between crosslinks. Test temperature: 23 °C. The density of the thermoplastic Phenoxy is indicated. The *diamond* represents polymer E.

Crosslinked polymers were denser than the thermoplastic. The densities of the polymers increased proportional to the number of crosslinks ($\sim \bar{M}_C^{-1}$) as shown by the two straight lines in Fig. 5.1. The volume occupied by the polymer was reduced by 0.008 nm^3 for each junction introduced in the network. The change of volume was deduced from the slope in Fig. 5.1. Likewise, small voids close to the ends of the molecular chains may well be responsible for the lower density of the Phenoxy resin ($\varrho = 1.1807$ Mg m^{-3}) as compared to polymer E.

5.2 Modulus and van der Waals' Forces

Young's moduli of the polymers were determined in tensile tests [53] using samples of 4 mm by 10 mm cross-section and a gauge length of 50 mm. The results of the tensile tests were regarded as reliable. In addition, we performed flexural tests [54] on small samples in order to show the effects of annealing. These samples measured merely 1.3 mm by 4 mm and were supported with a span length of 20 mm. Consequently, the results are not as accurate as the moduli from the tensile tests.

Young's moduli are given in Table 5.1 as well as in Fig. 5.2, plotted against $\bar{M}_C^{-1}$, the inverse molecular mass between crosslinks. Obviously, the moduli of the polymers increased as the number of crosslinks multiplied. Such an effect could be caused by the additional number of rigid covalent bonds or, just as well, by the increased density of the crosslinked polymers (Sect. 5.1).

The flexural tests on quenched and on annealed samples showed a striking gain in rigidity for all the slowly cooled samples. However, the difference between the two sets of results almost vanished when the moduli from the flexural tests were plotted against the corresponding densities in Fig. 5.3. A single smooth curve fits both sets of data. Apparently, moduli correlate well with the densities of the polymers. The crosslinks do not contribute noticeably to the rigidity of the glassy polymers. A similar observation is reported by Vogt [55].

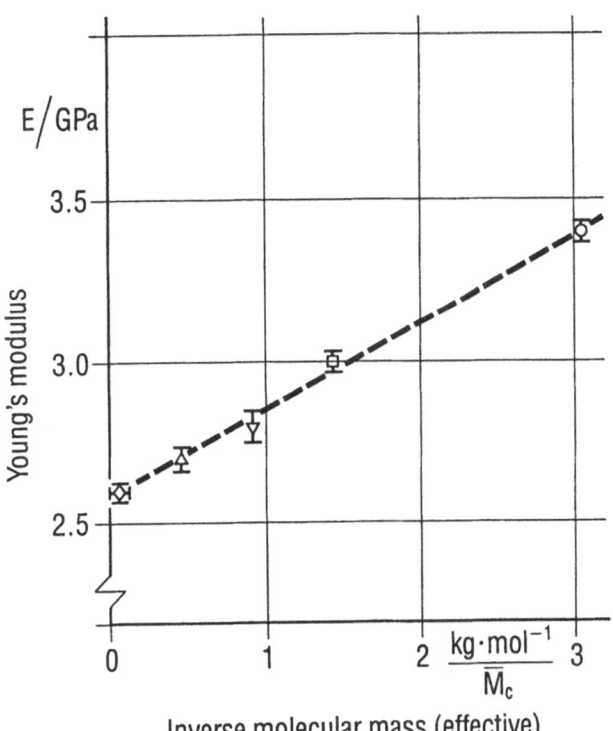

Fig. 5.2. Young's modulus E, as determined in tensile tests, is plotted against $1/\bar{M}_C$, that is the inverse molecular mass between crosslinks. Test temperature: 23 °C

Table 5.1. Some properties at ambient temperature (T = 296 K)

Polymer	Effective molecular mass between crosslinks $\bar{M}_C/kg \cdot mol^{-1}$	Density $\varrho/Mg \cdot m^{-3}$	Young's modulus E/GPa	Thermal expansion $\alpha/10^{-6}\,K^{-1}$	Product Barker [58] $E\alpha^2/PaK^{-2}$
A	0.327	1.1977	3.4 ± 0.03	62	13.1
B	0.693	1.1908	3.0 ± 0.03	66	13.1
C	1.090	1.1876	2.8 ± 0.05	68	12.9
D	2.170	1.1856	2.7 ± 0.05	70	13.2
E	·?	1.1837	2.6 ± 0.03	72	13.5

The denser packing of the molecular chains results in stronger interactions between segments and their near neighbours by means of van der Waals' forces which are responsible for the rigidity of glassy polymers just as in the solidification of high polymers [56]. The important features of van der Waals' forces are [57, 111]:
- always attractive,
- acting between every atom or molecule and its neighbours,

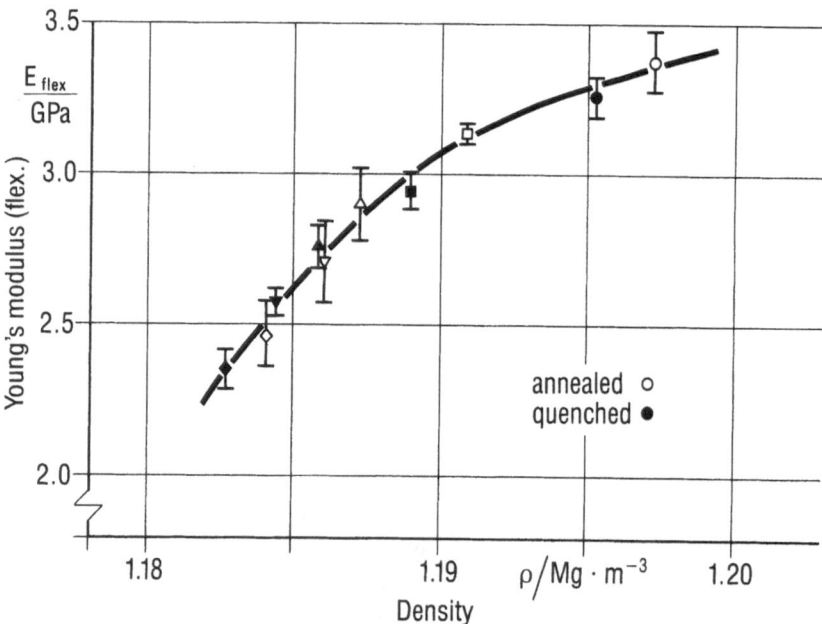

Fig. 5.3. Young's moduli E_{flex} as determined by flexural tests on small samples after thermal treatment are plotted against the densities of those samples. The dots are situated along a single line since the annealed samples are denser and more rigid than the quenched samples prepared from the same polymer

- not confined to fixed orientations or locations,
- simply additive,
- vary as r^{-7}, (r interatomic distance).

Although the energy of a single van der Waals' pair interaction is rather small compared to the energy of a covalent bond, the sum of such interactions can be remarkable large. Covalent bonds help the action of van der Waals' forces as the covalent bonds increase the density of the polymer and bring molecules close together. The increase of the glass transition temperature T_g by crosslinking, for instance, is partly due to such a dense packing of the molecular chains.

5.3 Thermal Expansion

Thermal expansion – as elasticity – depends directly upon the strength of the intermolecular forces in the material. Strongly bonded materials usually expand little when heated, whereas the expansion of weak materials may be a hundred times as large. This general trend is confirmed by Table 5.1. The coefficient of thermal expansion α was found to be lower in the crosslinked polymers and higher in the less crosslinked or thermoplastic materials as observed by Nielsen [1]. In addition, Table 5.1 presents the Young's moduli E of the polymers at ambient temperatures as well as the products $\alpha^2 E$. The values of $\alpha^2 E$ are all close to 13.1 Pa K^{-2} with a coefficient of variation of 1.6%.

A rather general, approximate relation between Young's modulus E and the coefficient α of linear thermal expansion was proposed by Barker [58]:

$$\alpha^2 E = 15\,\mathrm{Pa\,K^{-2}} = 15\,\mathrm{J\,m^{-3}\,K^{-2}} \tag{5.1}$$

Its validity at normal temperatures was shown for more than 60 materials, ranging from pure metals to glassy polymers. Obviously, the polymers of the present study are good examples for Barker's rule. The product $\alpha^2 E$ is linked to the difference of the two heat capacities c_σ and c_ε, measured under constant stress and under constant strain, respectively [58]. Also, $\alpha^2 E$ is linked to the difference of two Young's moduli E_s, and E_T measured adiabatically and isothermally [59].

A conclusion as to the effect of crosslinking on thermal expansion is not possible. Clearly, the polymers with many crosslinks and with short strands expand less than the uncrosslinked materials when heated. However, this effect cannot exclusively be attributed to the presence of crosslinks. It may just as well originate from the increased density of the crosslinked materials which was shown to be responsible for the increase in the moduli.

6 Yielding

When the load is high enough, a polymer yields and loses its resistance. The corresponding stress level is specific to the polymer and the actual temperature. Knowledge of the yield strength of a polymer is crucial in order to avoid the risk of failure in application. However, the polymer can fracture even at loads below the yielding level. Consequently, crack growth and yielding are both subjects of equal relevance with respect to failure in use.

The onset of plastic deformation in a material under load is called yielding [60]. In contrast to the experiments described in the previous sections, yielding causes a permanent deformation, i.e. a deformation that remains after the load is removed. The effects of crosslinks on the yield behaviour of polymers are demonstrated by three experiments:

— tensile tests,
— thermal history,
— strain rate.

6.1 Tensile Tests

Yield strength as determined in tensile tests [53] at ambient temperature was plotted in Fig. 6.1 against $\bar{M}_C^{-1}$, the inverse molecular mass between crosslinks. All the samples of polymer A (the most crosslinked polymer) failed before the polymer started to yield. Therefore, load-extension-curves were extrapolated up to a hypothetical yield strain in this case. The extrapolated tensile is marked by brackets (Table 6.1).

Yield strength was found to increase inversely proportional to the mass $\bar{M}_C$ of the chains (Figure 6.1) as was observed in the case of the glass transition temperature T_g (Fig. 4.1). Actually, the tensile yield strengths of the polymers correlate well with the corresponding glass transition temperatures as shown in Fig. 6.2. The increase $\Delta\sigma_y$ of the tensile yield strength amounts to

$$\Delta\sigma_y = \left(1.5\,\frac{\text{MPa}}{\text{K}}\right)\Delta T_g \tag{6.1}$$

ΔT_g is the change in the glass transition temperature T_g.

Table 6.1. Some results on polymer failure (T = 296 K)

	Effective molecular mass between crosslinks $\bar{M}_C$/kg mol^{-1}	Tensile yield strength σ_y/MPa	Energy release rate G_{IC}/Jm^{-2}	Half crack opening displacement $w = \delta/2$ $= G_{IC}/2\sigma_y$ w/µm	Chain contour length (Eq. 7.9) l_c/nm
A	0.327	[95]	160 ± 34	0.84	1.5
B	0.693	79	284 ± 56	1.8	3.2
C	1.090	73	486 ± 49	3.3	5.0
D	2.170	68	675 ± 64	5.0	9.9
E	?	65	1003 ± 152	7.7	
Phenoxy PKHJ	5.05*	60	1490 ± 143	12.4	23*

* value from M.D. Glad [37] p. 95

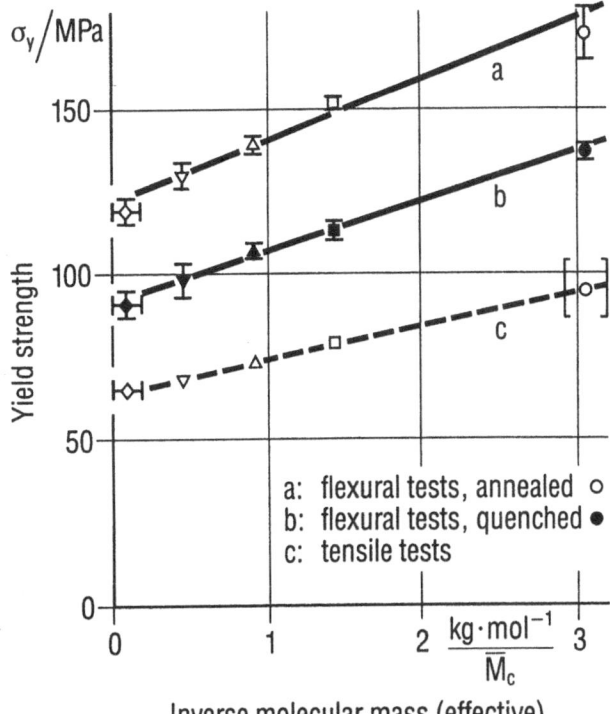

Inverse molecular mass (effective)
between crosslinks

Fig. 6.1. Yield strengths of the five polymers are plotted against $1/\bar{M}_C$ that is the inverse molecular mass between crosslinks. The *diamond* represents polymer E. Test temperature: 23 °C. *a* and *b* represent results of flexural tests on small samples (thickness 1.3 mm)
a: annealed,
b: quenched,
c: results of tensile tests, cross-section of the samples 4 mm by 10 mm; [ ] result of extrapolated stress-strain-curve

The simple relation (6.1) is quite useful. It allows an estimate of the gain or loss in yield strength if the glass transition temperature is changed, as for instance by the progress of the curing reaction, by radiation damage, or by absorption of water and of solvents. The ΔT_g is determined fairly easily by thermoanalytical measurements.

The increase $\Delta \tau_y$ of the shear yield strength is

$$\Delta \tau_y = p_\tau \Delta T_g \tag{6.2}$$

with $p_\tau = 0.75\,\text{MPa K}^{-1}$, since the maximum shear stress τ equals half the unidirectional tensile stress σ [61].

Necking occurred in samples of polymer E and of the thermoplastic Phenoxy. The other, more crosslinked polymers failed before necking started. The results on the post yield behavior are included in Table 2.1. Apparently, a process such

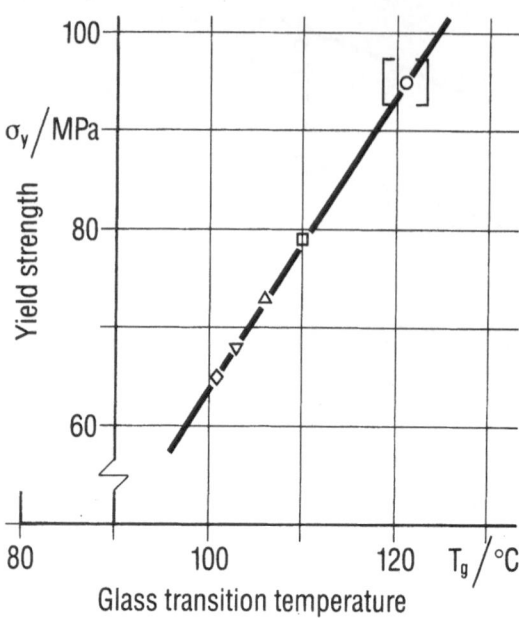

Fig. 6.2. Yield strengths from tensile tests at 23 °C are plotted against the glass transition temperatures ($T_{\lambda\,\text{max}}$) of the five polymers; [ ] result of extrapolated stress-strain-curve

as necking is prevented by a certain concentration of chemical crosslinks between the molecules. They impede the strong lateral contraction of the material that has to take place during necking.

6.2 Thermal History

The flexural strength of the annealed polymers proved to be consistently about 30% higher than the strength of the quenched polymers as shown in Fig. 6.1. Tests were evaluated in accordance with ISO 178 [54]. As the samples yielded, they deformed plastically. Therefore, the assumptions of the simple beam theory were no longer justified and consequently the yield strength was overestimated.

The results of the flexural tests from Fig. 6.1 were replotted against the densities of the samples in Fig. 6.3. The points aligned along two well separated curves representing annealed and quenched samples. In comparison to Fig. 6.1, the two curves approach each other but there is still a gap of 22%. Obviously, the annealed polymer became stronger than expected from the increase of density in comparison to the quenched material. The rigidity of the polymers exhibited a different behavior. According to Fig. 5.3, Young's moduli proved to be related to the density of the samples almost independently of the thermal history.

Haward [62] proposed that annealing leads to changes in the distribution of rotational isomers within the molecular chains. Such changes may be accompanied by closer packing and more extensive regularity. In glassy polycarbonate e.g. Turska et al. [63] observed by X-rays the straightening of macromolecules during annealing.

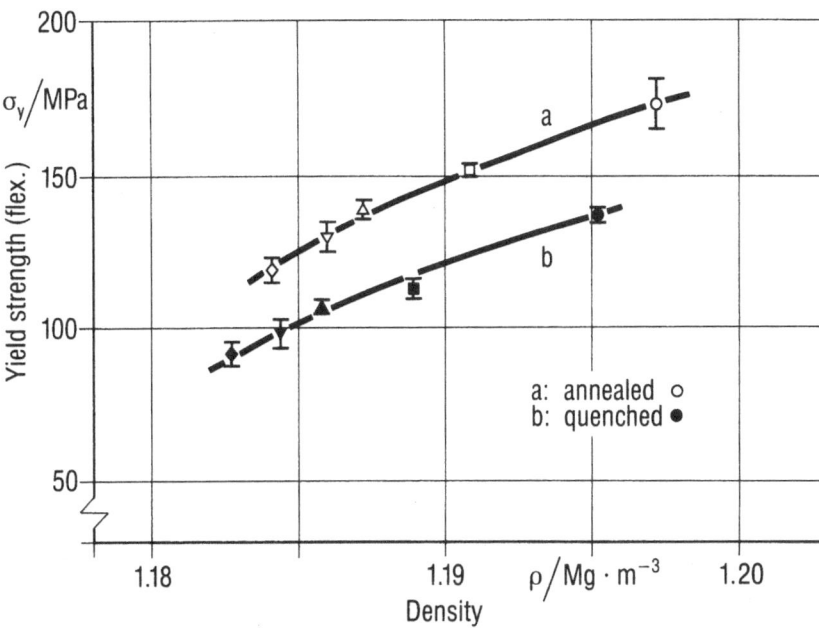

Fig. 6.3. Yield strengths from flexural tests are plotted against the densities of the polymers. The annealed samples were noticeably stronger than the quenched ones of similar density. Rigidity (Fig. 5.3.) was governed by the density of the polymer whereas yield strength seemed to depend mostly on molecular conformations

An increased yield stress is required [62] in order to reverse the unfavourable conformations of the molecular chains that develop during annealing. This explanation is supported by the energy changes observed in annealed polymers. The enthalpy difference, as determined by DSC was $\Delta H = 1.8 \text{ J/g}$ (Sect. 4.2), whereas the additional work required for yielding in an annealed sample was 1 J/g.

The annealing process rearranged the chain molecules apparently more efficiently than was expected by the gain of density. For instance, the density of PVC was increased up to 1% by cooling the polymer under pressure (200 MPa) from the rubbery to the glassy state [64]. However, the ultimate properties were scarcely changed [64, 65], although the densification of the polymers caused by pressure surmounted significantly the densification of the present polymers achieved by annealing.

The number of crosslinks had no significant effect on the outcome of the thermal history in the present materials.

6.3 Effect of Strain Rate

The yielding of polymers can be understood analogue to the transition-state theory [15, 45, 60, 66–71]. In viscous liquids the molecules pass each other with some

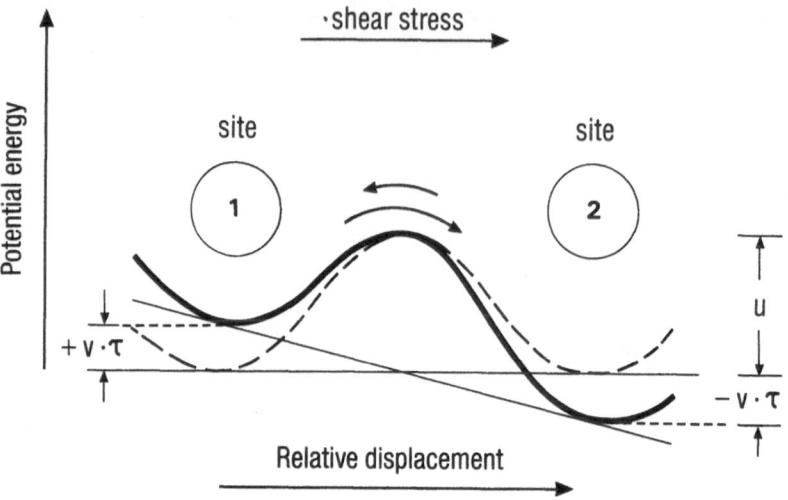

Fig. 6.4. Energy barrier between occupied and empty molecular sites; u: activation energy. The applied shear stress τ deforms the energy barrier analogous to Eyring's theory of viscosity; v: activation volume

hindrance when they are squeezed through the narrow gap between adjacent molecules. Eyring [72] modelled this passage by a potential barrier, which is crossed over by the molecules with a certain frequency. A shear stress τ that acts across the fluctuating molecules will favour the passage of molecules along its direction and will hinder the passage of molecules in the opposite direction. According to the model (Fig. 6.4), the height of the potential barrier will be diminished by τv for crossings in the direction of the shear stress τ and it will be increased by τv for crossings in the opposite direction. v stands for the activation volume. It includes all the molecular segments that participate in a discrete jump by giving way.

The shear rate $d\gamma/dt$ in the polymer will be proportional to the number of segments crossing the barrier, reduced by the number of segments crossing back in a given time interval. The exchange of segments amounts to the following balance, if Γ is a factor of proportionality:

$$\frac{d\gamma}{dt} = \Gamma\left(\exp\frac{-u + v\tau}{kT} - \exp\frac{-u - v\tau}{kT}\right) = \Gamma\frac{1}{2}\exp\frac{-u}{kt}\cdot\sinh\frac{v\tau}{kT}$$

$$(6.3)$$

u   potential barrier;
k   Boltzmann constant, $k = 1.38 \times 10^{-23}$ J/K;
T   absolute temperature;
τ shear stress;
v   activation volume.

In liquids, and in glasses at temperatures above their glass transition $(T > T_g)$, the shear stresses are usually very small. The equation for the viscosity η can be derived if $\tau v \ll kT$ is a reasonable assumption

$$\tau \Big/ \frac{d\gamma}{dt} = \eta = \frac{kT}{\Gamma v} \exp \frac{u}{kT}. \tag{6.4}$$

At temperatures below the glass transition, the polymers are fairly rigid and the yield strength τ_y is high. $v\tau_y \gg kT$ is plausible in the case of solid polymers. Eq. (6.3) can be rearranged to read:

$$\tau_{y1} - \tau_{y2} = \frac{kT}{v} \ln \left(\frac{d\gamma_1}{dt} \cdot \frac{dt}{d\gamma_2} \right) \tag{6.5}$$

τ_{y1} and τ_{y2} are two yield stresses measured at the shear rate $\dfrac{d\gamma_1}{dt}$ and $\dfrac{d\gamma_2}{dt}$, respectively .

Accordingly, the activation volume v can be determined from yield experiments performed at various shear rates.

The yield strengths of the polymers A, B and E from flexural tests are plotted in Fig. 6.5 against the strain rate on a logarithmic scale. The crosshead speed was

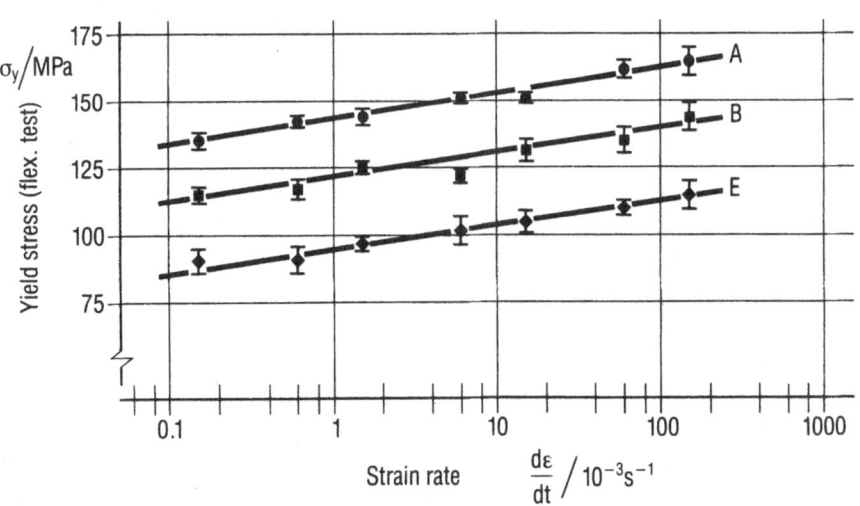

Fig. 6.5. Yield strengths from flexural tests are plotted against strain rates at the surface of the samples. Tests were performed on polymers A, B, and E; test temperature 23 °C. The slope of the three lines correspond to similar activation volumes: $v = 2 \pm 0.1 \text{ nm}^3$

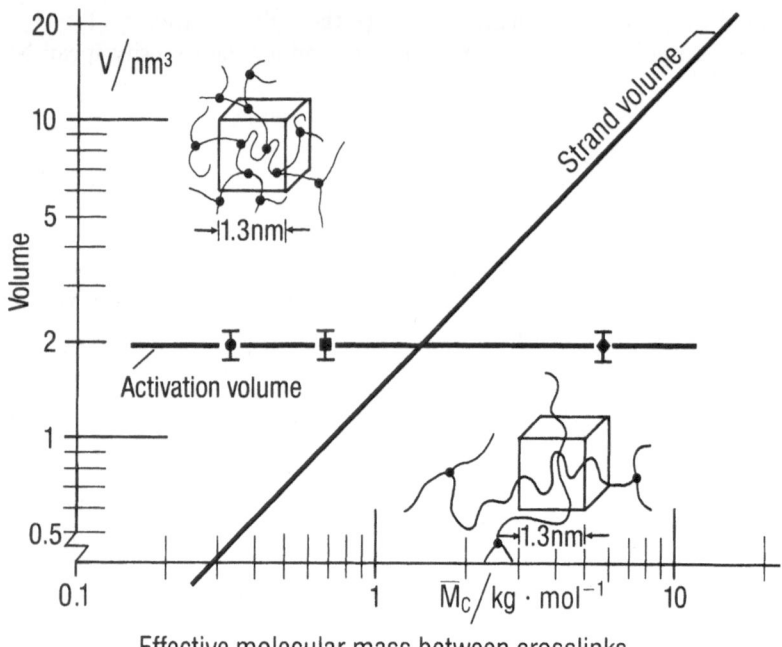

Effective molecular mass between crosslinks

Fig. 6.6. Comparison of activation volume and average volume of the strands between crosslinks. The effective molecular mass $\bar{M}_C$ between crosslinks is varied from 0.1 to 10 kg/mol. The *cubes* represent the activation volume of 2 nm³.
Left-hand side $\bar{M}_C < 1$ kg/mol: the activation volume contains several molecular junctions.
Right-hand side $\bar{M}_C > 1$ kg/mol: the volume of one strand is much larger than the activation volume

varied by three orders of magnitude. The general trend is obvious in spite of the scatter of the results: each polymer exhibited a linear dependence of the yield stress to the logarithm of the strain rate.

Activation volumes of the polymers according to Eq. (6.5) were calculated with $\tau_y = 0.5\sigma_y$ and plotted in Fig. 6.6 versus the average molecular mass $\bar{M}_C$ of network strands. The volume of a molecular strand is plotted also

$$V_C = \frac{\bar{M}_C}{N_L} \frac{1}{\varrho} \tag{6.6}$$

ϱ density
N_L Avogadro number.

The activation volume of the three polymers turned out to be $v \approx 2$ nm³, independent of their crosslink density. In the crosslinked polymer A the strands are short and about five of them fit into the activation volume. In contrast, one strand of polymer E requires a volume five times larger than the activation volume!

Apparently, crosslinks between the molecular chains do not interfere with the yield process in the activation volume. Yielding involves the cooperative movement of about 10 to 20 chain segments, if the average volume of one segment measures about 0.15 nm³. Such a volume seems to be reasonable according to "molecular parameters" [73–75].

The activation energy u is determined from the temperature dependence of the yield stress [45]. Fischer et al. [45] found $u/v = 248$ MPa for Phenoxy resins. Therefore, the activation energy is about $u = 0.5 \times 10^{-18}$ J, if the activation volume $v = 2$ nm³.

7 Fracture

7.1 Experiments and Results

Critical stress intensity factor K_{IC} and critical strain energy release rate G_{IC} quantify the stability of a polymer against the initiation and propagation of cracks. Stress intensity factor K_1 and energy release rate G_1 are not independent but they are related [76] by means of the appropriate modulus E*.

$$G_1 = \frac{K_1^2}{E*} \tag{7.1}$$

$$E* = \frac{E}{1 - \mu^2} \quad \text{in case of plane strain}$$

$$E* = E \quad \text{in case of plane stress}$$

μ Poisson's ratio.

Compact tension samples (Fig. 7.1) were prepared from cast plates. Pressure injection moulding was used for Phenoxy PKHJ. The toughness tests were performed at ambient temperature according to ASTM-E399 [77]. The results are given in Table 6.1.

Polymer A with $G_{IC} = 160$ J m^{-2} is typical for thermoset materials which are expected to be brittle [78]. At the other end of the series, polymer E and Phenoxy with $G_{IC} > 1$ kJ m^{-2} are tougher than several wellknown thermoplastics (PMMA, PS, PES). In contrast to the more crosslinked polymers, polymer E and Phenoxy PKHJ show necking after yielding in tensile tests with draw ratios $\Lambda = 1.7$ and $\Lambda = 2.1$, respectively (Table 2.1).

7.2 Analysis Based on the Dugdale-Barenblatt-Model

Good agreement is reported to exist between the Dugdale plastic zone model and optical interference experiments, performed at the tip of a crack. Morgan and Ward [79], Fraser and Ward [80] and more recently and extensively Döll and

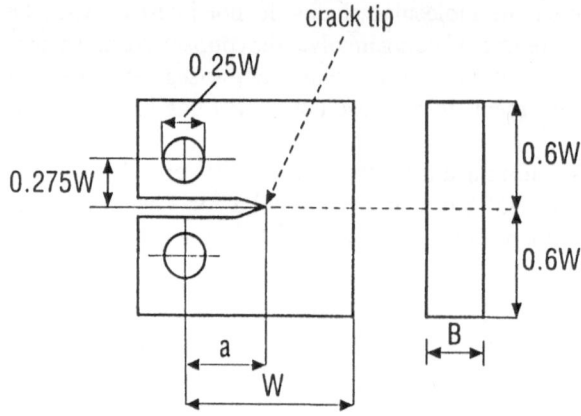

Fig. 7.1. Compact tension test-piece [77], W = 20 mm, B = 4 mm; the crack (length a) was introduced by tapping with a razor blade after cooling of the polymer with liquid nitrogen, if necessary

Könczöl [81] looked at PMMA, PC, PVC and other thermoplastic polymers. Their results support the plastic zone model.

A flat elliptical hole of length 2a is considered in an infinite plate which is loaded from a tensile stress σ normal to the ellipse. Plastic zones of length s will develop at the far ends of the ellipse as is shown in Fig. 7.2 [81]. The yielded zones at the crack tip are described by the plastic zone model of Dugdale [82] and the cohesive force model of Barenblatt [83, 84]. Rice [85] derived the following expression for the displacements $w_{(x)}$ of the plastic zone boundaries assuming small-scale yielding with stress $\sigma \ll \sigma_C$, the cohesive stress σ_C acting across the zone of length s

$$2w_{(x)} = \frac{8\sigma_C s}{\pi E^*}\left\{\xi - \frac{x}{2s}\ln\frac{1+\xi}{1-\xi}\right\} \tag{7.2}$$

$$\xi = \left(1 - \frac{x}{s}\right)^{1/2} \tag{7.3}$$

E^* is the appropriate modulus (see Eq. (7.1)).

The length s of the plastic zone for small-scale yielding can be expressed by

$$s = \frac{\pi}{8}\frac{K_I^2}{\sigma_C} = \frac{\pi}{8}\frac{G_I E^*}{\sigma_C^2} \tag{7.4}$$

K_I stress intensity factor
σ_C cohesive stress
G_I energy release rate.

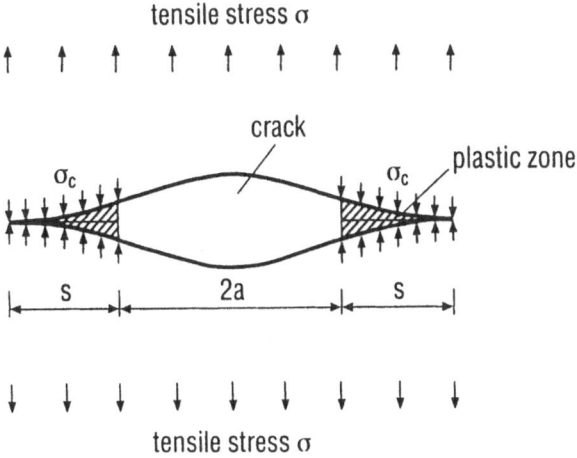

Fig. 7.2. Dugdale-Muskhelishvili-model of plastic zones at the ends of a loaded elliptical hole [81].
s: length of plastic zone
a: crack length
σ: tensile stress, remote from the defect
σ_c: cohesive stress

The boundary of the proposed deformation zone is shown in Fig. 7.3 according to Eq. (7.2). The displacement at the crack tip at $x = 0$ is called crack opening displacement δ or critical crack opening displacement δ_C if the crack is going to propagate.

$$\delta = 2w_{(0)} = \frac{G_I}{\sigma_C} \tag{7.5}$$

The work done in the deformation zone can be estimated assuming the deformation zone propagates together with the crack and maintains its steady state displacement profile [86]

$$G_{IC} = 2 \int_0^{w_{(0)}} \sigma_C \, dw = 2\sigma_C w_{(0)} = \delta_C \sigma_C. \tag{7.6}$$

The cohesive stress σ_C is assumed to be constant (Dugdale model) as in Eq. (7.5). Chan, Donald and Kramer [87] found a good agreement between the critical energy release rate G_{IC}, as estimated by the Dugdale model and G_{IC} as computed from the actual stress and displacement profiles in their experiments.

The deformation zones were calculated for the polymers of Table 5.1 and Table 6.1 according to the Dugdale-Barenblatt-model. Yield stress σ_y from tensile tests was used instead of the cohesive stress σ_C since a reasonable agreement of σ_y and σ_C

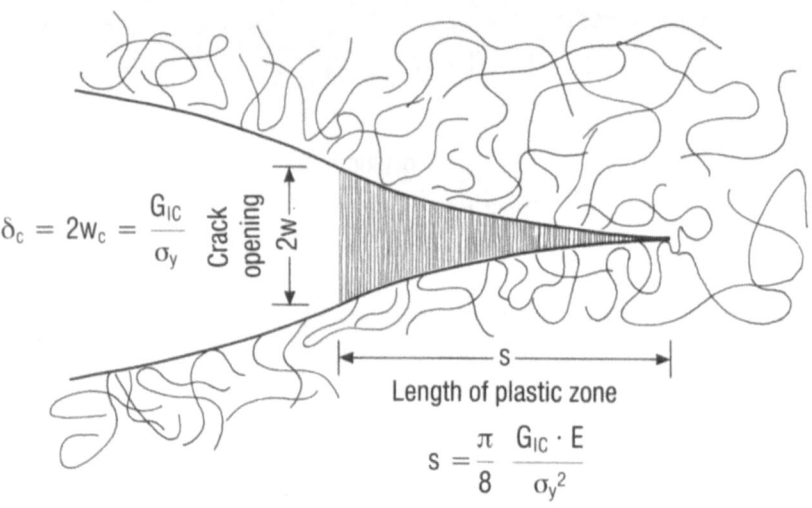

$$\delta_c = 2w_c = \frac{G_{IC}}{\sigma_y}$$

$$s = \frac{\pi}{8} \frac{G_{IC} \cdot E}{\sigma_y^2}$$

Length of plastic zone

Fig. 7.3. Deformation zone as calculated from the Dugdale-Barenblatt-model (Eq. 7.2). In order to magnify the displacements, $E/\sigma_y = 7$ was assumed for the diagram, whereas, in reality the ratio is about 38 (Tables 5.1 and 6.1)

has been observed, e.g. in PMMA [88]. The length s of the deformation zone is proportional to the crack opening displacement δ

$$s = \frac{\pi}{8} \frac{E^*}{\sigma_y} \delta . \qquad (7.7)$$

The ratio s/δ of length s and width $\delta = 2w$ does not depend on the toughness of the polymer but just on its modulus E and on its yield strength σ_y.

The zones look like thin wedges with a length fifteen times their width (Fig. 8.2). They all bear resemblance to each other although they are quite different in size. $\delta_c/2$ is below 1 μm in the most crosslinked polymer (A), but $\delta_c/2$ is larger than 10 μm in the thermoplastic polymers. Half of the crack opening displacement ($\delta_c/2$) is plotted in Fig. 7.4 against the effective molecular mass $\bar{M}_C$ of the network strands for the polymers A to E [11]. A straight line with slope 1 indicates the proportionality of $\delta_c/2$ and $\bar{M}_C$ in the double logarithmic plot:

$$\frac{\delta_c}{2} = \left(2.6 \frac{\mu m \, mol}{kg} \right) \bar{M}_C . \qquad (7.8)$$

Apparently, the width and the length of the deformation zone are simply proportional to the molecular mass of the network strands! The more highly crosslinked a polymer, the smaller is its deformation zone. On the other hand, polymers with few crosslinks will exhibit large deformation zones ahead of the growing crack. Since deformation zones pick up energy proportional to their crack

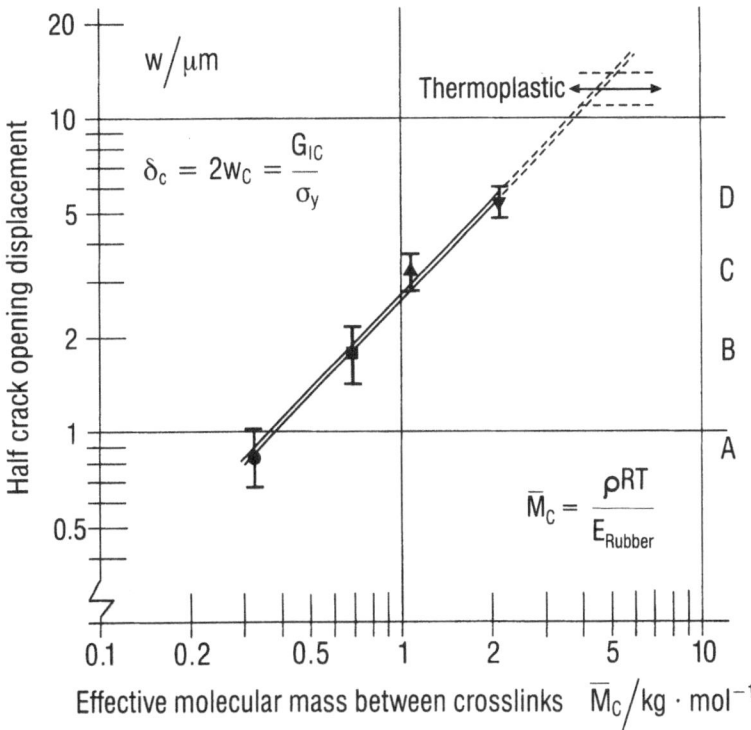

Fig. 7.4. Half the crack opening displacement δ_c, is plotted against the effective molecular mass $\bar{M}_C$ between crosslinks. $\delta_c = G_{IC}/\sigma_y$ was calculated from the results in Table 6.1, measured at 23 °C. $\bar{M}_C$ was determined from the moduli of the polymers in the rubbery state

opening displacements (Eq. 7.6), polymers with few crosslinks will be much tougher than highly crosslinked networks even if the yield strength of the latter is slightly higher (Sect. 6.1).

Usually, the molecular strands are coiled in the glassy polymer. They become stretched when a crack arrives and starts to build up the deformation zone. Presumably, strain softened polymer molecules from the bulk material are drawn into the deformation zone. This microscopic surface drawing mechanism may be considered to be analogous to that observed in lateral craze growth or in necking of thermoplastics. Chan, Donald and Kramer [87] observed by transmission electron microscopy how polymer chains were drawn into the fibrils at the craze-matrix-interface in PS films [92]. One explanation, the hypothesis of devitrification by Gent and Thomas [89] was set forth as early as 1972.

Of course, the network strands cannot be stretched completely. Stretching ratios of 1.4 for PC [31, 90] and of 1.3 for epoxy polymers [37] have been reported. The chain contour length of the strands is an appropriate measure for a simple estimation of the number of strands that are stretched across the deformation zone. The chain contour length of the strands is assumed to be proportional to the number of bond projections onto the backbone chain axis. A contribution of $\zeta = 0.093$ nm per bond was estimated using results on Phenoxy [37], PC [31] and PPO [31]. The contour lengths l_C in Table 6.1 were calculated [90] according to

$$l_C = \frac{\bar{M}_C}{M_1} \zeta \qquad (7.9)$$

$M_1 = 0.0203$ kg/mol.

M_1 equals the mass of a typical polymer chain devided by the number of atoms in its backbone. Slightly more than 1000 strands (length l_C) are needed to cross the deformation zone of width $\delta = 2w$ for all the polymers listed in Table 6.1. It is one of the essential findings of this report that the size of the deformation zone is scaled according to the length of the molecular strands.

Alle the deformation zones contain a finite and equal number of extended chains in their most highly stretched strands. This surprising conformity of the deformation zones may well be the consequence of the imposed plane-strain fracture condition which impedes lateral contraction of the material. However, no quantitative explanation has been presented as yet. A plausible explanation would be to assume that due to the hindered lateral contraction additional tensile stresses are transferred to the most extended strand with each additional chain pulled out of the matrix [112].

7.3 Strain Softening

The properties of glassy polymers such as density, thermal expansion, and small-strain deformation are mainly determined by the van der Waals' interaction of adjacent molecular segments. On the other hand, crack growth depends on the length of the molecular strands in the network as is deduced from the fracture experiments.

The fracture behavior can be attributed to strain softening [91] in the deformation zone [92, 93] or to stress-activated devitrification [89, 96]. The strands are comparatively free to move in the strain softened regions of the deformation zone. The van der Waals' interaction between adjacent strands is greatly reduced and the clearence between molecular segments is enlarged.

The behavior of the strain softened material resembles the behavior of rubberlike polymers. For instance, the Poisson's ratio of an ideally plastic material is also close to 0.5 [94, 95]. Proper understanding of crack propagation involves the microscopic level. Apparently, the load is transmitted by the molecular strands [97] from one crosslink to the next crosslink, exactly, as it is in rubberlike materials. However, two things are different in strain softened polymers as compared to rubberlike materials:

- The test-temperature is below the glass transition temperature
- The inability of the strain softened molecules to recover their random coil conformation when unloaded.

7.4 Results from the Literature

Broutman and McGarry [98] examined the effects of crosslinking on toughness as early as 1965. Bell [99] observed a threefold increase in notched impact strength as the molecular mass between crosslinks was increased. Schmid et al. [100] and Lohse et al. [101] pointed out the dominating effect of molecular strand length on the ultimate properties and the toughness of crosslinked polymers. Later, Batzer et al. [46], Schmid [44], and Fischer et al. [45] compared the behavior of various networks composed of epoxy resins.

An interesting and elegant complementary approach to study the fracture of crosslinked polymers was chosen by Nguyen, Kausch et al. [102] and Kausch et al. [103] who explored the healing of cracks in crosslinked styrene-co-acrylonitrile and other polymer alloys. Clearly, healing is a complementary process to fracture [104]. More recently, Le May and Kelley [105] studied the structure-fracture-relationship in simple epoxy systems whose structural variables were systematically controlled. For the first time, they developed a quantitative relation between the crack arrest fracture energy and the square root of the average molecular mass $\bar{M}_C$ of the network strands in the glassy material:

$$G_{arrest} = 193 \, \frac{J}{m^2} \, \sqrt{\frac{mol}{kg}} \, \sqrt{\bar{M}_C}. \tag{7.10}$$

Such a $\sqrt{\bar{M}_C}$ dependence has been observed for the *threshold* fracture energies in rubberlike materials [106, 107] in accordance with the model of Lake and Thomas [108]. However, Eq. (7.10) is based on arrest energies and the reason for the use of *arrest* energies is not clear.

Data on crack initiation were also presented by Le May and Kelley [105] but not further analysed. The authors were concerned that influences other than the network chain length could complicate or obscure simple relations. They suspected

Table 7.1. Le May's and Kelley's results [105, 110] on polymer failure (T = 296 K)

Polymer Network A/E = 1 [105]	Glass transition temperature	Molecular mass of strands (experimental)	Critical stress intensity factor (crack initiation)	Yield stress, taken from Fig. 9 [110]	Half crack opening displacement* $w = K_{IC}^2/2E\sigma_y$
	$T_g/°C$	$\bar{M}_C/kg\,mol^{-1}$	$K_{IC}/MPa\sqrt{m}$	σ_y/MPa	$w/\mu m$
1007F/DDS	105	3.33	1.8	63	8.6
1004F/DDS	113	1.51	1.3	65	4.3
1002F/DDS	121	1.09	1.1	70	2.9
1001F/DDS	132	0.824	1.0	75	2.2
828/DDS	212	0.360	0.75	100	0.94

* Young's modulus E = 3 GPa at 23 °C (approximately)

unidentified effects when they compared their results with the rather low fracture toughness values of Chang and Brittain [109] on similar, but poorly cured polymers. Nevertheless, the results of Le May and Kelley [105] on stoichiometric and properly cured DGEBA/DDS polymers prove the proportionality [11] between crack opening and $\bar{M}_C$, when plotted in a way similar to Fig. 7.4. To this end, the relevant results from the two papers [105, 110] are collected in Table 7.1 and plotted in Fig. 7.5.

Glad [37] studied the micro deformations of thin films prepared from DGEBA/MDA by electron microscopy. His results are also shown in Fig. 7.5. The deformation of the sample with high strand density was small and consequently its image in the EM rather blurred. Therefore, the result on $\bar{M}_C = 0.5\,\mathrm{kg/mol}$ should perhaps have been omitted.

Both sets of experiments seem to support the proportionality of crack opening displacement $\delta_C = 2w$ and molecular mass $\bar{M}_C$ between crosslinks as indicated by the slope 1 in the double logarithmic plot (Fig. 7.5). Even if $\bar{M}_C$ had to be adjusted due to doubts about the front factor in Eq. (4.3), the proportionality would stay unaffected. Consequently, the size of the deformation zone ahead of the crack is determined by the length of the molecular strands in the chemical network.

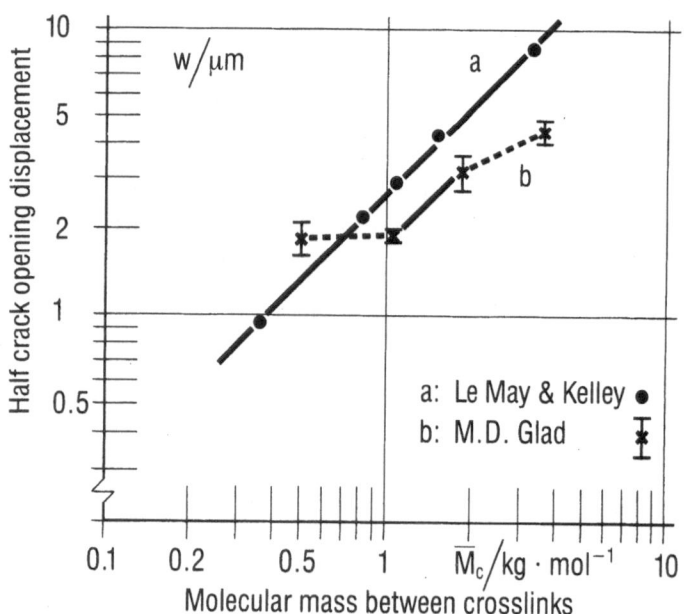

Fig. 7.5. Half the crack opening displacement δ_c is plotted against the effective molecular mass $\bar{M}_C$ between crosslinks.

a) Results from Le May and Kelley [105, 110] as listed in Table 7.1, were used to calculate $\delta_c = G_{IC}/\sigma_y$.

b) Glad's results [37] were derived from measurements on thin films in the transmission electron microscope

The same consideration can also be applied to physical networks. A thermo-plastic polymer consists of entangled molecular chains with an average molecular mass $\bar{M}_C$ between entanglements. If $\bar{M}_C$ were truly decisive, a constant crack opening displacement had to be expected (e.g. at various temperatures), as long as the degree of entanglements is not changed.

Morgan and Ward [79] as well as Fraser and Ward [80] looked at the temperature dependence of the craze shape in PMMA and PC at temperatures down to $-70\,°C$. They observed a constant crack opening displacement δ_C for a given grade of polymer, independent of temperature! However, Döll and Weidmann [88] tested PMMA also at higher temperatures than Morgan and Ward did. There, they discovered an enlarged crack opening displacement.

The repeat unit in the Phenoxy molecule is longer by two carbon atoms than that of PC. Otherwise the molecular structures of the two polymers are similar. Eq. (7.8) should, therefore, apply also to PC, at least approximately. $\delta_C = 8.9\ \mu m$ was measured for the crack opening displacement in a Makrolon sheet ($M_n = 9.5\ kg/mol$) by Fraser and Ward [80]. An estimate of the molecular mass between entanglements according to Eq. (7.8) yields $\bar{M}_C = 1.7\ kg/mol$. This result is not very different from the value of $2.5\ kg/mol$ for $\bar{M}_C$ in PC as given in Table 3.2.

The presented results and the additional information taken from various references indicate the direct relevance of the size of the network strands for the crack opening displacement and consequently for the toughness of the polymer. In polymers under load, the molecular chains at the tip of the crack break after the deformation zone ahead of the crack has grown to a critical width δ_C, that is the crack opening displacement. This value δ_C is proportional to the length of the molecular strands of the network and is linked in this way to the molecular structure of the polymer. However, the molecular mechanism for chain breakage in the deformation zone is not known at present.

8 Conclusions

a) There is no basic difference in the thermo-mechanical behavior of thermoplastics and thermoset polymers in spite of their entirely different processing tech-nologies. The properties of the glassy polymers are mainly determined by the particular chemical building blocks, the molecules are composed of.

b) At small strain levels, crosslinks have just a minor influence on the properties of the glassy polymers (Fig. 8.1). The behavior of a material is dominated by the near neighbours' interaction of molecular segments within the chains as well as between adjacent chains. The modest effects of crosslinking on various properties (X) are well described by Fox's equation

$$\frac{X - X_\infty}{X_\infty} = \frac{\phi_x}{\bar{M}_C} \tag{8.1}$$

350   M. Fischer

X   property of the polymer
X_∞ property of the thermoplastic polymer
$\bar{M}_C$ effective molecular mass of network strand
ϕ_x parameter; measure for the influence of crosslinks.

Results for the density, glass transition temperature, thermal expansion, rigidity, and yield strength are plotted according to Eq. (8.1) in Fig. 8.1. These properties remain proportional to each other.

c) The slippage of the molecular chains ($\bar{M}_C > 0.3$ kg/mol) is not noticeably hindered by the crosslinks. The activation volume proved to be $2\,\text{nm}^3$, irrespective of the strand length. Therefore, molecular flow takes place at the tip of any crack whether the polymer is crosslinked or not.

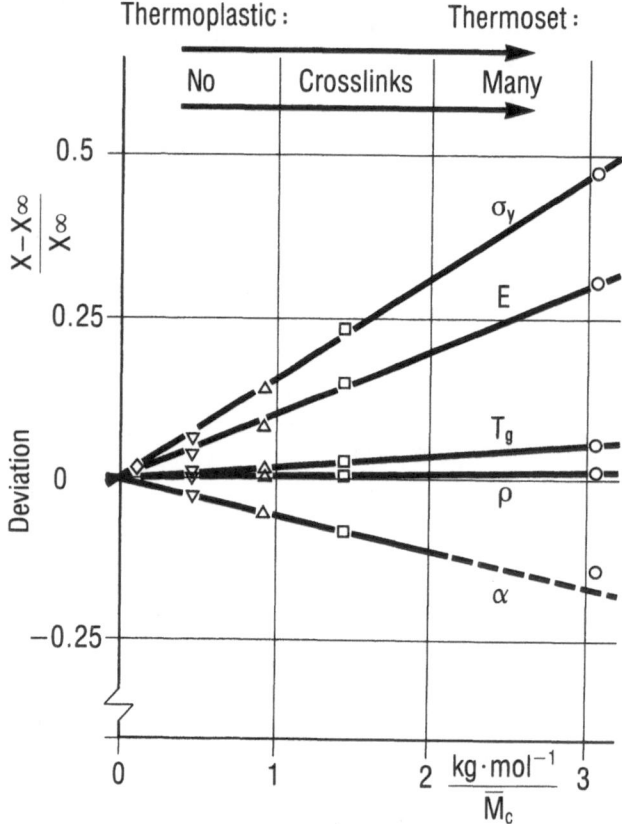

Fig. 8.1. Effects of crosslinking on various properties (X) of the polymer. The deviation $[X - X_\infty]/X_\infty$ from the thermoplastic property (X_∞) is plotted against M_C^{-1}, the inverse molecular mass between crosslinks.
ϱ: density; T_g: glass transition temperature; α: coefficient of thermal expansion; E: Young's modulus; σ_y: yield strength

Table 8.1. Effect of crosslinks on small strain behavior

Property: X	X_∞ (thermoplastic)	Influence of crosslinks
density	$\varrho_\infty = 1.1837\ \mathrm{Mg\ m^{-3}}$	$\phi_\varrho = 0.0033\ \mathrm{kg\ mol^{-1}}$
glass transition	$T_{g\infty} = 100\ ^\circ\mathrm{C}$	$\phi_{T_g} = 0.020\ \mathrm{kg\ mol^{-1}}$
thermal expansion	$\alpha_\infty = 72 \times 10^{-6}\ \mathrm{K^{-1}}$	$\phi_\alpha = -0.060\ \mathrm{kg\ mol^{-1}}$
Young's modulus	$E_\infty = 2.6\ \mathrm{GPa}$	$\phi_E = 0.103\ \mathrm{kg\ mol^{-1}}$
yield strength	$\sigma_\infty = 64\ \mathrm{MPa}$	$\phi_\sigma = 0.157\ \mathrm{kg\ mol^{-1}}$

$$\frac{X - X_\infty}{X_\infty} = \frac{\phi_x}{\bar{M}_C} \tag{8.1}$$

X   property of the polymer
X_∞ property of the thermoplastic polymer
$\bar{M}_C$ effective molecular mass of network strand
ϕ_x parameter; measure for the influence of crosslinks

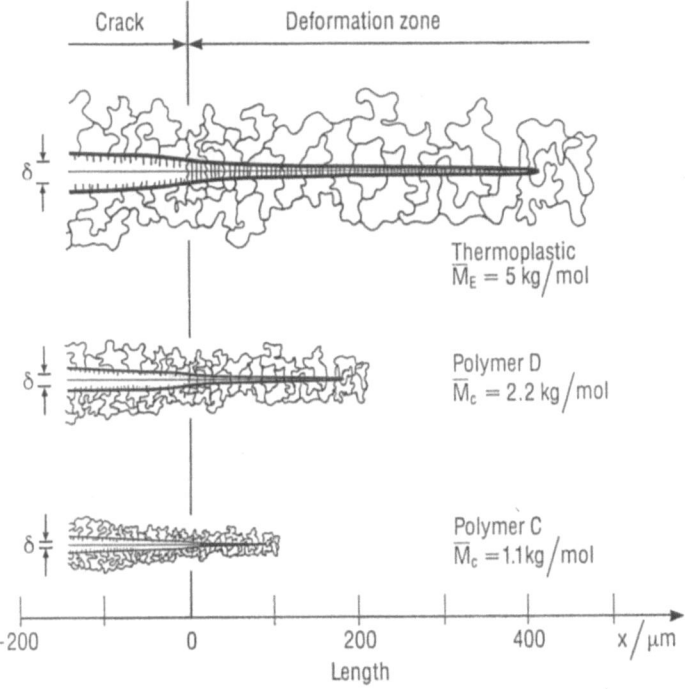

Fig. 8.2. Crack tip and deformation zone of the thermoplastic polymer with entanglements ($\bar{M}_E = 5$ kg/mol) and of two crosslinked polymers ($\bar{M}_C = 2.2$ kg/mol and 1.1 kg/mol).

network strands:	deformation zone:	fracture:
long	large	tough
short	small	brittle

d) The crack opening displacement δ_C at the crack tip was observed to be proportional to the mass $\bar{M}_C$ of the molecular strands. δ_C is approximately the length of 1000 molecular strands. Apparently, the strands are stretched in the deformation zone after the polymer has been locally strain softened. However, the crack does not propagate until the deformation zone has widened and is bridged by about 1000 strands. This picture is derived from the Dugdale-Barenblatt-model and from the fracture toughness of the polymers. It is confirmed by results from the literature.
The lengths of the molecular chains dominate large strain behavior and crack propagation in contrast to their minimal influence at small strain levels. Consequently, thermosets are characterized by small deformation zones (Fig. 8.2) and brittle fracture.

e) The average molecular mass $\bar{M}_R$ of the chains between crosslinks equals the total mass of the network devided by the number of strands. $\bar{M}_R$ agrees reasonably well with the effective molecular mass $\bar{M}_C$ of the strands that was derived via the theory of rubber elasticity from experiments on the Young's modulus of the rubbery polymers. Thermal equilibrium during the tests is required for quantitative evaluations and the functionality of the network junctions has to be accounted for (Eq. 3.1).

f) The yield strength of the polymers is more strongly affected by the rate of cooling than is the modulus. Annealing effects are probably caused by variations in the molecular conformations and by slightly different arrangements of the chains in the polymer (changes in density). No significant effect of crosslinking on the annealing behavior was noticed in the present experiments.

9 Appendix

9.1 Preparation of the Polymers

The diglycidyl ether (DGEBA) and the polyglycidyl compound (PGCBA) were mixed and heated to about 120 °C. The stoichiometric amount of bisphenol A was dissolved in the resin and, after cooling to about 60 °C, 0,1% by weight of 2-ethyl-4-methyl-imidazol was added to the mixture. The well stirred composition was degassed in a vacuum oven in order to remove trapped air. Afterwards, the reactive mixture was poured into preheated moulds and cured for two hours at 140 °C followed by two hours at 180 °C.

After this procedure the curing reaction was completed. This was demonstrated by the following:

— No epoxy groups were detectable in the cured polymer by infrared spectroscopy.
— The glass transition temperature T_g of the polymers remained constant and was not increased by extended heat treatment up to 240 °C.

Acknowledgements: The author appreciates the support and encouragement of Prof. Dr. P. Junod, Head of the Materials Science Research of CIBA-GEIGY.

The reported work was greatly inspired by Dr. R. Schmid, who brought the subject of crosslinked polymers to the author's attention.

The help of many collegues is gratefully acknowledged: The polyglycidyl compound was synthesized by Dr. Ch. Monnier. Ms. Ch. Irrgang and Dr. J. Vogt studied visco-elasticity and Dr. W. Sieber provided results on the density of the polymers. Mr. D. Martin and Mr. P. Rohrbasser assisted the experimental work.

The manuscript was improved by the advice of Dr. J. Huggins and Dr. B. Dobinson as well as by the friendly assistance of Ms. B. Doutaz. Last but not least, Prof. Dr. H. H. Kausch deserves special thanks for kindly having initiated this review.

10 References

1. Nielsen LE (1969) J Macromol Sci Rev Macromol Chem C3: 69
2. Fischer EW, Müller FH, Bonart R (eds) (1978) Physik der Duroplaste und anderer Polymerer. Prog Colloid & Polym Sci vol 64
3. Kinloch AJ, Young RJ (1983) Fracture behaviour of polymers. Applied Science Publishers, London
4. Batzer H (ed) (1985) Polymere Werkstoffe I–III, Georg Thieme, Stuttgart
5. Queslel JP, Mark JE (1984) Adv Polym Sci 65: 135
6. Dušek K (ed) (1986) Epoxy resins and composites I–IV. Adv Polym Sci, vols 72, 75, 78, 80
7. Pietralla M, Kilian H-G (eds) (1987) Permanent and transient networks. Prog Colloid & Polym Sci vol 75
8. Heinrich G, Straube E, Helmis G (1988) Adv Polym Sci 85: 33
9. Kaiser T (1989) Prog Polym Sci 14: 373
10. Lazár M, Rado R, Rychlý J (1990) Adv Polym Sci 95: 149
11. Fischer M (1988) 20th Europhysics conference on macromolecular physics. Lausanne, Preprint L 16
12. Zarzycki J (1982) Les verres et l'état vitreux. Masson, Paris, p 335
13. Bowden PB, Raha S (1970) Philos Mag 22: 463
14. Struik LCE (1978) Physical aging in amorphous polymers and other materials, Elsevier, Amsterdam
15. Brady TE, Yeh GSY (1971) J Appl Phys 42: 4622
16. Morgan RJ (1979) J Appl Polym Sci 23: 2711
17. Kreibich UT, Schmid R (1985) In: Ref [4] vol I, p 665
18. Meyer KH, Ferri C (1935) Helv Chim Acta XVIII: 570
19. Treloar LRG (1975) The physics of rubber elasticity. 3rd edn Clarendon, Oxford
20. Flory PJ (1976) Proc R Soc London A 351: 351
21. Burchard W (1985) Ber Bunsen — Ges Phys Chem 89: 1154
22. Edwards SF, Vilgis TA (1988) Rep Prog Phys 51: 243
23. Graessley WW (1975) Macromolecules 8: 186
24. Fischer M, Kreibich UT, Schmid R (1985) In: Ref [4] vol I, p 233
25. Oppermann W, Rennar N (1987) Prog Colloid & Polym Sci 75: 49
26. Brandrup J, Immergut EH (eds) Polym Handb 3rd edn (1989) John Wiley, New York, p V/8
27. Baur H (1985) In: Ref [4] vol I, p 151
28. Duiser JA, Staverman AJ (1965) In: Prins JA (ed) Physics of non-crystalline solids. North-Holland, Amsterdam, p 376
29. Langley NR (1968) Macromolecules 1: 348
30. Graessley WW, Edwards SF (1981) Polymer 22: 1329
31. Donald AM, Kramer EJ (1982) J Polym Sci, Polym Phys Ed 20: 899

32. Dettenmaier M (1983) Adv Polym Sci 52/53: 78
33. Kausch HH (1987) Mater Res Soc Symp Proc 79: 379
34. Potter DK, Rudin A (1991) Macromolecules 24: 213
35. Hoffmann M (1977) Angew Chem 89: 773
36. Graessley WW (1974) Adv Polym Sci 16: 1
37a. Glad MD (1986) Microdeformation and network structure in epoxies. Thesis, Cornell University, Ithaca, New York
37b. Glad MD, Kramer EJ (1991) J Mater Sci 26: 2273
38. ISO 537: International Standard: Plastics — Testing with the torsion pendulum
39. Fox TG, Flory PJ (1950) J Appl Phys 21: 581
40. Fox TG, Losheak S (1955) J Polym Sci XV: 371
41. Nielsen LE (1974) Mechanical properties of polymers and composites. Marcel Dekker, Inc, New York, vol 1, p 23
42. Lau CH, Hodd KA, Wright WW (1986) Br Polym J 18: 316
43. Schroeder JA, Madsen PA, Foister RT (1986) Polymer 28: 929
44. Schmid R (1978) Prog Colloid & Polym Sci 64: 17
45. Fischer M, Lohse F, Schmid R (1980) Markomol Chem 181: 1251 (Library Translation 2067, R.A.E. Farnborough, UK)
46. Batzer H, Lohse, F, Schmid R (1973) Angew Makromol Chem 29: 349
47. Rietsch F, Daveloose D, Froelich D (1976) Polymer 17: 859
48. Kong ESW (1986) Adv Polym Sci 80: 145
49. Kreibich UT, Schmid R (1985) In: Ref [4] vol 1, p 665
50. ASTM-D 1505–85: Standard test method "Density of plastics by the density-gradient technique"
51. Sieber W (to be published) Results were kindly provided by Dr W Sieber
52. Bero CA, Plazek DJ (1991) J Polym Sci, Polym Phys Ed 29: 39
53. ISO R 527: Recommendation: Plastics — Determination of tensile properties
54. ISO 178: International Standard: Plastics — Determination of flexural properties of rigid plastics
55. Vogt J (1987) J Adhesion 22: 139
56. Kausch HH (1987) Polymer Fracture, 2nd edn, Springer, Berlin Heidelberg New York p 117
57. Stuart HA (1952) Die Physik der Hochpolymeren. Springer, Berlin Göttingen Heidelberg, vol 1, p 37
58. Barker RE (1963) J Appl Phys 34: 107 and (1967) J Appl Phys 38: 4234
59. Baur H (1985) In: Ref [4] vol 1, p 152
60. Bowden PB (1973) In: Haward RN (ed) The physics of glassy polymers, Applied Science Publishers, London, p 279
61. Timoshenko SP, Goodier JN (1970) Theory of elasticity. 3rd edn, McGraw-Hill, New York, p 22
62. Haward RN (1980) Colloid & Polym Sci 258: 643
63. Turska E, Hurek J, Zmudzinski L (1979) Polymer 20: 231
64. Wetton ER, Moneypenny HG (1975) Br Polym J 7: 51
65. Bree HW, Heijboer J, Struik LCE, Tak AGM (1974) J Polym Sci, Polym Phys Ed 12: 1857
66. Ward IM (1971) J Mater Sci 6: 1397
67. Ward IM (1984) Polym Eng Sci 24: 724
68. Ward IM (1983) Mechanical properties of solid polymers. 2nd edn, John Wiley, Chichester, p 329
69. Kambour RP, Robertson RE (1972) In: Jenkins AD (ed) Polymer Science, North-Holland, Amsterdam, p 786
70. Hansen AC, Baker-Jarvis J (1990) Int J Fract 44: 221
71. Schnell HF, Göritz D, Schmid E (1991) J Mater Sci 26: 661
72. Eyring H (1936) J Chem Phys 4: 283
73. Argon AS (1973) Philos Mag 28: 839

74. Argon AS, Bessonov MI (1977) Polym Eng Sci 17: 174
75. Lee SM (1988) In: Dickie RA, Labana SS, Bauer RS (eds) Cross-linked polymers, ACS Symposium Series 367, p 136
76. Williams JG (1984) Fracture mechanics of polymers. Ellis Horwood, Chichester, p 46
77. ASTM-E 399-81: Standard test method: Plane-strain fracture toughness of metallic materials
78. Williams JG (1984) In: Ref [76], p 162
79. Morgan GP, Ward IM (1977) Polymer 18: 87
80. Fraser RAW, Ward IM (1978) Polymer 19: 220
81. Döll W, Könczöl L (1990) Adv Polym Sci 91/92: 137
82. Dugdale DS (1960) J Mech Phys Solids 8: 100
83. Landau LD, Lifschitz EM (1975) Elastizitätstheorie, Akademie-Verlag, Berlin, p 157
84. Barenblatt GI (1962) Adv Appl Mech 7: 55
85. Rice JR (1968) In: Liebowitz H (ed) Fracture − An advanced treatise, Academic, New York, p 191
86. Kramer EJ (1978) J Mater Sci 14: 1381
87. Chan T, Donald AM, Kramer EJ (1981) J Mater Sci 16: 676
88. Döll W, Weidmann GW (1979) Prog Colloid & Polym Sci 66: 291
89. Gent AN, Thomas AG (1972) J Polym Sci, Polym Phys Ed A-2, 10: 571
90. Kramer EJ (1983) Adv Polym Sci 52/53: 1
91. Haward RN (1973) In: Haward RN (ed) The physics of glassy polymers. Applied Science, London, p 340
92. Kramer EJ, Berger LL (1990) Adv Polym Sci 91/92: 1
93. Plummer CJG, Donald AM (1991) Polymer 32: 409
94. MacKenzie P, McKelvie J, McDonach A, Walker CA (1986) Strain 1986: 13
95. Vincent PI (1965) In: Ritchie PD (ed) Physics of plastics, van Nostrand, Princeton, p 110
96. Gent AN (1970) J Mater Sci 5: 925
97. Kausch HH (1987) In: Ref [56], p 105 and (1991) Makromol Chem, Macromol Symp 41: 1
98. Broutman LJ, McGarry FJ (1965) J Appl Polym Sci 9: 609
99. Bell JP (1970) J Appl Polym Sci 14: 1901
100. Schmid R, Lohse F, Fisch W, Batzer H (1970) J Polym Sci Part C: Polym Symp 30: 339
101. Lohse F, Schmid R, Batzer H, Fisch W (1969) Br Polym J 1: 110
102. Nguyen TQ, Kausch HH, Jud K, Dettenmaier M (1982) Polymer 23: 1305
103. Kausch HH, Petrovska D, Landel RF, Monnerie L (1987) Polym Eng Sci 27: 149
104. Kausch HH (1987) In: Ref [56], p 416
105. Le May JD, Kelley FN (1986) Adv Polym Sci 78: 115
106. Gent AN, Tobias RH (1982) J Polym Sci, Polym Phys Ed 20: 2051
107. Ahagon A, Gent AN (1975) J Polym Sci, Polym Phys Ed 13: 1903
108. Lake GJ, Thomas AG (1967) Proc R Soc London A 300: 108
109. Chang TD, Brittain JO (1982) Polym Eng Sci 22: 1228
110. Le May JD, Swetlin BJ, Kelley FN (1984) In: Labana SS, Dickie RA (eds) Characterization of highly cross-linked polymers, ACS Symposium Series 243, p 165
111. Moore WJ (1972) Physical chemistry, Longman, London p 913
112. Kausch HH (1991) private communication

Editor: H.-H. Kausch
Received July 2, 1991

Polymer Interfaces on a Molecular Scale: Comparison of Techniques and Some Examples

Manfred Stamm
Max-Planck-Institut für Polymerforschung, 6500 Mainz, FRG

There has been considerable progress in experimental techniques for analyzing polymer surfaces and interfaces at a microscopic level. In particular, it is possible to obtain molecular or even atomic resolution of polymer molecules at interfaces. Some of the most common techniques are briefly described and discussed with respect to their use for the analysis of the polymer free surface or polymeric "buried" interfaces well within material. They may be roughly divided into optical, ion beam, reflectivity or scattering techniques. The information gain is quite different for the various methods ranging from chemical composition, interfacial width and profile to structural arrangement of polymers at the interface. With many techniques, deuterated components and extremely smooth and homogeneous thin films are required.

Specific examples of polymer surface and thin film problems are briefly discussed. These include measurements of surface roughness, enrichment of components at the surface and surface-induced ordering effects. Special emphasis is given to the interdiffusion and welding of polymers investigated on a nanometer scale and to initial stages of interfacial mixing. While interesting first experiments with molecular resolution are available in many areas, analytical techniques in particular for structural and conformational investigations of polymers at surfaces and interfaces still need to be improved and further experiments will be necessary for a detailed understanding of the influence of surfaces and interfaces on the physical properties of polymers.

Advances in Polymer Science, Vol. 100
© Springer-Verlag Berlin Heidelberg 1992

List of Abbreviations

A-b-B	diblock copolymer of polymer A and B
AFM	atomic force microscopy
D	deuterium
DSIMS	dynamic secondary ion mass spectroscopy
EELS	electron energy loss spectroscopy
ELLI	ellipsometry
ERD	elastic recoil detection
ESCA	electron spectroscopy for chemical analysis
FRS	forward recoil spectroscopy
H	hydrogen
HREELS	high-resolution electron energy loss spectroscopy
IR-ATR	infrared atternated total reflection spectroscopy
IR-D	infrared densitometry
IR-GIR	grazing incidence infrared spectroscopy
LB	Langmuir-Blodgett
NR	neutron reflectometry
NRA	nuclear reaction analysis
PBrS	poly(para-bromo styrene)
PEP	poly(ethylene propylene)
PMIM	phase measurement interference microscopy
PMMA	poly methylmethacrylate
PMS	poly(para-methyl styrene)
PS	polystyrene
PVC	polyvinylchloride
PVP	poly(2-vinylpyridine)
RBS	Rutherford backscattering spectroscopy
SANS	small angle neutron scattering
SAXS	small angle X-ray scattering
SEM	scanning electron microscopy
STEM	scanning transmission electron microscopy
SP	surface plasmons
SSIMS	static secondary ion mass spectroscopy
ST	surface tension
STM	scanning tunneling microscopy
TEM	transmission electron microscopy
XPS	X-ray photoelectron spectroscopy
XR	X-ray reflectometry

1 Introduction

Polymer surfaces and interfaces have attracted increasing interest during recent years because of their practical and commercial importance as well as their high potential for future applications. The use of polymers in various fields including medicine, biotechnology, optics, microelectronics is discussed for, among others, artificial organs, protective or surface active coatings, insulating or conducting thin films, composites or laminates, as well as improved polymer blends and adhesive joints. Of course, the bulk properties of polymeric materials are very important, but in all these areas polymer molecules and their properties at surfaces and interfaces greatly influence material performance. One of the problems, for instance, is adhesion or the quality of contact between different materials, but also surface composition, roughness or interfacial mixing can be decisive factors for specific applications.

It is thus necessary to analyze polymer surfaces and interfaces on a molecular level corresponding to a nanometer scale. Techniques well suited for characterization of polymers in the bulk will generally not be successful for this problem since they should be especially surface or interface sensitive while the "background" from the bulk should be mostly suppressed. Our knowledge of polymers at interfaces was largely limited by the analytical techniques available and to a great extent restricted to some model systems. During recent years however, there has been significant development and improvement in surface analytical techniques and it is now possible to analyze polymer surfaces and interfaces on a molecular scale in much more detail.

One might try to start with the apparently simple problem of a free polymer surface (Fig. 1a) which on a closer view is, however, not simple at all. Due to the influence of the surface as a boundary condition the chain conformation is modified as compared to the bulk and for instance chain ends may be preferentially located in the vicinity of the surface. Despite many theoretical treatments there are hardly any experimental investigations of those effects. A polymer blend might also behave quite differently in the vicinity of a surface or interface as compared to the bulk (Fig. 1b). Due to the different surface energies of the components a surface enrichment of one species might be expected which may depend largely on sample preparation and treatment. Another largely unresolved problem is the intermixing and interdiffusion of polymers at the interface (Fig. 1c). In a typical welding experiment, two pieces of polymeric materials are put together and annealed for a certain period of time. Depending on temperature, time and compatibility of the materials, one ends up with a zone of intermixing of varying depth. The interdiffusion is determined by the particular starting situation where chains do not cross the interface, by the initial chain conformation and by details of the mechanism of the interdiffusion process. Since none of those points is rigorously known, the welding process is far from being understood.

Even from those first remarks it is evident that our knowledge of polymers at surfaces and interfaces depends largely on analytical techniques. They should yield information on chemical composition, density, roughness, chain conformation, end distribution etc. across the interface with subnanometer resolution. In Sect. 2

we will discuss some of the most common surface and interface analysis techniques for polymers with respect to these requirements. The investigation of "buried" interfaces *between* polymers is a particular challenge for the experimentalist since one is trying to resolve a possibly nanometer thick interfacial layer hidden within a bulk sample. Astonishingly enough, several techniques have been developed in recent years which are capable of tackling this problem. Suitable sample preparation techniques are a prerequisite for those investigations and will also be discussed briefly. Examples of the use of those techniques will be given in Sect. 3.

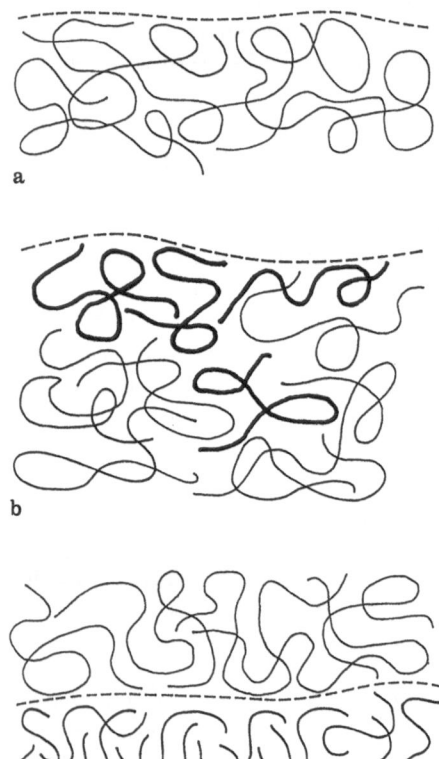

Fig. 1 a–c. Schematic drawing of some specific examples of polymer molecules at an interface: (a) the free surface of a homopolymer, (b) the surface enrichment of one component in a miscible polymer blend, and (c) the interface between polymers of different molecular weight and/or chemical composition

This article is not intended to give a comprehensive overview over the entire field, but will concentrate on some recent developments and highlights as perceived by the author. Only most recent references will be given where most of the previous work will be found. There are several books, proceedings and review articles available on earlier work and on specific aspects of polymers at interfaces [1–13]. In particular the area of tethered chains in solution or melt is covered by another article in this book [14].

2 Techniques for the Investigation of Polymer Interfaces

2.1 General Considerations

From solid state physics there are many surface and interface sensitive techniques available which in most cases can also be used in the area of polymer surfaces and interfaces. There are, however, limitations imposed by the special nature of polymers. Thus polymers are mostly non-conductive, consist of low mass elements and are generally sensitive to high energy radiation. They do offer on the other hand specific tagging possibilities through the exchange of hydrogen by deuterium which turn out to offer unique surface and interface analysis possibilities for specific problems. Thus for instance a contrast between otherwise indistinguishable materials can be generated by the deuteration of one component and single molecules (or parts of them) can be made visible against the background of the protonated ones. An example is the welding of polystyrene (PS) mentioned earlier where a deuterated and protonated piece of material, PS(D) and PS(H), can be put together. The interface between the two materials can be analysed to a high accuracy with techniques which are sensitive on the specific hydrogen or deuterium concentration profile. Using the deuterium tagging technique one should keep in mind that while the thermodynamic difference between hydrogen and deuterium is small it is not always negligible. It has been demonstrated in neutron small angle scattering experiments on PS(H)/PS(D) blends [15] that at high molecular weights a phase separation can occur. Similarly the spinodal decomposition temperature T_C between polymers is reported to shift significantly (up to 40 K) through the replacement of e.g. protonated by deuterated polybutadiene in a blend of PB/PS [16] or of deuterated by protonated polystyrene in a blend of PS/poly vinyl methyl ether [17]. In a thin film of a PS(H)/PS(D) blend on the other hand one component will be enriched at the surface (see Sect. 3.2). Those effects are most pronounced for high molecular weights and in the vicinity of a phase transition. They can be neglected to a large extent in most other cases.

The most common surface and interface analysis techniques for polymers are summarized in Table 1 and 2 and will be discussed in detail in Section 2.2 and 2.3. Some of them offer a resolution on the molecular scale. Analysis techniques may be roughly divided into those which can only be applied to polymer surfaces and those which can be used both for surfaces and buried interfaces within the sample. The sensitivity of the surface techniques for a region near to a surface ($\lesssim 5$ nm) is due to either the limited penetration (or escape) depth of the radiation used or interaction, or the purposeful restriction of the interaction of the radiation to the surface using e.g. evanescent wave (total) reflection. For the investigation of buried interfaces, higher energetic ion beams for instance may be utilized with a typical penetration depth of some micrometers or the upper layer may be sputtered away. The resolution of the described techniques goes down to the Ångstrom level and in some cases it is mostly the sample quality which restricts the available depth or lateral resolution. Each technique has its pecularities and restrictions. It is generally best to apply different approaches to identical samples

Table 1. Common techniques for the investigation of polymer surfaces

	radiation in	radiation out	sampling depth/ depth resolution	depth profiling	chemical information	lateral structural information	lateral sampling width	typical information
surface tension/ contact angle (ST)			$\lesssim 1$ nm	–			mm	hydrophobicity
X-ray photoelectron spectroscopy (XPS) electron spectroscopy for chemical analysis (ESCA)	X-rays	electrons	$\lesssim 5$ nm	yes	quantitative	10 μm (scanning)	$\gtrsim 10$ μm	surface composition
high-resolution electron energy loss spectroscopy (HREELS)	electrons	electrons	~1 nm	–	–	–	μm	vibrational spectrum
infra-red attenuated total reflection spectroscopy (IR-ATR)	light	light	μm	yes	quantitative	–	mm	vibrational spectrum
static secondary ion mass spectroscopy (SSIMS)	ions (Ar, Xe, Ga)	ions (fragments)	~1 nm	–	semi-quantitative	1 μm (scanning)	$\gtrsim 1$ μm	surface composition
phase measurement interference microscopy (PMIM)	light	light	0.6 nm	–	–	μm	μm	roughness, homogeneity
atomic force microscopy (AFM)			0.1 nm	–	possible	0.5 nm	1 nm	topography of surface
scanning tunneling microscopy (STM)		electrons	0.1 nm	–	possible	0.1 nm	0.1 nm	topography of surface
scanning electron microscopy (SEM)	electrons	electrons	>10 nm	–	quantitative	5 nm	10 nm	structure, composition
X-ray reflectometry/ scattering (XR)	X-rays	X-rays	0.2 nm	yes	quantitative	0.1 nm	mm	roughness, composition profile

Table 2. Common techniques for the investigation of polymeric "buried" interfaces

	radiation		depth resolution (FWHM)		contrast by	typical depth of interface (scan width)	lateral structural resolution	typical information
	in	out	at surface	at 100 nm				
elastic recoil detection (ERD) forward recoil spectroscopy (FRS)	^{4}He	^{1}H(H), ^{2}H(D)		80 nm (TOF: 20 nm)	H/D	≲1 μm	—	interface width/profile
Rutherford backscattering (RBS)	^{4}He	^{4}He		30 nm	heavy atoms	μm	—	movement of markers
nuclear reaction analysis (NRA)	^{3}He ^{15}N	^{4}He γ(4.4 MeV)	14 nm 4 nm	30 nm 12 nm	H/D H/D	μm	—	interface width/profile, element specific
dynamic secondary ion mass spectroscopy (DSIMS)	ions (Ar, Xe, Ga)	ions (fragments)		12 nm	H/D or others	μm	possible	interface width/profile, chemical analysis
X-ray reflectometry (XR)	X-rays	X-rays		0.1 nm	heavy atoms	300 nm	—	interface width/profile, marker movement
neutron reflectometry (NR)	neutrons	neutrons		0.2 nm	H/D	300 nm	—	interface width/profile
transmission electron microscopy (TEM)	electrons	electrons		3 nm	staining	(μm)	3 nm	interface width/profile
infra-red densitometry (IR-D)	light	light		10 μm	H/D or others	(mm)	10 μm	interface width/profile

Table 2. (continued)

	radiation		depth resolution (FWHM)		contrast by	typical depth of interface (scan width)	lateral structural resolution	typical information
	in	out	at surface	at 100 nm				
ellipsometry (ELLI)	light	light		0.2 nm	refractive index	mm	μm	thickness of layer
surface plasmons (SP)	light	light		0.2 nm	refractive index	μm	μm	thickness of layer
small angle X-ray scattering (SAXS)	X-rays	X-rays		1 nm	electron density	mm	–	interface width, interfacial area
small angle neutron scattering (SANS)	neutrons	neutrons		1 nm	H/D	mm	–	interface width

to obtain unique results. Surface and interface analysis on the molecular scale is not a trivial task and for a specific problem the most reasonable techniques and preparation procedures have to be chosen and optimized. Then it will be possible in most analytical cases to obtain molecular resolution with available surface and interface analysis techniques.

2.2 Polymer Surfaces

The free surface between a polymer and air (or vacuum) can be investigated by different techniques (Table 1) to obtain information on chemical composition, roughness and structure. Depending on the method used, either the actual surface region ($\lesssim 1$ nm) or a region near a surface is analysed. Some techniques offer the possibility of depth profiling in a region near a surface and the capability of obtaining information on lateral structure.

The simplest technique introduced by Young as early as 1805 [18] is the measurement of the *contact angle* as a measure of surface tension and surface energy [1, 19, 20, 21]. In many cases this gives an indication of surface composition and can be used to observe changes in composition, structure and/or roughness at the surface during a particular surface treatment. A quantitative description or distinction between different parameters is hardly possible in most cases.

Other techniques utilize various types of radiation for the investigation of polymer surfaces (Fig. 2). *X-ray photoelectron spectroscopy* (XPS) has been known in surface analysis for approximately 23 years and is widely applied for the analysis of the chemical composition of polymer surfaces. It is more commonly referred to as *electron spectroscopy for chemical analysis* (ESCA) [22]. It is a very widespread technique for surface analysis since a wide range of information can be obtained. The surface is exposed to monochromatic X-rays from e.g. a rotating anode generator or a synchrotron source and the energy spectrum of electrons emitted

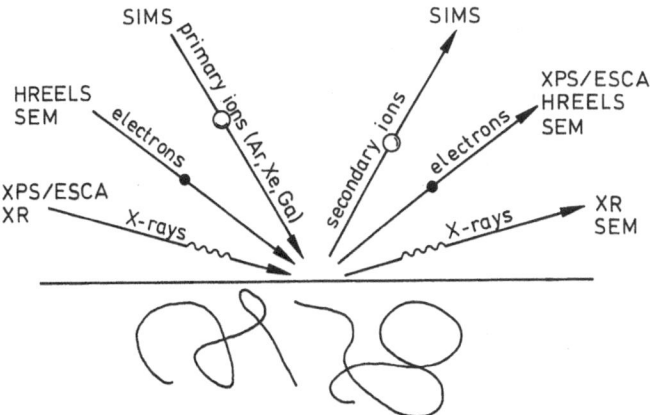

Fig. 2. Schematic diagram of some surface analysis techniques. Abbreviations are explained in Table 1 and in the text

via the photoelectric effect is analyzed. The kinetic energy of photoelectrons is directly related to the binding energy of electrons of different electron levels within atoms. This energy spectrum of electrons is specific with respect to chemical composition, environment and binding character of atoms and oxidation level of elements at the surface. From the peak heights, the surface concentration of species can be deduced with an accuracy of approximately 10 per cent. Light elements (H, He) cannot be analyzed and the sampling depth of the technique depends on the penetration depth of the exciting X-ray radiation as well as on the escape depth of the electrons. Depending on electron energy and the materials investigated it is quite variable and ranges from typically 1 to 10 nm. By rotation of the sample with regard to the energy analyzer assembly the effective electron flight path in the sample is varied and some depth profiling is achieved. The position of peaks in the energy spectrum yields information on the environment of atoms and on their oxidation state.

Recently, surface two dimensional imaging with a lateral resolution in the micrometer range has become possible with XPS [23] to obtain the distribution of elements within the surface. With most commercial machines this lateral resolution is, however, limited to 10 μm and ion beam techniques (SSIMS, Auger electron spectroscopy) are better suited for imaging due to a smaller spot size. Charging of non-conducting surfaces causes problems but can be overcome by specific experimental techniques (low-energy flood gun, calibration by additives or internal standards etc.). XPS has been applied to many polymer surface analysis problems including surface enrichment of components in polymer blends, chemical changes at the surface due to surface treatments or the distribution of low molecular weight additives in polymers. Complete valence band spectra are used to "fingerprint" materials when e.g. an unknown surface contamination is investigated. Several reviews on the application of XPS to polymer surface analysis are available [3, 6, 22–25].

High resolution electron energy loss spectroscopy (EELS/HREELS) is a less common technique for polymer surface analysis and has only recently been used mostly for polymer surface vibrational spectroscopy [26]. The energy loss of low energy electrons after interaction with the surface is investigated. The electrons deposit some of their energy at the surface by the excitation of vibrational modes and at higher incident energies by the excitation of electrons of atoms into higher electron bands. The energy loss spectrum at low energies is thus characteristic of the vibrational spectrum and at higher energy to electronic levels of atoms at the surface and reveals similar information to Raman, IR or XPS techniques. The resolution of characteristic peaks and peak shifts is better than that of XPS, especially for light elements (H/D), while the radiation damage at the surface and charging effects are more severe. Due to the strong interaction of electrons with matter, the penetration depth is typically of the order of 1 nm depending on the energy of the electrons. Information on vibrational spectra is thus quite surface specific while the resolution is worse than by Raman or IR-techniques ($\gtrsim 20$ cm^{-1} versus 1 cm^{-1}). Both Raman and IR-active modes can, however, be detected, and sub-monolayer coverages may be analyzed over a wide spectral range. Results are not quantitative, but peak intensities can be compared on a relative scale (e.g. surface versus bulk). There are only few reports on the application of this technique to polymer surface analysis [26–28].

Static secondary ion mass spectroscopy (SSIMS) has also recently been applied to polymer materials [3, 6, 22, 29, 30] and reveals information on the mass spectrum of polymer fragments from the uppermost surface region (typically 1 nm). It is an interesting supplementary technique to XPS since it overcomes some of the problems of XPS where for instance, because of small chemical shifts or peak overlap, a detailed chemical surface analysis sometimes may not be possible. A primary ion beam (Ar, Xe, Cs, Ga etc.) is accelerated towards a surface (energy $\lesssim 4\,keV$, current $\lesssim 1\,nA\,cm^{-2}$) and secondary particles are emitted from the surface. Due to the impinging ions, bonds will break (and reform) and mostly neutral particles will leave the surface. Typically 1% of positive and negative ions will be formed and these can be extracted into a mass spectrometer to yield a characteristic mass spectrum of the material under investigation. The ion yield will depend on various parameters and is different for different elements but also for the same elements in different environments (matrix effect). A quantitative analysis of surface composition is thus difficult while SSIMS is extremely sensitive to small surface contaminations (as low as 10^{-6} atom%). Different isotopes are also distinguished (e.g. H versus D) and this can be used to identify organic species. SSIMS spectra are very complex and are sometimes only used as "fingerprints" in the comparison with spectra of known materials. Since ion beams can be accurately focused, two-dimensional lateral chemical imaging of surfaces with a scanning beam is possible. The lateral resolution is in the micrometer range [23]. Radiation damage of the surface can be minimized utilizing low acceleration voltage, low currents and a sensitive mass detector. Recently, very sensitive time-of-flight detectors have become available and they also allow the detection of high atomic masses [30]. During ion bombardment, a nonconducting polymer surface will be charged which, however, may be carefully compensated by low energy electron "neutralization" to avoid changes in secondary ion yield during the experiment. SIMS is also used in a dynamic mode (DSIMS) where the sample surface is sputtered away to obtain depth resolution. This technique will be described in Sect. 2.3. There are several reviews available on the application of SSIMS to polymer surface analysis [3, 6, 22, 23, 29, 30] and with the improvement of experimental facilities this technique is gaining ground against XPS for polymer surface analysis.

While electron or ion beam techniques can only be applied under ultra-high vacuum, optical techniques have no specific requirements concerning sample environment and are generally easier to use. The surface information which can be obtained is, however, quite different and mostly does not contain direct chemical information. While with infra-red attenuated total reflection spectroscopy (IR-ATR) a deep surface area with a typical depth of some micrometers is investigated, other techniques like phase-measurement interference microscopy (PMIM) have, due to interference effects, a much better surface sensitivity. PMIM is a very quick technique for surface roughness and homogeneity inspection with subnanometer resolution.

In *infra-red attenuated total reflection spectroscopy* (IR-ATR) and *grazing incidence reflection IR spectroscopy* (IR-GIR) the evanescent wave of a totally reflected beam at the interface between a high-refractive-index-glass and a polymer layer or at the surface of a thin film, respectively, is utilized to illuminate a thin interfacial region [31, 32]. Depending on the ratio of the indices of refraction of the materials, the angle of incidence and the wavelength, an interfacial area of typically some micrometers in the polymer layer is illuminated. The infra-red absorption from this layer is obtained after one or more reflections at the glass or polymer surface to yield information on the vibrational spectrum of the polymer layer. Because of the large penetration depth of IR-radiation ($\gtrsim 0.5$ μm depending on wavelength and difference in refractive index) this technique is not particulary surface or interface sensitive. IR-ATR is also limited in application since specific high-refractive-index-glasses have to be used. Also other IR techniques like diffuse reflectance IR spectroscopy are applied to surface analysis. By the development of special detectors and Fourier transform techniques the established IR methods have gained much in sensitivity and resolution. Monolayer thin films on a substrate may be investigated [32].

The application of interference techniques overcomes the limitations exerted by the large optical wavelengths. With commercial *phase-measurement interference microscopes* (PMIM), a surface resolution of the order of 0.6 nm can be achieved [33, 34]. In a microscope a laser beam is both reflected from the sample surface and from a semitransparent smooth reference surface (Fig. 3). The interference pattern is recorded on an area detector and modulated via the piezo-electric driven reference surface. The modulated interference pattern is fed into a computer to generate a two-dimensional phase map which is converted into a height level contour map of the sample surface. While the lateral resolution (typically of the

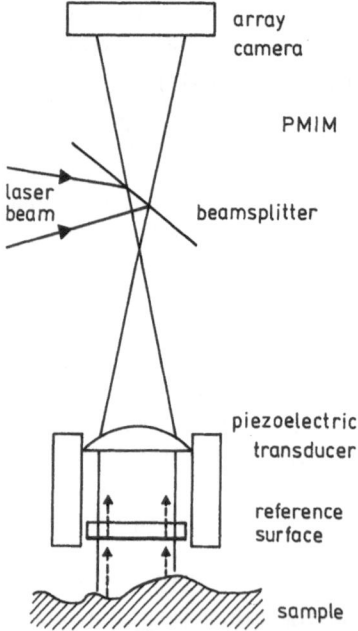

array
camera

PMIM

laser
beam

beamsplitter

piezoelectric
transducer

reference
surface

sample

Fig. 3. Schematic beam path of a phase-measurement interference microscope (PMIM, Fizeau optics). The beam partially reflected at the reference plane and at the sample surface interfere with each other while the reference plane is moved by the piezoelectric transducer for automatic phase determination. A reflectivity of at least 1% is required for the sample surface

order of 1 μm) is limited by the laser wavelength the sample height can be measured by suitable averaging and normalization procedures with respect to a high quality reference surface with a resolution as low as 0.1 nm (typically 0.6 nm). The measurement is very fast and easy to perform and provides a good measure of sample quality with respect to lateral sample homogeneity (on a μm- to mm-scale) and roughness (on an Ångstrom- to μm-scale). Since the computer software considers a simple reflecting sample surface, the calculated height pattern might be erroneus in cases of laterally inhomogeneous and in particular absorbing materials as well as multilayer samples with different reflecting layers. Also steep jumps in height may cause phase problems depending on magnification. With simple polymer films on glass substrates one obtains very good results. Specific examples will be discussed in Sect. 3.1.

With *scanning tunneling microscopy* (STM) a much better lateral resolution as compared to PMIM is achieved by a small scanning needle which is very accurately controlled by piezoelectric drives in three directions. The resolution is mostly limited by the quality of the needle. A "single atom"-like top of the needle is sometimes achieved for STM where the tunneling current between the needle and the conducting surface is measured. In the constant current mode this tunneling current is kept constant while the needle is scanned at high speed over the surface. Thus the surface topography is obtained. Since a conducting surface (graphite, gold, conducting polymer) is needed the application of STM to polymers is limited to specific cases. Either a monolayer-type film of an insulating polymer on a conducting surface or single adsorbed molecules on a conducting substrate can be investigated. Quite spectacular pictures of single polymer molecules adsorbed on graphite [35, 36] or epitaxial structures of molecules on the substrate [37] have been reported (see Sect. 3.2). These results have to be seen, however, in the light of specific conformations due to the interaction of the molecules with the surface which thus might not be representative for the bulk. The technique offers, however, extreme surface sensitivity and atomic resolution at the surface which is hardly met by any other technique. Not all molecules on a surface will be seen but only those which are fixed to the surface on the timescale of an STM scan (milliseconds). The technique has atomic resolution in all directions which is, however, lost with large aggregates or thicker films. It can be applied under atmospheric conditions and even in organic solutions or water.

A very similar technique is *atomic force microscope* (AFM) [38] where the force between the tip and the surface is measured. The interaction is usually much less localized and the lateral resolution with polymers is mostly of the order of 0.5 nm or worse. In some cases of polymer crystals atomic resolution is reported [39]. The big advantage for polymers is, however, that non-conducting surfaces can be investigated. Chemical recognition by the use of specific tips is possible and by dynamic techniques a distinction between forces of different types (van der Waals, electrostatic, magnetic etc.) can be made. The resolution of AFM does not, at this moment, reach the atomic resolution of STM and, in particular, defects and localized structures on the atomic scale are difficult to see by AFM. The technique, however, will be developed further and one can expect a large potential for polymer applications.

A more conventional technique is *scanning electron microscopy* (SEM) which can also be used to obtain the surface topography and chemical composition on molecular to micrometer scale [40, 41]. With modern electron microscopes it is not only possible to record the backscattered or secondary electrons, but also to perform an energy analysis of Auger electrons for chemical surface analysis or to detect the element-characteristic X-ray fluorescence (with μm-resolution). The electron beam can be focused down to the nanometer scale. Either high or low energy electrons are used to obtain either high spatial or improved depth resolution. With polymers, beam damage can result in structural changes, loss of light elements and modifications at the surface, and thus high resolution with high accelerating voltages is mostly not possible for polymers. To reduce beam damage, polymers are coated e.g. with platinum. A recent development is environmental SEM where specimens can be observed at high pressure (20 Torr) and in different environments including wet surfaces [41].

X-ray reflectometry (XR) is again a technique which gains its high depth resolution from interference effects [42]. The incident X-ray beam is reflected from the surface following optical principles. The ideal Fresnel reflectivity curve of a sharp interface is damped exponentially by surface roughness. This effect has been used to measure the roughness of a water surface to 0.32 nm [43] and of a solid polystyrene film to 0.3 nm [44]. A composition profile at the surface can also be detected if sufficient X-ray contrast between components is present. The refractive index of materials which determines the X-ray reflectivity can be expressed in terms of electron density. If the difference in electron density between materials is sufficient, surface enrichment for instance can be detected from an analysis of the X-ray reflectometry curve. Since most polymers contain only carbon, hydrogen, oxygen etc. but no heavy elements, contrast between polymers is usually not sufficient and neutron reflectometry may be advantageous (see Sect. 2.3). While the resolution of XR is in the subnanometer range perpendicular to the surface, there is no lateral resolution in conventional X-ray reflectivity experiments. Lateral surface structures can, however, be resolved by scattering experiments in eva- nescent wave geometry [45, 46], which preferentially have to be performed with synchrotron radiation for intensity reasons. Examples of XR will be discussed in Sect. 3.1.

There are still other surface analysis techniques including ellipsometry, surface enhanced Raman scattering, light scattering, nano-hardness measurements etc. which are used for specific investigations. It is, however, already evident from this discussion that many new and powerful techniques now are available which offer the capability of investigating various aspects of polymer surfaces on a molecular level. Some of those techniques are surface specific while others can be used for the analysis of buried interfaces, too.

2.3 "Buried" Interfaces

To investigate the interface between polymeric materials, i.e. a so-called buried interfaces, several techniques are available schematically shown in Fig. 4 and listed in Table 2. They have quite different characteristics and depth resolution depending on the type and energy of radiation used. The interface studied can be located at a typical depth of some nanometers to some centimeters within the material. The polymer surface is in most cases also accessible as the particular interface between polymer and air/vacuum. With some techniques, in situ experiments e.g. at high–temperatures or during exposure to a swelling agent are possible and a suitable layer system can be investigated without further preparation. Only with TEM and IR-D the sample has to be cut perpendicular to the interface. Most techniques based on ion bombardment will destroy the sample during the analysis and only XR, NR, ELLI, SP, SAXS and SANS are non-destructive techniques. In comparison with surface analytical techniques (Table 1) less techniques are available which offer a depth resolution on a molecular scale, i.e. below 10 nm. Those are in particular XR, NR, TEM, ELLI, SP, SAXS and SANS where again some techniques are largely restricted with respect to available information or necessary sample preparation as will be discussed below. Also two-dimensional imaging of buried interfaces on a molecular scale is presently not possible. Most techniques use the contrast between hydrogen and deuterium when applied to polymers while with others (RBS, XR, TEM) heavy markers are needed to generate a contrast between components.

There are several techniques where ions are used for the analysis of polymer interfaces. For *elastic recoil detection* (ERD) (also called forward recoil spectroscopy, FRS) [47, 48] helium ions are accelerated to a typical energy of 3 MeV and impinge on a thin polymer layer system of deuterated and protonated polymer films shown schematically in Fig. 4a. The energy spectrum of elastic recoil particles, hydrogen (H) and deuterium (D), is recorded in the forward direction to obtain a depth profile of the deuterated and protonated polymer with typically 80 nm depth resolution. The energy of the recoil particles is a direct measure of the location of their production since ions lose energy proportional to the length of their path through the sample. Depth resolution is largely limited by the introduction of a stopper foil before the detector to eliminate elastically scattered ^{4}He particles. It has recently significantly been improved by the introduction of a time-of-flight technique which brings the resolution down to 35 nm [49] or even 20 nm [50]. With polymers mostly ^{4}He is used as a projectile to keep radiation damage to a minimum. Even if molecular resolution has only recently been approached by this technique, it has been applied very successfully in the past to the analysis of polymer diffusion [51, 47]. It is well suited for the analysis of light elements.

Heavy element depth profiles are analyzed by *Rutherford backscattering* (RBS) [47, 52]. RBS can be utilized both for composition and for depth profile analysis of heavy atoms. The principle is quite similar to ERD but because of the mass difference of impinging and target atoms the ^{4}He ions are scattered back to a certain extent and can be analysed with respect to their energy distribution (Fig. 4a). The depth resolution can be of the order of 30 nm or better and depends largely on the system studied. This technique has also been applied to the problem of polymer diffusion, but its application is largely limited in polymer science due to the need for heavy atoms. By RBS, movement of gold markers at the interface between polystyrenes of different molecular weights has been followed [52] and

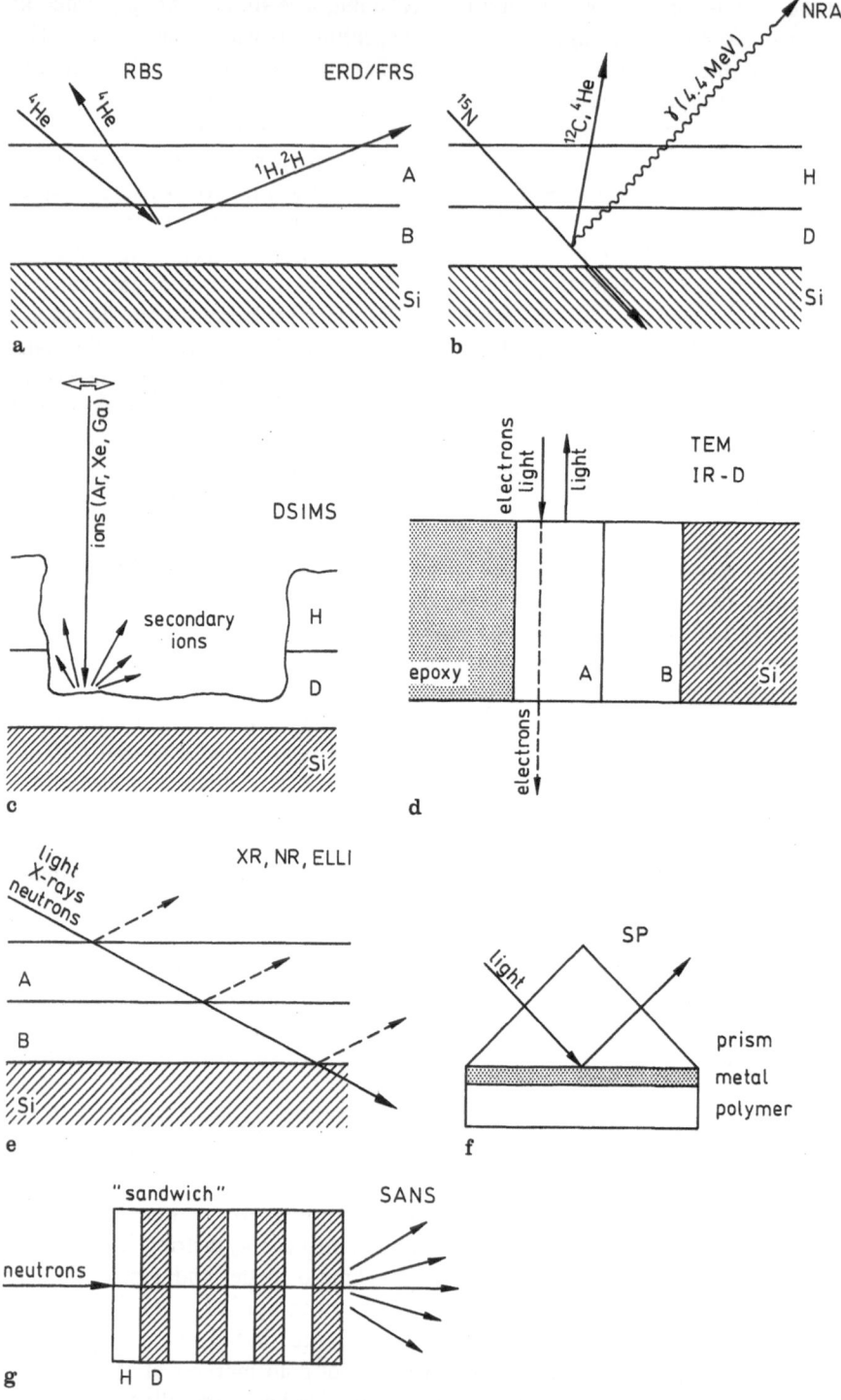

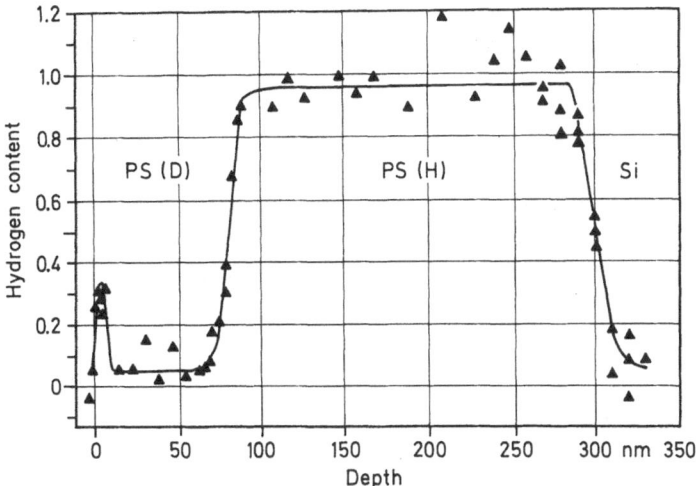

Fig. 5. Hydrogen depth profile of a deuterated polystyrene PS(D) film deposited on a protonated polystyrene PS(H) film on top of a silicon wafer as obtained by ^{15}N-nuclear reaction analysis (^{15}N-NRA). The small hydrogen peak at the surface is due to contamination (probably water) of the surface. The sharp interface between PS(D) and PS(H) is smeared by the experimental resolution (approx. 10 nm at a depth of 80 nm) [57]. The *solid line* is a guide for the eye

the non-Fickian diffusion of low molecular weight solvents in glassy polymers investigated [53].

Using the *nuclear reaction analysis* (NRA) technique one is slowly approaching molecular resolution. One can use either the D (^{3}He, ^{4}He) H-reaction for deuterium (D) profiling [54, 55] or the H(^{15}N, $\alpha\gamma$)^{12}C-reaction for hydrogen (H) profiling [56, 57]. With the ^{3}He-NRA technique a 0.7 MeV ^{3}He-beam is incident on the sample. Those ions will react with deuterium at a certain depth to produce a proton. The energy loss of that proton which is proportional to the length of the flight pass through the sample is used to determine the location of the reaction [54]. Alternatively, the energy of ^{4}He ions may be analysed for depth profiling [55]. In both cases the depth resolution is approximately 14 nm at the surface and 30 nm at a depth of 100 nm mainly due to straggling. With the ^{15}N-NRA technique (Fig. 4b) a resonant reaction is used which takes place between ^{15}N-ions and hydrogen at a sharp resonance energy of 6.4 MeV [56]. Incident ^{15}N-particles lose energy on their path through the sample until they reach the resonant energy. If there are hydrogen atoms present at that depth the reaction will take place and the emitted γ-radiation of 4.4 MeV is detected. Scanning the incident energy an accurate hydrogen depth profile is obtained. The depth resolution at the surface

---

◀ **Fig. 4 a – g.** Some interface analysis techniques applied to polymers. Mostly the interface of a double layer system of polymer A/polymer B on a substrate (silicon wafer (Si) or glass) is investigated. In some cases deuterated (D) versus protonated (H) polymer films are used. Techniques and abbreviations are explained in the text and in Table 2

is 4 nm and at a depth of 100 nm approximately 12 nm again mostly limited by straggling effects [57]. Quantitative hydrogen concentrations well below 1 atom% can be detected. The application of this technique to polymers is quite new [57], and an example is given in Fig. 5. The hydrogen density profile of a layer sample PS(D) on PS(H) on a silicon wafer has been recorded showing the resolution and sensitivity of the ^{15}N-NRA technique. A small H-surface contamination is observed. The sharp interface between PS(H) and PS(D) at a depth of approximately 80 nm is reproduced within resolution limits of the technique. While radiation damage of polymers by ions is generally not negligible, run times are kept short enough to ensure that radiation damage does not influence experimental results. Ion beam techniques thus provide the concentration profile of hydrogen or deuterium in a quite straightforward and quantitative way which is a big advantage over some of the other interface analysis methods. They are now also capable of achieving molecular resolution. Another example will be discussed in Sect. 3.2.

Also *dynamic secondary ion mass spectroscopy* (DSIMS) is used for depth profiling of polymers [22, 58]. In contrast to SSIMS applied to surface analysis a much higher primary ion energy and current is utilised. As a consequence surface material is sputtered away and slowly a crater develops in the sample (Fig. 4c). Scanning the surface several times with a small beam results in a progressive removal of sample material. The sputtering speed is kept constant as far as possible to obtain a depth profile by monitoring secondary ions as a function of time. Depth resolution is mostly limited by atomic mixing due to impinging primary ions and may be as low as 12 nm for optimized conditions. Problems are of course similar to SSIMS and in particular charging and matrix effects may influence results. A constant sputtering rate is obtained only after some initial sputtering time and one should chose those conditions under which matrix effects do not change with sputtering depth. The sputtering time scale is converted into a depth scale by a measurement of the depth of the crater after the experiment. Charging of the surface can be largely avoided by the deposition of a thin gold layer on top of the polymer film. The development of textures or surface roughness during the sputtering process sometimes makes a detailed analysis difficult. The application of DSIMS to polymers is again mostly based on the use of deuterated polymers where hydrogen and deuterium profiles are obtained. Thus for instance the surface induced order of partially deuterated diblock copolymers (PS-*b*-PMMA) [59] or the interface width during the interdiffusion of PS(H) and PS(D) [60] has been obtained. It is generally, however, difficult to obtain quantitative results with DSIMS due to the uncertainties mentioned above. They are difficult to recognize and to correct during DSIMS experiments.

X-ray reflectometry (XR) has already been described in Sect. 2.1 as a technique for polymer surface investigations. If a suitable contrast between components is present buried interfaces may also be investigated (Fig. 4d) [44, 61, 62]. The contrast is determined by the difference in electron density between materials. It is, in the case of interfaces between polymers, only achieved if one component contains heavy atoms (chlorine, bromine, metals, etc.). Alternatively the location of the interface may be determined by the deposition of heavy markers at the interface.

In these cases a difference in the refractive index for X-rays at the interface gives rise to partial reflection of the incident beam and thus to interference effects [42]. A detailed analysis of the reflection curve reveals information on film thickness, interface width and profile with subnanometer resolution. Investigated films should not be significantly thicker than 300 nm since interference fringes of thicker films will be difficult to resolve. Also the interface width should not exceed 20 nm and a relatively large, smooth and homogeneous sample of some cm^2 is needed for the experiments. Surface and interface roughness of the sample due to the preparation limits the obtainable resolution in many cases and special preparation techniques are needed to meet the requirements for this technique. Some examples of the application of XR to polymer interface analysis will be discussed in Sect. 3. The general use of this technique is, however, limited by the lack of a suitable contrast between most polymeric materials and thus only very specific problems can be resolved.

This limitation is largely overcome by the use of *neutron reflectometry* (NR) where a contrast between polymers can be generated by deuteration of one component while the subnanometer resolution is still retained [61–63]. This technique has only recently been applied to polymer problems but is ideally suited for the study of buried polymer interfaces with subnanometer resolution. The principle of the technique is similar to XR (Fig. 4d) with the exception that the index of refraction of neutrons is determined by the scattering length density of materials. Thus hydrogen- and deuterium-containing materials show a large contrast for neutron reflection and individual film thicknesses, interface width and profile between materials can be investigated. There are, however, plenty other possibilities with NR where (partial) deuteration can be utilized. Thus the organisation of partially deuterated diblock copolymers or the compatibalising effect of the addition of a blockcopolymer to two homopolymers may be envisaged where alternatively one of the three components may be deuterated. This technique truly has resolution on a molecular scale perpendicular to the interface while, however, similarly to XR it averages laterally over a relatively large area. A typical sample dimension is 5×10 cm^2 and the sample has to be homogeneous and smooth over such a large area. Thus again sample preparation and quality is what mostly limits resolution of NR. In addition there are limitations similar to those of XR with respect to film thickness ($\lesssim 300$ nm) and interface width ($\lesssim 20$ nm). Because of the negligible absorption of neutrons in most materials, the solid/liquid interface through a 10 cm thick quartz crystal may be investigated under special conditions [61]. Quite similar to scattering experiments, the interpretation of NR data is not unique. For complicated scattering density profiles other techniques (ERD, NRA etc.) with lower resolution but more direct information on H/D-concentrations should be utilized to obtain complementary information. Otherwise NR proves to be ideally suited for the investigation of buried polymer interfaces on a molecular scale and some examples will be discussed in Sect. 3.

Also a good interface resolution is obtained with *transmission electron microscopy* (TEM), where, however, a dedicated sample preparation and treatment are necessary to achieve nanometer resolution and suitable contrast [64]. Thus the thin film has to be embedded in some matrix material, cut perpendicularly to the interface with a microtome and in most cases stained by chemical treatment (e.g. OsO_4) to achieve a contrast between components (Fig. 4d). From this specimen which then is typically 50 nm thick a TEM picture is taken to obtain an image of the interfacial region. In the scanning mode (STEM) or with SEM a depth profile may be obtained by an element specific X-ray fluorescence or electron energy loss scan through the interface which, however, is generally only qualitatively due to uncertainties in the previous sample treatment. The depth resolution is typically in the range of some nanometer for TEM and 100 nm for SEM. There are some examples for the study of polymer interdiffusion by TEM or SEM [1, 65, 66]. The technique is restricted to the interdiffusion of chemically different polymers which can be contrasted or distinguished. A nice demonstration of the potential of this technique is the localization of a specially marked triblockcopolymer at the interface of a mixture of incompatible homopolymers [67]. Because of the tedious preparation procedures and limited quantitative information not very many polymer interface investigations have been reported. This is quite different for inorganic crystals, where through the use of high resolution techniques, atomic resolution within interfacial regions has also become possible [68]. Radiation damage of polymers still hinders the application of these techniques to polymers.

The resolution of *infra-red densitometry* (IR-D) is on the other hand more in the region of some micrometers even with the use of IR-microscopes. The interface is also viewed from the side (Fig. 4d) and the density profile is obtained mostly between deuterated and protonated polymers. The strength of specific IR-bands is monitored during a scan across the interface to yield a concentration profile of species. While in the initial experiments on polyethylene diffusion the resolution was of the order of 60 μm [69] it has been improved e.g. in polystyrene diffusion experiments [70] to 10 μm by the application of a Fourier transform-IR-microscope. This technique is nicely suited to measure profiles on a micrometer scale as well as interdiffusion coefficients of polymers but it is far from reaching molecular resolution.

In *ellipsometry* (ELLI) the change of the state of polarization during the reflection of light from an interface or thin film is analyzed to obtain accurate information on the index of refraction and film thickness (Fig. 4e) [71 – 75]. The correlation of measured ellipsometric angles and molecular parameters is generally not unique and model dependent. Good results are obtained for metallic surfaces where the difference in refractive index between components and the reflectivity are high [71]. For a polymer layer on solid substrate the mean refractive index and layer thickness may be obtained with high precision. In favourable cases, information on the density profile by spectroscopic or angle dependent ellipsometry may also be obtained [72, 73]. With ultra-thin films or monolayers in connection with dielectric substrates the situation is much worse. Mostly only one quantity, e.g. the total adsorbed amount of polymer material, can be deduced from ellipsometric data [74, 75]. Density profiles or film thickness cannot be obtained in those cases. Ellipsometry, however, still proves to be an extremely sensitive technique for detecting minor changes e.g. during the adsorption process of polymers from solution [74] or during the lateral organisation of ultra-thin Langmuir films on water [76]. The technique is frequently applied to the analysis of adsorbed polymer layers on metallic substrates [71], but is generally restricted with respect to the available molecular information since mostly only two ellipsometric angles are measured at fixed incident angle and wavelength. A model has to be assumed for their interpretation. Other similar techniques are sometimes used including internal reflection interferometry [77] or dynamic scanning angle reflectometry [78].

Very good resolution is also obtained through *surface plasmons* (SP) (Fig. 4f) [79–82]. Surface plasmons are electromagnetic surface waves propagating for instance from a thin metal film into a dielectric medium. Plasma oscillations in the metal film may be excited by polarized light from a laser which couples into the film at a particular angle. Since the resonance condition for a given system depends on dielectric constants of components, the measurement of the location of the resonance by variation of the incident angle yields information on e.g. an adsorbed polymer film on the metal surface. The penetration depth of the surface plasmon into the dielectric medium (e.g. polymer solution) corresponds approximately to the wavelength (micrometers) and the location of the resonance is extremely sensitive on changes of the system over this distance. SP can thus be used to determine the variation of the surface excess in a kinetic polymer adsorption experiment [80]. Recently, however, surface plasmon microscopy has also been demonstrated [81] to yield a surface image of resonant and non-resonant regions which differ in thickness. It can be used to obtain a thickness pattern with subnanometer resolution. The lateral resolution is of the order of some micrometers limited by the propagation length of plasma oscillations. SP can be combined with Raman scattering and imaging to obtain information on vibrational modes and inelastic processes [82]. This technique thus offers some unique possibilities for polymer thin film investigations even though it is restricted with respect to specific substrates which need to be metals. Density profiles cannot be obtained since the surface plasmon integrates over typically one micrometer. With the knowledge of the refractive index of the polymer film the film thickness can be accurately obtained.

Also *small angle X-ray scattering* (SAXS) or *small angle neutron scattering* (SANS) [83, 84] may be used in various ways to extract information about "buried" surfaces. The most conventional technique is the Porod-analysis [83, 84] where the decrease of the small angle intensity at larger angles is analysed. Assuming a two-phase system with sharp boundaries, a q^{-4}-behavior of the intensity is expected. The electron scattering density difference between phases, the volume content of the phases and/or the total interfacial area can be extracted. Deviations from the q^{-4}-behavior are attributed to the influence of an interfacial region. In some cases it was possible to obtain, from a careful analysis, both the interface width as well as an indication of its functional form. For microphase separated polymer blends and diblock copolymers, the interface width was determined to be of the order of 2 nm in accordance with theory [85, 86]. It is only 1 nm for segmented polyurethanes [87], where phase-separated amorphous and crystalline regions are formed. For many polymers there is enough electron density difference between components to perform SAXS experiments. By deuteration of one component, a large neutron scattering density difference can generally be generated, which helps in the extraction of the small scattering signal at large angles against the background needed for the Porod-analysis. Thus it was possible by SANS to distinguish between different interfacial profiles in a spinodal system of deuterated and protonated polybutadiene [88]. The big advantage of this technique against the other ones listed in Table 2 may be seen from the fact that no special sample geometry is needed and bulk samples typically 1 mm thick may be investigated. On the other hand a very tedious analysis of the tail of the small angle scattering intensity is necessary where according to Porod's law it has fallen close to zero on top of a constant background.

Recently another SAXS or SANS technique has been developed [89–91] which is especially sensitive on the interfacial region where molecules are intermixed. Because of the special geometry of the experiment (see Fig. 4g) where thin films of components are alternatively stacked to a sandwich-like sample, only those regions contribute to small angle scattering where the two components are mixed. In a typical real-time SANS experiment [89] 1800 films of PS(H) and PS(D), each 0.7 μm thick, are heated above the glass transition temperature and the SANS intensity is monitored as a function of time. The region of intermixing and accordingly the SANS intensity increases due to interdiffusion. The resolution with respect to interface width is better than 5 nm and diffusion coefficients of the order of 10^{-21} m^2 s^{-1} may be measured [89]. Information on the detailed form of the interfacial profile cannot be obtained with this technique. It may also be used with SAXS when a suitable X-ray contrast is present between components [91].

Thus, for the investigation of buried polymer interfaces, several techniques with molecular resolution are also available. Recently NMR spin diffusion experiments [92] have also been applied to the analysis of a transition zone in polymer blends or crystals and even the diffusion and mobility of chains within this layer may be analyzed. There are still several other techniques used, such as radioactive tracer detection, forced Rayleigh scattering or fluorescence quenching, which also yield valuable information on specific aspects of buried interfaces. They all depend very critically on sample preparation and quality, and we will discuss this important aspect in the next section.

2.4 Sample Preparation

The aspect of sample preparation and characterization is usually hidden in the smallprint of articles and many details are often not mentioned at all. It is, however, a very crucial point, especially with surface and interface investigations since there might be many unknown parameters with respect to surface contaminations, surface conformations, built-in stresses, lateral sample inhomogeneities, roughness, interfacial contact etc. This is in particular important when surfaces and interfaces are investigated on a molecular scale where those effects may be quite pronounced. Thus special care has to be taken to prepare well defined and artifact free specimens, which is of course not always simple to check. Many of these points are areas of research in themselves and cannot be answered easily at the present state of our knowledge.

Sample preparation and requirements are quite different for different techniques and materials. Some techniques are very sensitive to surface contamination (XPS, SSIMS, NRA, ELLI, SP) or require a microscopically very smooth surface (AFM, STM). For some techniques a large homogeneous film is needed (XR, NR) while others require specific sample treatment like cutting (IR-D) and staining (TEM) or the deposition of a conductive metal film on top (DSIMS, SEM). Thus a variety of different sample preparation techniques is needed and a particular sample can hardly be investigated by two techniques under identical conditions. Sample preparation even for a given technique will of course also depend very much on the specific problem which one would like to solve under optimized conditions, and on the polymer under investigation. We, therefore mention here just some of the most common techniques for preparing thin polymer films. Specimens then might be further treated in different ways to meet the requirements of the specific surface or interface analysis technique. We will not mention the problem of the substrate and its chemical modification [93, 94] which can of course also influence interfacial properties significantly.

Probably the easiest method of thin film preparation is *solution casting*. A polymer solution of suitable concentration is spread on a flat substrate. The solvent should evaporate slowly to obtain a homogeneous smooth film. The quality of the films depends largely on solvent and substrate used as well as on spreading and evaporation conditions. It is easily possible to obtain e.g. a mostly amorphous or largely crystalline film from the same material by simply changing solvent and evaporation conditions. The surface roughness of solution cast films can be kept in the region of 2 nm and fairly large films may be prepared. It may be possible to float films off on water and pick them up e.g. on top of another film to produce a sandwich sample. They may also be picked up by a frame to obtain a free standing film. For SANS experiments [89] we have thus prepared 30×30 cm^2 free-standing films of PS approximately 700 nm thick. Solution casting may be used for thin film preparation, when no extreme requirements on surface roughness or film homogeneity are demanded, when little material is available or when relatively thick films have to be prepared.

Much better quality films are prepared by *spin coating*. For this technique the substrate is rotated at high speed (typically 3000 rpm) while the polymer solution is deposited on it. Alternatively the solution is first put on the substrate before the spin coater is switched on. Important for a good quality film is a high acceleration and rotation connected with favourable viscosity, evaporation rate etc. of the solution. If parameters are well chosen, the central part of the film shows excellent homogeneity and surface roughness [44] while there is generally a small hump at the corner. High quality films of various thickness may be obtained changing the concentration of the solution. A large portion of the polymer is, however, lost since most of the solution is flung away. Films for XR or NR of a typical size of 10×10 cm^2 and typical surface roughness of less than 1 nm [95] are prepared in this way. Layered samples can again be prepared by the floatation technique described previously.

Ultra-thin films or monolayers may be obtained by the *Langmuir-Blodgett* (LB) technique. Many polymers and in particular low molecular weight organic materials [94, 96] can be spread from organic solution on the surface of a liquid (often water). Under suitable conditions in a Langmuir through sometimes a well-organized monolayer is formed which can be transferred onto a solid support by dipping. With polymers an orientation of this film may be observed [97] and by repeated dipping, a large number of layers can be transferred. The formation of LB films depends critically on various parameters and films may be not always homogeneous over larger areas.

Very thin films may be also obtained through *adsorption* of a thin layer from solution [11, 71, 74] or *chemical grafting* [98] which is achieved by a polymerization reaction at the surface. A polymer film may also be deposited on the surface by plasma polymerization [99]. It is then, however, usually crosslinked and chemically not well-defined.

During sample preparation one needs simple techniques to characterize the prepared films with respect to thickness, roughness and lateral homogeneity. This can be achieved by standard techniques like ST, ELLI, PMIM or XR which are commercially available for laboratory use and which can be applied with relative ease. Examples of polymer films and their parameters as well as various applications of the described techniques to polymeric surface and interface problems will be described in the following section.

3 Experimental Results

3.1 Polymer Surface

Polymer surfaces have been extensively investigated by several techniques (see Table 1) to yield information on chemical composition, structure, roughness, homogeneity etc. One can distinguish four main areas of interest: (i) surface chemistry including surface modification and functionalization, (ii) surface composition, where e.g. the enrichment of one component of a polymer blend is investigated, (iii) surface structure including lateral organization, roughness, order and conformation of chains near the surface and (iv) dynamics of chains at the surface.

We will only briefly discuss the area of surface chemistry. There are essentially two ways of introducing specific functional groups at a surface [93, 94], i.e. first, by chemical reaction including e.g. plasma treatment [99] or second, by adsorption of organic low molecular weight compounds bearing specific surface active groups at one end and the desired functional group at the other end [100, 101]. In this way, metal and inorganic surfaces (glass, mica etc.) as well as polymer surfaces may be modified for further adsorption or reaction with other chains. For polymers, plasma treatment is often used. Functional groups at the surface are mostly investigated by XPS or simply by contact angle experiments.

By chemical treatment for instance, the hydrophobicity of the surface may be changed to facilitate the deposition of a specific layer on the surface. For a good contact and homogeneous film formation, the wetting of substrate and deposited material will of course be advantageous. In some cases chemical bonding of the molecules to the substrate is achieved. While there are already many surface functionalised substrate materials commercially available, this field of specifically designed and functionalized polymer and substrate surfaces is still growing due to its importance in various technical areas including composites, filler materials, fiber processing etc.

Additives of polymers may also play an important role in surface modification. In some cases small contaminations of low molecular weight additives may migrate to the surface and completely change surface properties and composition [3]. Therefore, in polymer blends, one will generally not observe the same composition of components at the surface as compared to the bulk. One component will be enriched at the surface to minimize the total free energy of the system (Fig. 1 b). This should be the component with the lowest surface free energy and the surface excess will depend on the free energy difference. With largely immiscible polymers essentially one component will separate at the surface, while in a miscible blend one component will only be enriched at the surface. Surface excess and concentration profile can be calculated on the bases of mean field arguments [102, 103]. Several experimental investigations on homopolymer blends have been performed mostly with XPS, FRS, SIMS and IR techniques to obtain the surface excess in different blends. Examples include poly(caprolactone)/poly(vinylchloride) (PVC) [104], poly(methylmethacrylate)/PVC [105], poly(ethyleneoxide)/PS [106], poly(dimethylsiloxane)/polycarbonate [107], poly(vinylmethylether)/PS [106] and PS(H)/PS(D) [108]. In the last case, the small difference due to isotopic exchange between PS(H) and PS(D) of approximately equal molecular weight is sufficient for a significant surface enrichment of PS(D). By NR, a detailed concentration profile is obtained showing deviations to the expected mean field profile [108]. In an XR study of poly(vinylidene fluoride)/PMMA [109] the profile could not be accurately resolved because the X-ray contrast was not large enough. Recently surface-directed spinodal decomposition was observed with poly(ethylenepropylene) (PEP) isotopic mixtures, PEP(H)/PEP(D), by FRS experiments [110]. Spinodal composition modulations in the two phase region near the critical temperature T_C are shown to be directed to the surface resulting in an enrichment of PEP(D) at the surface followed by a depletion layer beneath.

With diblock copolymers, similar behavior is also observed. One component is enriched at the surface and depending on miscibility and composition a surface-induced ordered lamellar structure normal to the surface may be formed. Recent investigations include poly(urethanes) [111], poly(methoxy poly(ethyleneglycol) methacrylate)/PS [112] and PS/PMMA [113, 114]. In particular the last case has been extensively studied by various techniques including XPS, SIMS, NR and optical interferometry. PS is enriched at the surface depending on blockcopolymer composition and temperature. A well ordered lamellar structure normal to the surface is found under favourable conditions. Another example is shown in Fig. 6 where the enrichment of poly(paramethylstyrene), PMS(H), in a thin film of a di-

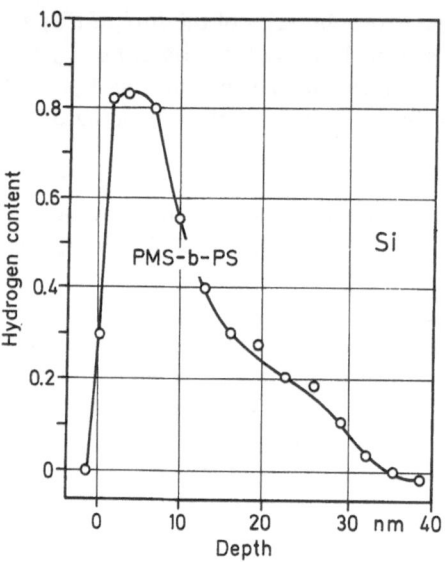

Fig. 6. Hydrogen depth profile of a thin film of poly(p-methylstyrene)(H)/PS(D) diblock copolymer, PMS(H)-b-PS(D), on a silicon wafer as obtained by the ^{15}N-NRA technique [57]. The sample has been annealed for 1 h at 140 °C. PMS(H) is largely enriched at the surface. The *solid line* is a guide to the eye

block copolymer of PMS(H)-b-PS(D) (50:50 composition, $M_w \approx 230000$ g/mol) is shown [57]. The quantitative concentration profile normal to the surface is obtained from ^{15}N-NRA and the material is expected to be phase separated at this composition and temperature also in the bulk. The in the bulk randomly oriented domains are, however, ordered parallel to the surface in the thin film due to the surface interaction of PMS(H). Since the copolymer has been annealed at 140 °C for 1 h not very far from its critical phase separation temperature one still observes a significant amount of the other component in the two phases and the transition region is relatively wide.

A very specific surface structure is observed after the annealing of a PS/polybutadiene (PB) diblock copolymer, PS-b-PB, shown in Fig. 7b. The surface is very smooth directly after preparation of the film from solution (similar to Fig. 7a). By annealing at 120 °C the surface structure shown in Fig. 7b evolves, which we believe is due to the formation of layers of PS and PB parallel to the surface. The outermost layer might not be completely filled due to lack of material leading to steps at the surface. Similar behavior is observed with other diblock copolymers such as PS-b-PMMA [61]. Enrichment of one component is also observed at the surface of a polymer solution [115, 116] by X-ray fluorescene and evanescent wave techniques.

Another characteristic of a polymer surface is the surface structure and topography. With amorphous polymers it is possible to prepare very smooth and flat surfaces (see Sect. 2.4). One example is the PMIM-picture shown in Fig. 7a where the root-mean-square roughness is better than 0.8 nm. Similar values are obtained from XR-measurements of polymer surfaces [44, 61, 62]. Those values compare quite well with observed roughnesses of low molecular weight materials. Thus for instance, the roughness of a water surface is determined by XR to 0.32 nm

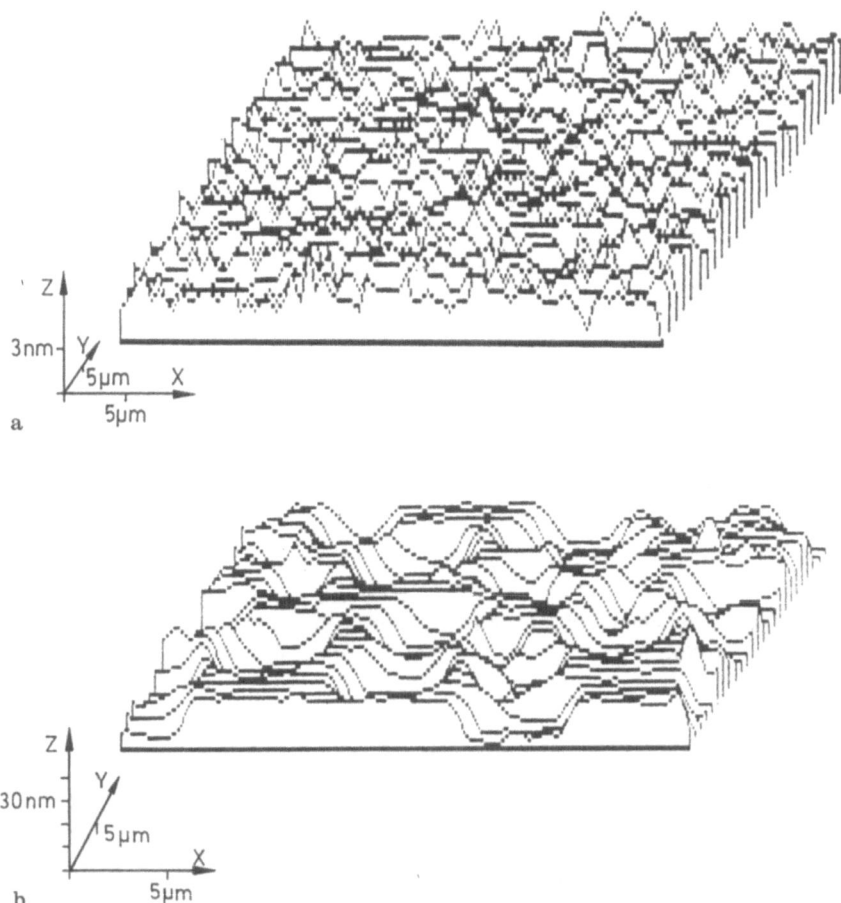

Fig. 7a, b. PMIM-image of (**a**) a poly-*p*-bromostyrene surface [118], (**b**) a PS/polybutadiene diblock copolymer, PS-*b*-PB, at approximately 100 fold magnification. The lateral resolution is of the order of 1 μm while the height resolution is of the order of 0.6 nm. The root-mean-square roughness averaged over the area shown is 0.8 nm in (a) close to the resolution limit of the technique. It is much larger (10 nm) in (b) due to the formation of steps after annealing. The scale in z-direction in (a) and (b) is different by a factor of 7

[117] largely dictated by capillary surface waves. Polished silica glass shows complicated surface density behavior which extends over tens of nanometer [119]. Similar behavior is observed with float glass in NR [62] which might be due to a particular hydrogen distribution at the surface and is seen differently with XR (see Sect. 3.2). A monolayer of behenic acid on water again shows a surface roughness of 0.3 to 0.4 nm and phase transitions in the layer are also reflected in changes of the surface roughness [120]. This is also observed with liquid crystalline polymer surfaces at the smectic-nematic phase transition [121] where the surface roughness changes from 0.6 nm to 1.1 nm during the transition. Thus the surface

roughness of polymers is quite comparable to low molecular weight compounds and reflects the bulk structure.

The structure near the surface will of course be influenced by the presence of the surface leading to e.g. surface-induced ordering effects well known for inorganic compounds. Since only a very small region near the surface might be involved, evanescent wave techniques should be used. While the theory of evanescent wave scattering has recently been discussed [122, 123], only few experimental investigations on polymers are reported. The ordering of phase separated regions in diblock copolymers has already been mentioned (see Fig. 6 and [57]) which leads, in the case of PS-*b*-PMMA, to a well ordered lamellar morphology parallel to the surface [124]. Similarly with side-chain liquid crystalline polymers smectic-A layers are organized parallel to the surface as observed by XR [121]. The order within a monolayer of 1-heneicosanol on water changes with surface pressure and phase transitions are detected by X-ray scattering at grazing incidene [125]. Both lateral order and tilting of molecules is obtained by additional so-called Bragg-rod scans [126]. A surface structural investigation on a polymer is performed with an aromatic polyimide [46] where the enhancement of crystalline order is detected over a region near the surface extending approximately 9 nm into the bulk. With decreasing penetration depth of X-rays at small angles of incidence, Bragg reflections are recorded. They sharpen and shift to smaller scattering vectors in the vicinity of the surface. This observation is interpreted in terms of the surface-induced crystalline order of the aromatic polyimide.

Besides crystalline order and structure, the chain conformation and segment orientation of polymer molecules in the vicinity of the surface are also expected to be modified due to the specific interaction and boundary condition at the surface between polymers and air (Fig. 1a). According to detailed computer simulations [127, 128], the chain conformation at the free polymer surface is disturbed over a distance corresponding approximately to the radius of gyration of one chain. The chain segments in the outermost layers are expected to be oriented parallel to the surface and chain ends will be enriched at the surface. Experiments on the chain conformation in this region are not available, but might be feasible with evanescent wave techniques described previously. Surface structure on a micrometer scale is observed with IR-ATR techniques [129].

Another largely unexplored area is the change of dynamics due to the influence of the surface. The dynamic behavior of a latex suspension as a model system for Brownian particles is determined by photon correlation spectroscopy in evanescent wave geometry [130] and reported to differ strongly from the bulk. Little information is available on surface motion and relaxation phenomena of polymers [10, 131]. The softening at the surface of polymer thin films is measured by a mechanical nano-indentation technique [132], where the applied force and the path during the penetration of a thin needle into the surface is carefully determined. Thus the structure, conformation and dynamics of polymer molecules at the free surface is still very much unexplored and only few specific examples have been reported in the literature.

3.2 Polymer Thin Films

In this section we will describe some experiments on polymer thin films, where the film thickness is of comparable or smaller magnitude than the molecular extension. Since the end-to-end distance of polymer molecules of high molecular weights can be of the order of 50 nm, a typical film thickness should range between 1 nm and 200 nm. Very smooth and homogeneous polymer films of this thickness are prepared by spin coating as described in Sect. 2.4. An example of a thin polystyrene film on float glass as investigated by XR is shown in Fig. 8 [160]. One obtains an excellent model fit to reflectivity data assuming a homogeneous film of thickness 59.1 ± 0.1 nm and surface roughness 0.6 ± 0.1 nm. The size of this sample is 7×2.5 cm^2 and the film is prepared by spin coating from toluene solution. In deuterated PMMA-films on the other hand in some cases (scattering) density gradients are observed by NR [133]. The density in a surface layer appears to be up to 5.7% higher than the bulk value. A similar effect is observed in per-fluoropolyether polymer thin films on carbon by XR [134] which is explained by preferential adsorption of end groups. The electron scattering density near the surface is as much as 11% higher than in the rest of the film. In this case the film has, however, only a total thickness of 2.5 nm. Dealing with glass substrates we generally have to introduce into NR data analysis an interface layer of low scattering density between glass and polymer [62] which might be due to some hydrogen-containing thin layer of e.g. water. It is similarly observed in ^{15}N-NRA experiments on the surface of polymers (see e.g. Fig. 5 and Ref. [57]). Careful XR experiments on PMMA films on polished quartz substrates [135] similar to those used in NR experiments [133] reveal slight distortions of the glass-polymer profile from the error-functional behaviour which we believe is due to the polishing process. Other experiments on thin polymer films are reported [61] which, however, also mostly show that very homogeneous and smooth amorphous polymer films can be prepared and characterized. The preparation of semi-crystalline polymer thin films causes many more problems since those polymers mostly need solvation at high temperature and smooth films by spin coating are hard to obtain.

While thin polymer films may be very smooth and homogeneous, the chain conformation may be largely distorted due to the influence of the interfaces. Since the size of the polymer molecules is comparable to the film thickness those effects may play a significant role with ultra-thin polymer films. Several recent theoretical treatments are available [136–144, 127, 128] based on Monte Carlo [137–141, 127, 128], molecular dynamics [142], variable density [143], cooperative motion [144], and bond fluctuation [136] model calculations. The distortion of the chain conformation near the interface, the segment orientation distribution, end distribution etc. are calculated as a function of film thickness and distance from the surface. In the limit of two-dimensional systems chains segregate and specific power laws are predicted [136, 137]. In 2D-blends of polymers a particular microdomain morphology may be expected [139]. Experiments on polymers in this area are presently, however, not available on a molecular level. Indications of order on an

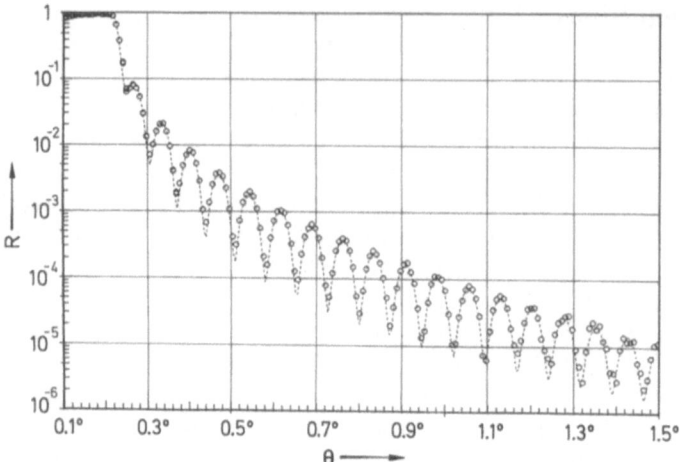

Fig. 8. X-ray reflection diagram of a thin polystyrene film on float glass [160]. The reflectivity R is plotted against the glancing angle Θ. The film is spin coated from solution. A model fit *(dashed line)* to the reflectivity data is also shown where the following parameters are obtained: film thickness = 59.1 ± 0.1 nm, interface roughness glass-polymer = 0.4 ± 0.1 nm, surface roughness polymer-air = 0.6 ± 1 nm, mean polymer density = 1.05 ± 0.01 g/cm^{-3}. The X-ray wavelength is 0.154 nm

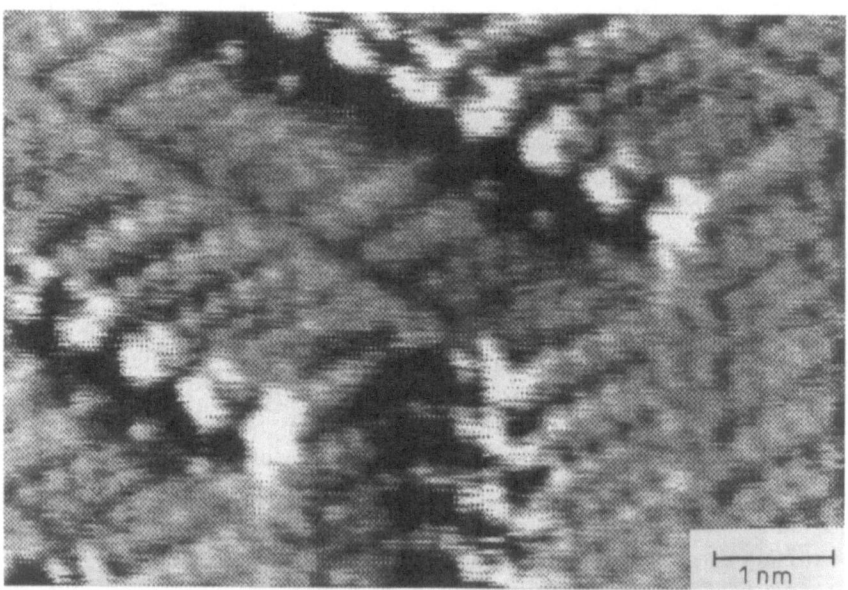

Fig. 9. STM image of a monolayer of didodecylbenzene adsorbed on pyrolytic graphite from a phenyloctane solution. The area shown corresponds to 7.2×4.7 nm^2 [37]. The strong features represent the benzene rings and the ordered arrangement of side groups is clearly resolved

atomic level are obtained from force balance experiments from melts of low molecular weight compounds where the order induced by the mica surface is deduced from periodic force oscillations [145–148]. Individual polymer molecules can also be resolved by STM where investigations are, however, limited to specific cases of ultra-thin polymer films on a conducting substrate. A very recent example is shown in Fig. 9 where in an STM picture of the surface of a didodecylbenzene monolayer adsorbed on pyrolytic graphite from phenyloctane solution, atomic resolution is achieved [37]. The lamellar structure with the alkyl side groups and phenyl center is clearly resolved and even the movement of domain boundaries with time can be followed. Similarly ordered structures of rodlike polyglutamate monolayers [149], electrically conducting polypyrrole with polymeric anions [150] and polyimide crystalline monolayers [151] are other examples where well ordered structures of polymers on graphite are obseved with atomic resolution by STM.

Thin polymer films may also be investigated by TEM and high resolution images are obtained for e.g. thin films of liquid crystalline polymers [64]. Usually thin microtome cuts from bulk samples are investigated, but also epitaxial growth of polyoxymethylene on NaCl [152], chain folding of polyethylene crystals [153], epitaxial crystallization of polypropylene on polystyrene [154] or monomolecular polystyrene particles [155] are observed. The resolution is, however, in most cases not comparable to STM.

AFM images of PET film surfaces have also recently been measured [156] and they showed a microroughness of PET films of the order of 1 nm averaged over a surface area of 200×200 nm^2. In some cases atomic lateral resolution is achieved [39] when e.g. dialkylamonium (C$_{16}$) layers adsorbed on mica from cyclohexane are examined [157]. The interpretation of AFM data is at present not always clear and further advances will be made with improved instrumentation. Up till now, only very specific examples of polymeric structure have been investigated.

When dealing with polymer blends or blockcopolymers, surface enrichment or microstructures may be observed as already discussed in Sect. 3.1. Quite similar effects may be expected for buried interfaces e.g. between polymer and substrate where one component may be preferentially enriched. In a system of PS, PVP and diblock copolymer PS-b-PVP it has been shown by FRS that the copolymer enrichment is strongly concentration dependent [158]. In a mixed film of PS(D) and end-functionalized PS on a silicon wafer the end-functionalized chains will be attached to the silicon interface and can be detected by NR [159].

Also wetting and dewetting phenomena may influence the stability of thin polymer films on a substrate and this has been discussed by several authors theoretically [20, 161, 162]. The wetting process of a polydimethylsiloxane drop on a silicon wafer is nicely followed by several techniques including XR and ellipsometry [163 – 165]. Dewetting on the other hand may roughen and destabilize thin polymer films. This phenomenon is e.g. observed by PMIM for a thin PS layer on a silicon wafer [166]. In a 10 nm-thick film for instance, during annealing above the glass transition temperature small holes develop which grow in size with time to finally form droplets. The surface becomes significantly rougher and at later stages this can be seen directly by optical inspection. A theoretical description of the kinetics of dewetting has been given recently [167, 168] which predicts thickness and time dependence.

Several aspects of polymer thin films have thus been investigated while many others are still unexplored. These include structural and conformational aspects where polymer thin film properties are theoretically well-treated but experimental data are generally missing. However, with further development of experimental techniques this area might become accessible in the near future.

3.3 Polymer Interdiffusion

3.3.1 The Welding Problem

The welding of polymers is schematically shown in Fig. 1c. Two pieces of material are brought together in the melt and the mechanical connection of the two sides is achieved by interdiffusion of chains from one side to the other. If both components are compatible, molecules will freely diffuse across the interface and the region of the interface will be indistinguishable from the bulk after a time τ_d when molecules have completely crossed the interface. During the initial stages of interdiffusion, however, only chain segments will cross the interface and the chain conformation might still resemble the distorted chain conformation of the initial surfaces which will typically relax after a time τ_r. Thus the initial stages of interdiffusion of chains at the interface are expected to be different from later stages when essentially the bulk behavior is achieved.

Mechanical properties, in particular the fracture energy, will also change in those initial stages significantly with interdiffusion time and temperature [169] and it is still not clear, how far a chain has to diffuse until a certain mechanical strength of the joining interface is achieved. Several models have been discussed where the density of crossing chains [170], the diffusion over an average interpenetration depth [171] or the formation of "effective" crossings [172, 173] are taken as the essential factors which determine the strength of an interface. An excellent review on the welding problem is given in [51] and other reviews on polymer interdiffusion in general have recently been published [47, 174, 175]. As a main conclusion the polymer interdiffusion in the bulk is still not completely understood and in particular the question of "fast" or "slow" mode mutual diffusion lacks a rigorous theoretical basis [176]. Several experiments favour the "fast" mode theory, where the faster species in a diffusion couple determines the mutual diffusion coefficient [89, 177, 178]. Others favour the "slow" mode theory [179–181], while it is argued on the basis of Monte Carlo calculations in the bulk that neither of these theories can be applied [182]. The situation is not easier at the interface where different interdiffusion times correspond to the movement of segments of molecules of different length over the interface with lateral chain relaxation superimposed. The experimental situation might be, however, to some extent advantageous since sensitive techniques are available to follow segmental movement across the interface. We will discuss the welding and interdiffusion of polymers with emphasis on some recent results of initial stages of interdiffusion in Sect. 3.3.3.

The situation is quite different with incompatible materials where the welding process comes to a standstill after some initial interdiffusion. Only little intermixing at the interface is achieved which is within the mean field picture determined by the Flory-Huggins interaction parameter χ between monomers. For weak incompatibility χ is small and the interface width is large, while for strong incompatibility χ is large and the interface width is very narow. Several theories describe this functional behavior in the weak and strong segregation limit [183–187]. Because of little chain interpretation in the strong segregation limit, the mechanical strength of the interface will be weak. There have been attempts, however, to strengthen the interface by the introduction of agents which make the two materials compatible (compatibalizers) examples of which are diblock copolymers where each block is soluble in one component and a large number of effective crossings are introduced at the interface in this way [188]. Experiments on incompatible polymers will be described in Sect. 3.3.2.

For both investigations it will be necessary to utilize experimental techniques which have submolecular resolution since it is necessary to determine the movement of chain segments across the interface.

3.3.2 The Interface Between Incompatible Materials

Blends of polymers have a significant practical application range. Since most polymers are immiscible with each other, in most cases phase segregation between components is observed [189, 190]. The thermodynamics of polymer blends is treated within mean field theory [184] and the phase behavior is in most cases well-understood. The interfacial region between phases largely determines materials properties and there are various approaches to determine the interfacial width and functional form both theoretically [183–187] and experimentally (see below). In a three-dimensionally phase-separated system this is achieved e.g. by careful analysis of the angular dependence of the SAXS or SANS intensity (Porod's law) [83, 84] or by quantitative densitometry through a TEM picture where a thin cut through the sample has been stained in a suitable way [65, 66]. Since both techniques are not easy to apply there have not been very many of these investigations reported. A nice example is the spinodal system of deuterated and protonated polybutadiene [88]. From a careful SANS analysis an interface width of the order of 25 nm and details of the profile were resolved. Another example is a SAXS investigation of PS/polyisoprene [85] or PS/polybutadiene [86] where the interface width was determined to be approximately 2 nm. With TEM the sample treatment is very crucial and quantitative data are difficult to obtain. Results are reported for instance for PS/PMMA [1], PMMA/polyvinylidene fluoride [65] and polyphenylene ether with PS and different copolymers [66].

A very good resolution of interfacial mixing is achieved with XR or NR. A very specific sample geometry is needed. Two thin homogeneous, smooth films are deposited on each other and the development of an interfacial zone during annealing may be followed. An example is shown in Fig. 10 for the strongly immiscible polymers PS/poly-p-bromostyrene (PBrS) [191]. Due to the bromination a sufficient X-ray contrast between components is present and the interfacial mixing due to interdiffusion of chain segments can be followed by XR as a function

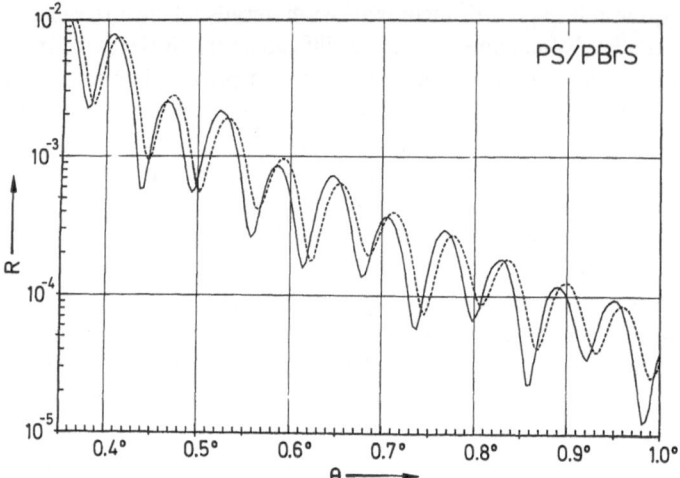

Fig. 10. X-ray reflectivity curves of polystyrene (PS)/poly-*p*-bromostyrene (PBrS) on a glass substrate before *(solid line)* and after annealing for 13 h at 130 °C *(dashed line)* [191]. The width of the interface changes from 1.3 nm to 2.0 nm due to interfacial mixing of components. The X-ray wavelength is 0.154 nm and films have a thickness of 37.8 nm (PS) and 45.0 nm (PBrS), respectively

of annealing temperature. Since the glass transition temperatures of both materials are significantly different, 103 °C and 142 °C, respectively, the mobile component (PS) swells the immobile component (PBrS) at annealing temperatures in between. This swelling proceeds into the immobile component lowering the glass transition temperature in the mixing zone, until a concentration is reached where no further movement is possible. This concentration is temperature dependent and the width of the interface reaches a maximum when both components are mobile. Then the mean interaction parameter χ solely determines the interface width, which is hardly temperature dependent. Thus the interface width increases from 1.3 nm to 2.2 nm between 103 °C and 139 °C and stays then approximately constant with temperature and time. The initial broadening of 1.3 nm is attributed to interface roughness and corresponds approximately to the surface roughnesses of the original films put together.

Another example is the investigation of PS(D)/PMMA by NR [192, 193] where an equilibrium interface width of (2.0 ± 0.5) nm is reported. It is worth mentioning that in two studies with different samples and experimental procedures the same result was obtained independently to within ±0.1 nm demonstrating nicely the reproducibility of the NR technique. The interface width can be correlated with mechanical strength of the interface between PS/PMMA [194] determined from a wedge cleavage fracture test. Based on simple scaling arguments an interface width is deduced from the measured fracture strength of the interface which is comparable to the NR results. The system PMMA(D)/solution chlorinated polyethylene (SCPE) is in the weak segregation limit when NR experiments are performed just above the critical temperature T_C within the two phase region

[195]. The interface is much wider (up to 9 nm) as compered to PS/PBrS and a movement of the position of the interface at larger annealing times is observed. Scattering density profiles are asymmetric and change with time. A specific time evolution of interfacial growth is determined for PS(H)/PS(D) of high molecular weight in the weak segregation regime by ^{3}He-NRA [55]. This particular time dependence can be explained on the basis of mean field theory [187].

Several studies have been performed to investigate the compatibalizing effect of blockcopolymers [67, 158, 188, 196–200]. It is generally shown that the diblock copolymer concentration is enhanced at the interface between incompatible components when suitable materials are chosen. Micell formation and extremely slow kinetics make these studies difficult and specific non-equilibrium starting situations are sometimes used. Diblock copolymers are tethered to the interface and this aspect is reviewed in another article in this book [14].

3.3.3 Initial Stages of Interdiffusion

The interdiffusion or welding of miscible materials is shown schematically in Fig. 1c. Both components are well separated from each other in the initial stages and chains do not cross the interface. The specific conformation of chains near the interface has already been mentioned. When this system is heated above the glass transition or melting temperature chains start moving across the interface. According to reptation theory movement proceeds along a tube formed by entanglements [201, 202]. Translational segmental movement takes place essentially at the chain ends. This picture is qualitatively consistent with molecular dynamics calculations on long chains [203] where chain ends can move much more freely than the central parts of the chains which are confined in their movement to a "tube". Following this picture one might expect different time regimes of segmental movement across the interface, which resemble segment dynamics in the bulk as well as modifications due to specific surface conformations. Several theoretical treatments of this problem of initial stages of polymer interdiffusion at an interface have been published [170–173, 204–207]. There have been, however, only a few experimental investigations since very good resolution and sample preparation are needed to tackle this problem.

Most investigations have been performed on PS(H)/PS(D) which are compatible at lower molecular weights [15]. Initial experiments using FRS [47] did not have the resolution to resolve the initial stages of interdiffusion. Using ^{3}He-NRA [55, 208], DSIMS [60] and TOF-FRS [50] this system has been subsequently investigated and the resolution has been considerably improved, but only with recent NR experiments [61, 95, 209, 210] are the initial stages of interdiffusion resolved where the movement of chain segments across the interface is followed. An example is given in Fig. 11 [95]. A bilayer sample of two thin films of PS(D) and PS(H) on a float glass substrate (see insert in Fig. 11) is investigated by NR at different stages of chain interdiffusion. The upper film is deposited on the lower film by the floatation technique (see Sect. 2.4). The interface shows an initial roughness of 1 nm due to the surface roughness of individual films. In this study all interfaces are assumed to be Gaussian, i.e. simple error function type profiles are fitted to reflectivity curves. The variance σ_{exp} of the error functions is taken

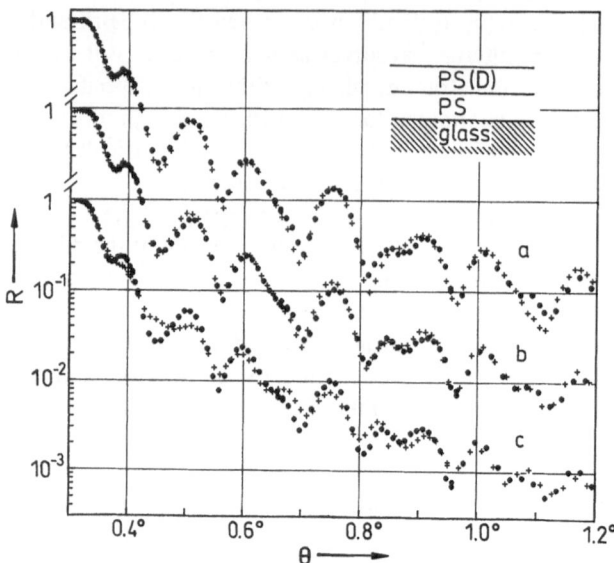

Fig. 11. Neutron reflectivity curves of PS(D)/PS(H) on a glass substrate during the initial stages of interdiffusion [95]. Different times of interdiffusion τ after annealing are compared: (a) unannealed (+) and 2 min at 120 °C ($\bullet$), (b) 2 min (+) and 910 min ($\bullet$), and (c) 910 min ($\bullet$) and 11750 min (+). Annealing times are reduced to 120 °C. The neutron wavelength is 0.43 nm

as a measure of interface width. The sample is subsequently heated above the glass transition temperature for different annealing times. From these times and temperatures a reduced time τ at the reference temperature 120°C is calculated [211] which ranges for the different curves of Fig. 11 from 0 (not annealed) to 11750 min.

Even by a simple comparison of curves at different times, some conclusions on the initial stages of interdiffusion can be drawn. There is a significant change in the reflectivity curve essentially at larger angles after 2 min of annealing. Thus a fast diffusion or relaxation process takes place which causes chain intermixing on a small distance scale (corresponding to large angles). Later, no significant change of the reflectivity curves over a considerable time interval (2 to 910 min) is recorded. With still longer times the reflectivity curves also change at small angles indicating segment diffusion over larger distances.

Those features are shown in Fig. 12 where the interfacial broadening due to segment interdiffusion, σ, is shown over a wide time range. σ has been corrected for interface roughness by: $\sigma = (\sigma_{exp}^2 - \sigma_0^2)^{1/2}$; σ_0 is the initial interface roughness also indicated in the lower left corner. The observed time dependence is clearly no simple center-of-mass diffusion of molecules and $t^{1/2}$-behavior is only achieved outside the range shown [212]. It may be explained on the basis of reptation theory taking into account the particular starting situation (Fig. 1c) and the modifications induced by the surface of the films [95]. Segmental motion can only be resolved in those NR experiments because of the particular starting

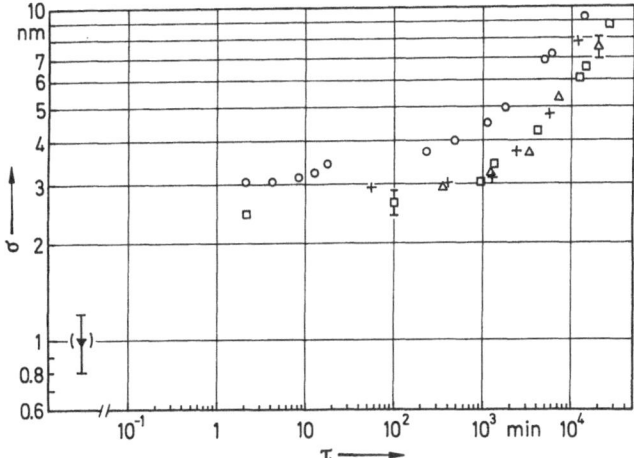

Fig. 12. Interface width σ as a function of annealing time τ during initial stages of interdiffusion of PS(D)/PS(H) [95]. Data points are obtained by a fit with error function profiles of neutron reflectivity curves as shown in Fig. 11. Different symbols correspond to different samples. The interface width σ_0 prior to annealing is also indicated ($\blacktriangledown$) and is subtracted quadratically from the data ($\sigma = [\sigma_{exp}^2 - \sigma_0^2]^{1/2}$)

situation where protonated and deuterated chains are separated from each other at the sharp interface (Fig. 1c). At very initial stages, segments may diffuse fairly quickly across the interface and cover a distance corresponding to the tube diameter in reptation theory before entanglements slow segmental motion down. Super-imposed on this diffusional motion will be the relaxation of chains within individual films which will tend to adapt an equilibrium conformation. Later, segmental motion over longer distances takes place and chains can disengage partially from the tube leading to a different power law in time. In addition to those time dependences, changes in the profile are also predicted [173, 206, 213] and these are detected by a careful analysis of NR data [212]. There are indications of the presence of a discontinuity in the profile at the interface which is, however, broadened by other effects and which disappears at later annealing times. Sample preparation plays a crucial role in those investigations where subnanometer resolution is achieved. While it was possible in those first experiments to monitor segmental motions of polymer molecules, details of the initial stages of interdiffusion are far from being understood and further experiments will be necessary.

In another type of experiment the mass flow across the interface during the initial stages of interdiffusion is determined via the movement of gold markers at the location of the interface [214]. Very small gold particles will be dragged over the interface, if a net mass flow takes place. This is observed by XR at large interdiffusion times, when two PS films with different molecular weights are put together. At small times, first a contraction of the films and then an incubation period with no marker movement is observed. Depending on molecular weight M_w of the smaller component marker movement starts at a well defined time.

Those observations can again be explained by segmental movement which at the beginning takes place at equal velocity from both sides. Thus the net segmental flux across the interface is zero. Later, however, complete chains of the lower molecular weight material move across the interface while on the other side the large chains and segments are restricted in their movement. Due to the difference in M_w of the components the low M_w component is much more mobile than the high M_w component leading to swelling and mass flow across the interface. The marker movement in this regime is determined by the diffusion coefficient of the faster species as was recognised previously from RBS experiments [215]. With NR, however, the early stages of interdiffusion are also resolved where this particular time behavior is observed which is characteristic of polymers of different molecular weights.

It has thus been demonstrated that the initial stages of interdiffusion at a polymer interface show a distinct diffusion behavior which is determined by the particular starting situation. There might be the possibility to get much more information on chain motion and interdiffusion by a careful analysis based on the now available extremely sensitive interfacial analysis techniques. This might facilitate our understanding of some of the underlying general processes. In particular, details of the welding and healing process might be resolved and it should be possible to understand e.g. fracture mechanical experiments [169] in much more detail. Experiments of this sort will, however, also help us to get a better insight into segmental motion and chain dynamics, an area which is still not completely understood even in the bulk. The application of different techniques would be very advisable and for instance a combination of mechanical, dynamical and structural techniques would be very helpful in this respect.

4 Conclusion

In the analysis of polymer surfaces and interfaces there has been tremendous progress in recent years. This is to a large extent due to the development of surface- and interface-sensitive analytical techniques which previously had not been applied to polymers. It is thus possible to achieve molecular resolution both for the free polymer surface and for buried interfaces between polymers. In addition, suitable sample preparation techniques are available and extremely homogeneous and smooth polymer thin films can be prepared. They may be put together to investigate the interface between polymers.

This area of research is still at its beginning and many aspects are not resolved. This includes in particular the structure and conformation of polymers at an interface as well as the modification of polymer dynamics by the interface. We have given several examples of the potential of surface and interface analytical techniques. They provide information on surface roughness, surface composition, lateral structure, depth profiles, surface-induced order and interfacial mixing of polymers on a molecular and sometimes subnanometer scale. They thus offer a large variety of possible surface and interface studies which will help in the understanding of polymer structure and dynamics as it is modified by the influence of the surface or interface. It will, however, also help in a purposeful modification of interfacial properties with a tremendous potential for future applications in particular also in so-called "high-tech" areas where smaller and smaller components are designed.

In this review we did not pay very much attention to the area of surface chemistry which particularly in practical applications is gaining increasing importance. There is a large potential for chemistry to tailor and modify interfaces. Similarly by physical means such as blockcopolymer adsorption, interfacial properties can be significantly affected. Many areas in polymer physics are influenced by interfacial properties and some examples have been given. We have shown, however, that our knowledge of interfacial properties is largely based on available analytical techniques and a further development in this area will be based on a further development in the sensitivity and potential of these techniques. It will be, however, necessary to perform many more experiments to get a better understanding of interfacial problems also on a molecular level.

Acknowledgement: I acknowledge the very stimulating and helpful discussions with Prof. Dr. E. W. Fischer, Dr. G. Reiter and members of the scattering group at Mainz. I am very grateful for the support from BMFT, the Forschungsanlage Jülich, and in particular from Prof. Dr. T. Springer at Jülich.

5 References

1. Wu S (1982) Polymer interfaces and adhesion, M Dekker, New York
2. Ishida H (ed) (1988) Interfaces in polymer, ceramic and metal matrix composites, Elsevier, New York
3. Briggs D, Rance DG, Briscoe BJ (1989) Surface properties. In: Allen J, Bevington JC (eds) Comprehensive polymer science, vol 2, Pergamon, Oxford, p 707
4. Koberstein JT (1987) Interfacial properties. In: Encyclopedia of polymer science and engineering, J Wiley, vol 8, p 237
5. Robb ID (1989) Adsorption. In: Allen J, Bevington JC (eds) Comprehensive polymer science, vol 2, Pergamon, Oxford, p 733
6. Briggs D (1989) Characterisation of surfaces. In: Allen J, Bevington JC (eds) Comprehensive polymer science, vol 1, Pergamon, Oxford, p 543
7. Jaycock MJ, Parfitt GD (1981) Chemistry at Interfaces, Ellis Horwood, Chichester
8. Mittal KL (ed) (1981) Physicochemical aspect of polymer surfaces, vol 1 and 2, Plenum, New York
9. Feast WJ, Munro HS (eds) (1987) Polymer surfaces and interfaces, J Wiley, Chichester
10. Andrade JD (ed) (1988) Polymer surface dynamics, Plenum, New York
11. Lipotov YS, Sergeeva LM (1974) Adsorption of polymers, J Wiley, New York
12. Tong HM, Nguyen LT (eds) (1990) New characterisation techniques for thin polymer films, J Wiley, New York
13. Paul S (1986) Surface coatings, J Wiley, New York
14. Halperin A, Tirrell M, Lodge TP (1991) Tethered chains in polymer microstructure. In: Adv Polym Sci, this volume
15. Bates FS, Wignall GD (1986) Phys Rev Lett 57: 1429
16. Atkin EL, Kleintjens LA, Koningsveld R, Fetters LJ (1982) Pol Bull 8: 347
17. Yang H, Shibayama M, Stein RS, Shimizu N, Hashimoto T (1986) Macromolecules 19: 1667

18. Young T (1985) In: Peacock G (ed) Miscellaneous works, vol 1, J Murray, London
19. Neumann AW, Good RJ (1979) Techniques of Measuring Contact Angles. In: Good RJ, Stromberg RR (eds) Surf and Coll Sci, vol 1, Plenum, p 31
20. de Gennes PG (1985) Wetting: Statics and dynamics, Rev Mod Phys 57: 827
21. Sacher E (1988) The determination of the surface tensions of solid films. In: Ratner BD (ed) Surface characterisation of biomaterials, Elsevier, Amsterdam
22. Chou NJ (1990) XPS/SIMS/AES for surface and interface characterisation of thin polymer films. In: Tong HM, Nguyen LT (eds) New characterisation techniques for thin polymer films, J Wiley, New York
23. Kelley MJ (1991) Materials Res Soc Bulletin 16 (3): 46
24. Pireaux JJ, Riga J, Boulanger P, Snauwaert P. Novis Y, Chtaib M, Gregoire C, Fally F, Beelen E, Caudano R, Verbist J (1990) J Electron Spectr Rel Phen 52: 423
25. Seki K (1989) Photoelectron spectroscopy. In: Bässler H (ed) Optical technique to characterize polymer systems, Elsevier, Amsterdam p 115
26. Wandass JH, Gardella JA (1985) Surf Sci 150: L 107
27. Gardella JA, Pireaux JJ (1990) Analyt Chem 62 (11): 103
28. Pireaux JJ, Thirty PA, Caudano R, Pfluger P (1986) J Chem Phys 84: 6452; Pireaux JJ, Vermeersch M, Gregoire C, Thiry PA, Candano R, Clarke TC (1988) J Chem Phys 88: 3353
29. Briggs D (1986) Surf Interf Analysis 9: 391
30. Niehuis E, van Velzen PNT, Lub J, Heller T, Benninghoven A (1989) Surf Interface Anal 14: 135
31. Hsu SL (1989) IR Spectroscopy. In: Allen J, Bevington JC (eds) Comprehensive polymer science, vol 1, Pergamon Press, p 429
32. Arndt T, Wegner G (1989) In: Bässler H (ed) Optical techniques to characterize polymer systems, Elsevier, Amsterdam, p 41; Bubeck C, Höltkamp D (1991) Adv Mater 3: 32
33. Biegen JF, Smythe RA (1988) Proc SPIE, Int Soc Opt Engin 897: 207
34. White HS, Earl DJ, Norton JD, Kragt HJ (1990) Anal Chem 62: 1130
35. Rabe JP (1989) Angew Chem Int Ed 28: 117 and 1127
36. McMaster TJ, Carr HJ, Miles MJ, Cairns P, Morris VJ (1991) Macromol 24: 1428
37. Rabe JP, Buchholz S (1991) Phys Rev Lett 66: 2096
38. Hansma PK, Elings VB, Marti O, Bracker CE (1988) Science 242: 209
39. Meyer E, Heinzelmann H, Rudin H, Güntherodt HJ (1990) Z Phys B79: 3; Lotz B, Wittmann JC, Stocker Y, Magonov S. N., Cantow HJ (1991) Pol Bull 26 209; 215; 223
40. Goldstein JI, Newbury DE, Ecklin P, Joy DG, Fiori G, Lipshin E (1981) Scanning electron microscopy and X-ray microanalysis, Plenum, New York
41. Sujata K, Jennings HM (1991) Mat Res Soc Bull 16 (3): 41
42. Lekner J (1987) Theory of reflection, Martinus Nijhoff, Dordrecht
43. Braslaw A, Deutsch M, Pershan PS, Weiss AH, Als-Nielsen J, Bohr J (1985) Phys Rev Lett 54: 114
44. Foster M, Stamm M, Reiter G, Hüttenbach S (1990) Vacuum 41: 1441
45. Mailänder L, Dosch H, Peisl J, Johnson RL (1990) Phys Rev Lett 64: 2527
46. Factor BJ, Russell TP, Toney MF (1991) Phys Rev Lett 66: 1181
47. Green PF, Doyle BL (1990) In: Tong HM, Nguyen LT (eds) New characterisation techniques for thin polymer films, J Wiley, new York, p 139
48. Mills PJ, Green PF, Palmstrom CJ, Mayer JW, Kramer EJ (1984) Appl Phys Lett 45: 958
49. Sokolow J, Rafailovich MH, Jones RAL, Kramer EJ (1989) Appl Phys Lett 54: 590
50. Bruder F, Straub W, Brenn R (1991) paper presented at DPG Frühjahrstagung Mainz
51. Kausch HH, Tirell M (1989) Ann Rev Mat Sci 19: 341
52. Green PF, Palmstrom CJ, Mayer JW, Kramer EJ (1985) Macromolecules 18: 501
53. Mills PJ, Palmstrom CJ, Mayer JW (1986) J Mater Sci 21: 1479
54. Payne RS, Clough AS, Murphy P, Mills PJ (1989) Nucl Instr Meth B42: 130
55. Chaturvedi UK, Steiner U, Zak O, Krausch G, Klein J (1989) Phys Rev Lett 54: 590
56. Lanford WA (1978) Nucl Instr Meth 149: 1
57. Endisch D, Rauch F, Götzelmann A, Reiter G, Stamm M Nucl Instr Meth B (in press)

58. Sykes DE (1989) Dynamic secondary ion mass spectrometry. In: Wells JM (ed) Methods of surface analysis, Cambridge University Press, Cambridge
59. Coulon G, Russell TP, Deline VR, Green PF (1989) Macromolecules 22: 2581
60. Whitlow SJ, Wool RP (1989) Macromolecules 22: 2648
61. Russell TP (1990) Materials Sci Rep 5: 173
62. Stamm M, Reiter G, Kunz K (1991) Physica B 173: 35
63. Stamm M, Majkrzak CF (1987) ACS Pol Prepr 28 (2): 18
64. Voigt-Martin I (1985) Adv Pol Sci 67: 194
65. Wu S, Chuang HK (1986) J Pol Sci, Phys 24: 143
66. Machate C (1989) PhD Thesis, University of Münster
67. Fayt R, Jerome R, Teyssier P (1986) J Pol Sci, Lett 24: 25
68. Gibson JM (1991) MRS Bulletin, 16 (3): 27
69. Klein J, Briscoe BJ (1979) Proc Roy Soc Lon 365: 53
70. Seggern JU, Klotz S, Cantow HJ (1989) Macromolecules 22: 3328
71. Takahashi A, Kawagushi M (1982) Adv Pol Sci 46: 1
72. Kim MW, Pfeiffer DG, Chen W, Hsiung H, Rasing T, Shen YR (1989) Macromolecules 22: 2682
73. Arwin H, Aspnes DE (1986) Thin Solid Films 138: 195
74. Motschmann H, Stamm M, Toprakcioglu C (1991) Macromolecules 24: 3681
75. Reiter R, Motschmann H, Lawall R, Stamm M, Knoll W (to be published) J Opt Soc Am
76. Motschmann H, Reiter R, Lawall R, Duda G, Stamm M, Wegner G, Knoll W (in press) Langmuir
77. Munch MR, Gast AP (1990) J Chem Soc Faraday Trans 86: 1341
78. Leermakers FAM, Gast AP (1991) Macromolecules 24: 718
79. Knoll W (1991) (1991) MRS Bulletin 16 (7): 29
80. Tassin JF, Siemens RL, Tang WT, Hadziioannou G, Swalen JD, Smith BA (1989) J Phys Chem 93: 2106
81. Sawodny M, Stumpe J, Knoll W (1991) J Appl Phys 69: 1927
82. Knobloch H, Knoll W (1991) J Chem Phys 94: 835
83. Glatter D, Kratky O (1982) Small angle X-ray scattering, Academic, London
84. Stamm M (in press) J Appl Cryst
85. Hashimoto T, Shibayama M, Kawai H (1980) Macromolecules 13: 1237; ibid, 1660
86. Bates FS, Berney CN, Cohen RE (1983) Macromolecules 16: 1101
87. Koberstein JT, Stein RS (1983) J Pol Sci, Phys 21: 2181
88. Bates FS, Dierker SB, Wignall GD (1986) Macromolecules 19: 1983
89. Stamm M (1983) ACS Polymer Prepr 24 (2): 380; Brautmeier D, Stamm M, Lindner P (in press) J Appl Cryst
90. Bartels CR, Crist B, Graessley WW (1983) J Pol Sci, Lett 21: 459
91. Garbella RW (1987) PhD Thesis, TH Darmstadt
92. Schmidt-Rohr K, Clauss J, Blümich B, Spiess HW (1990) Magnet Res in Chem 28: 3; Spiess HW (1991) Ann Rev Mater Sci 21: 131; Schmidt-Rohr K, Spiess HW (1991) Macromolecules 19: 5288
93. Ward WJ, McCarthy TJ (1989) Surface modification. In: Encyclopedia of Polym Sci and Engin, supplement I. J Wiley, New York, p 674
94. Ratner BD (1989) Biomedical applications of synthetic polymers. In: Allen J, Bevington JC (eds) Comprehensive polymer science, Pergamon, Oxford, p 201
95. Stamm M, Hüttenbach S, Reiter G, Springer T (1991) Europhys Lett 14: 1441
96. Roberts G (ed) (1990) Langmuir-Blodgett films, Plenum, New York
97. Duda G, Schouten AJ, Arndt T, Lieser G, Schmidt GF, Bubeck C, Wegner G (1988) Thin Solid Films 159: 221
98. Boven G, Folkersma R, Chella G, Schouten AJ (1991) Pol Commun 32: 50
99. Biederman H, Osada Y (1990) Adv Pol Sci 95: 57
100. Whitesides GM, Feruson GS (1988) Chemtracts Organ Chem 1: 171; Bain CD, Whitesides GM (1989) Adv Mater 4: 110; Whitesides GM, Laibinis PE (1990) Langmuir 6: 87
101. Kessel CR, Granick S (1991) Langmuir 7: 532

102. Schmidt I, Binder K (1985) J Phys (Paris) 46: 1631
103. Nakanshi H, Pincus P (1983) J Chem Phys 79: 997
104. Clark MB, Burkhardt CA, Gardella JA (1991) Macromolecules 24: 799; (1989) ibid
     22: 4495
105. Schmidt JJ, Gardella JA, Salvati L (1989) Macromolecules 22: 4489
106. Quamardeep SB, Pan DH, Koberstein JT (1988) Macromolecules 21: 2166
107. Schmitt RL, Gardella JA, Salvati L (1986) Macromolecules 19: 648
108. Jones RAL, Norton LJ, Kramer EJ, Composto RJ, Stein RS, Russell TP, Mansour A,
     Karim A, Felcher GP, Rafailovich MH, Sokolov J, Schwarz SA (1990) Europhys Lett
     12: 41
109. Russell TP, Jark W, Comelli G, Stöhr J (1989) Mat Res Soc Symp Proc 143: 265
110. Jones RAL, Norton LJ, Kramer EJ, Bates FS, Wiltzius P (1991) Phys Rev Lett 66: 1326
111. Yoon SC, Sung YK, Ratner BD (1990) Macromolecules 23: 4351
112. Teraya T, Takahara A, Kajiyama T (1990) Polymer 31: 1149
113. Green PF, Christensen TM (1990) J Chem Phys 92: 1478
114. Green PF, Christensen TM, Russell TP (1991) Macromolecules 24: 252
115. Bloch JM, Sansone M, Rondelez F, Pfeiffer DG, Pincus P, Kim MW, Eisenberger PM
     (1985) Phys Rev Lett 54: 1039
116. Rondelez F, Ausserré D, Hervet H (1987) Ann Rev Phys Chem 38: 317
117. Braslau A, Deutsch M, Pershan PS, Weiss AH (1985) Phys Rev Lett 54: 114
118. Hüttenbach S, Stamm M, Reiter G, Foster M (in press) Langmuir
119. Névot L, Croce P (1980) Rev Phys Appl (Paris) 15: 761
120. Daillant J, Borsio L, Benattar JJ, Meunier J (1989) Europhys Lett 8: 453
121. Mensinger H, Stamm M, Boeffel C, J Chem Phys (in press)
122. Sinha SK, Sirota EB, Garoff S (1988) Phys Rev B: 2297
123. Pynn R (to be published)
124. Ausserré D, Chatenay D, Coulon G, Collin B (1990) J Phys Frace 51: 257
125. Batow SW, Thomas BN, Flom EB, Rice SA, Lin B, Peng JB, Ketterson JB, Dutta P
     (1988) J Chem Phys 89: 2257
126. Jacquemain D, Grayer-Wolf S, Leveiller F, Lahav M, Leiserowitz L, Deutsch M,
     Kjaer K, Als-Nielsen J (to be published)
127. Madden WG (1987) J Chem Phys 87: 1405
128. Mansfield KF, Theodorou DN (1990) Macromolecules 23: 4430
129. Sung NH, Lee NY, Yuang P, Sung CSP (1989) Pol Engin Sci 29: 791
130. Lan KH, Ostrowsky N, Sornette D (1986) Phys Rev Lett 57: 17
131. Andrade JD, Gregonis DE, Smith LH (1985) In: Andrade JD (ed) Surface and interfacial
     aspects of biomedical polymers. vol 1, Plenum, New York, p 15
132. Ion R, Pollock HH, Roques-Carmes C (1990) J Mater Sci 25: 1444
133. Fernandez ML, Higgins JS, Penfold J, Shackleton CS (1990) Polymer Commun
     31: 124
134. Toney MF, Thompson C (1990) J Chem Phys 92: 3781
135. Reiter G, Stamm M (unpublished results)
136. Carmesin I, Kremer K (1990) J Phys France 51: 915
137. Bishop M, Clarke JHR (1990) J Chem Phys 94: 3936
138. Reiter J, Zifferer G, Olaj OF (1990) Macromolecules 23: 3120
139. van Vliet JH, ten Brinke G (1990) Macromolecules 23: 2797
140. Kumar SK, Vacatello M, Yoon DY (1990) Macromolecules 23: 2189
141. Vacatello M, Yoon DY (1990) J Chem Phys 93: 779
142. Bitsanis I, Hadziioannou G (1990) J Chem Phys 92: 3827
143. Theodorou DN (1989) Macromolecules 22: 4578; 4589
144. Pakula T (to be published)
145. Horn RG, Hirz SJ, Hadziioannou G, Frank CW, Catala JM (1989) J Chem Phys
     90: 6767
146. Montfort JP, Hadziioannou G (1988) J Chem Phys 88: 7187
147. van Alsten J, Granick S (1988) Phys Rev Lett 61: 2570
148. Horn RG, Israelachvili JN (1988) Macromolecules 21: 2836

149. McMaster TJ, Carr HJ, Miles MJ, Cairns P, Morris VJ (1991) Macromolecules 24: 1428
150. Yang E, Naoi K, Evans DF, Smyrl WH, Hendrickson WA (1991) Langmuir 7: 556
151. Sotobayashi H, Schilling T, Tesche B (1990) Langmuir 6: 1246
152. Sato Y (1990) J Pol Sci, Phys 28: 1163
153. Wittmann JC, Lotz B (1985) J Pol Sci, Phys 23: 205
154. Petermann J, Xu Y (1990) Pol Comm 31: 428
155. Kumaki J (1990) J Pol Sci, Phys 28: 105
156. Yorkgitis E (private communication)
157. Evans DF et al. (5/1991) paper presented at Tethered Chain I meeting, University of Minnesota
158. Shull KR, Kramer EJ, Hadziioannou G, Tang W (1990) Macromolecules 23: 4780
159. Jones RAL, Kramer EJ, Norton LJ, Shull KR, Felcher GP, Karim A, Fetters LJ (to be published)
160. Reiter G, Hüttenbach S, Foster M, Stamm M (in press) Fresenius
161. Bascom WD (1988) Adv Pol Sci 85: 89
162. Dietrich S (1988) Wetting phenomena. In: Phase Transitions and Critical Phenomena, Domb C, Lebowitz JL (eds.), vol 12, Academic London, p 1
163. Silberzahn P, Leger L (1991) Phys Rev Lett 66: 185
164. Daillant J, Benattar JJ, Leger L (1990) Phys Rev A41: 1963
165. Daillant J, Benattar JJ, Bosio L, Leger L (1988) Europhys Lett 6: 431
166. Reiter G (to be published)
167. Brochard F, Redon C, Rondelez F (1988) CR Acad Sci (Paris) 306 II:1143
168. Brochard-Wyart F, Daillant F (1990) Can J Phys 68: 1084
169. Kausch HH (1987) Polymer fracture, Springer Verlag, Berlin
170. Prager S, Tirrell M (1981) J Chem Phys 75: 5194
171. Jud K, Kausch HH, Williams JG (1981) J Mat Sci 16: 204
172. Mikos AG, Pappas NA (1988) J Chem Phys 88: 1137
173. Tirrell M, Adolf D, Prager S (1984) Springer Lect Notes Appl Math 1063: 37
174. Binder K, Sillescu H (1988) Polymer-Polymer diffusion. In: Encyl Pol Sci Engin, vol 2, Wiley, New York, p 1
175. Klein J (1990) Science 250: 640
176. Kehr KW, Binder K, Renlein SM (1989) Phys Rev B 39: 4891
177. Composto RJ, Mayer JW, Kramer EJ, White D (1986) Phys Rev Lett 57: 1312
178. Jordan AE, Bell RG, Donald AM, Fetters LJ, Jones RAL, Klein J (1988) Macromolecules 21: 235
179. Garbella RW, Wendorff JH (1988) Makromol Chem 189: 2459
180. Murshall U, Fischer EW, Herkt-Maetzky C, Fytas G (1986) J Pol Sci, Pol Lett 24: 191
181. Brereton MG, Fischer EW, Fytas G, Murhsall U (1987) J Chem Phys 87: 5048
182. Jilge W, Carmesin I, Kremer K, Binder K (1990) Macromolecules 23: 5001
183. Leibler L (1982) Macromolecules 15: 1283
184. Binder K (1983) J Chem Phys 79: 6387
185. Binder K, Frisch HL (1984) Macromolecules 17: 2930
186. Helfand E, Bhattacharjee SM, Fredrickson GH (1989) J Chem Phys 91: 7200
187. Harden JL (1990) J Phys Frace 51: 1777
188. Fayt R, Jerome R, Teyssie P (1989) Multiphase polymers: Blends and ionomers. In: Utracki LA, Weiss RA (eds) ACS Sympos Series 395, Washington, p 38
189. Olabisi O, Roberson LM, Shaw MT (1979) Polymer-Polymer miscibility, Academic, New York
190. Utracki LA, Weiss RA (eds) (1989) Multiphase polymers: Blends and ionomers, ACS Sympos Series 395, Washington
191. Hüttenbach S, Stamm M, Reiter G, Foster M (in press) Langmuir
192. Fernandez ML, Higgins JS, Penfold J, Ward RC, Shackleton C, Walsh DJ (1988) Polymer 29: 1923
193. Anastasiadis SH, Russell TP, Satija SK, Majkrzak CF (1990) J Chem Phys 92: 5677
194. Foster KL, Wool RP (1991) Macromolecules 24: 1397

195. Fernandez ML, Higgins JS, Penfold J, Shackleton C, Walsh DJ (1990) Polymer 31:2146
196. Brown HR, Char K, Deline VR (1990) Macromolecules 23: 3383
197. Cho K, Brown HR, Miller DC (1990) J Pol Sci, Phys 28: 1699
198. Russell TP, Anastasiadis SH, Menelle A (1991) Macromolecules 24: 1575
199. Creton C, Kramer EJ, Hadziioannou G (1991) Macromolecules 24: 1846
200. Götzelmann A, Reiter G, Stamm M (to be published)
201. Doi M, Edwards SF (1986) Theory of polymer dynamics, Oxford University Press, Oxford
202. de Gennes PG (1979) Scaling concepts in polymer physics, Cornell University Press, Ithaca, New York
203. Kremer K, Grest GS (1990) J Chem Phys 92: 5057
204. Brochard-Wyart F, de Gennes PG (1990) Makromol Chem, Makromol Sympos 40: 167
205. Kim YH, Wool RP (1983) Macromolecules 16: 1115
206. Zhang H, Wool RP (1989) Macromolecules 22: 3018
207. de Gennes PG (1989) CR Acad Sci (Paris) 308 II: 13
208. Steiner U, Krausch G, Schatz G, Klein J (1990) Phys Rev Lett 64: 1119
209. Karim A, Mansour A, Felcher GP (1990) Phys Rev B 42: 6846
210. Stamm M, Reiter G, Hüttenbach S, Foster M (1990) ACS Pol Prepr 31 (2): 73
211. Tassin JF, Monnerie L, Fetters JL (1988) Macromolecules 21: 2404
212. Reiter G, Steiner U (1991) J Phys France II 1: 659
213. de Gennes PG (1980) CR Acad Sci Paris 291 B: 219; (1989) 308 II: 13
214. Reiter G, Hüttenbach S, Foster M, Stamm M (1991) Macromolecules 24: 1179
215. Green PF, Palmstrom CJ, Mayer JW, Kramer EJ (1985) Macromolecules 18: 501

Editor: H.-H. Kausch
Received June 17, 1991

Author Index Volumes 1–100

Subject Index